Fish Vaccination

Fish Vaccination

Editor:

Vikas Mishra, Ph.D.

Delve Publishing
www.delvepublishing.com

Fish Vaccination

Editor: Vikas Mishra, Ph.D.

708 3rd Avenue, 6th Floor
New York, NY 10017
United States of America
www.delvepublishing.com

ISBN: 978-1-68095-849-2

Library of Congress Control Number: 2016951602

About the Editor

Vikas Mishra, Ph.D.

Vikas obtained his Ph.D. from King George's Medical University, India in 2010. His interests span from bio-resources, fermentation technology to medical virology and neuroscience research. He is currently working as a research associate at the University of British Columbia.

Contents

List of Contributors

Annelies Maria Declercq
Department of Morphology, Faculty of Veterinary Medicine, Ghent University, Merelbeke, Belgium

Freddy Haesebrouck
Department of Pathology, Bacteriology and Avian Diseases, Faculty of Veterinary Medicine, Ghent University, Merelbeke, Belgium

Wim Van den Broeck
Department of Morphology, Faculty of Veterinary Medicine, Ghent University, Merelbeke, Belgium

Peter Bossier
Laboratory of Aquaculture and Artemia Reference Center, Ghent University, Ghent, Belgium

Annemie Decostere
Department of Morphology, Faculty of Veterinary Medicine, Ghent University, Merelbeke, Belgium

Linn B Hølvold
UiT - The Arctic University of Norway, Faculty of Biosciences, Fisheries & Economics, Norwegian College of Fishery Science, 9037 Tromsø, Norway

Anne I Myhr
Genøk – Centre for Biosafety, The Science Park, Postbox 6418, 9294 Tromsø, Norway

Roy A Dalmo
UiT - The Arctic University of Norway, Faculty of Biosciences, Fisheries & Economics, Norwegian College of Fishery Science, 9037 Tromsø, Norway

Stephen Mutoloki
Department of Basic Sciences and Aquatic Medicine, Faculty of Veterinary Medicine and Biosciences, Norwegian University of Life Sciences, Oslo, Norway

Hetron Mweemba Munang'andu
Department of Basic Sciences and Aquatic Medicine, Faculty of Veterinary Medicine and Biosciences, Norwegian University of Life Sciences, Oslo, Norway

Øystein Evensen
Department of Basic Sciences and Aquatic Medicine, Faculty of Veterinary Medicine and Biosciences, Norwegian University of Life Sciences, Oslo, Norway

Hanna L. Thim
Norwegian College of Fisheries Science, UiT The Arctic University of Norway, Tromsø N-9037, Norway

Stéphane Villoing
MSD Animal Health Norway, Thormøhlensgate 55, Bergen N-5008, Norway

Marian McLoughlin
Aquatic Veterinary Services, 35 Cherryvalley Park, Belfast BT5 6PN, UK

Karen Elina Christie
MSD Animal Health Norway, Thormøhlensgate 55, Bergen N-5008, Norway

Søren Grove
Department of Laboratory Services, National Veterinary Institute, Oslo N-0106, Norway

Petter Frost
MSD Animal Health Norway, Thormøhlensgate 55, Bergen N-5008, Norway

Jorunn B. Jørgensen
Norwegian College of Fisheries Science, UiT The Arctic University of Norway, Tromsø N-9037, Norway

Brian Dixon
University of Waterloo, Department of Biology, Waterloo, ON, Canada

Francesca Andreoni
Department of Biomolecular Science, Biotechnology Section, University of Urbino "Carlo Bo", Via Arco d'Augusto 2, 61032

Fano, Italy

Mauro Magnani
Department of Biomolecular Science, Biotechnology Section, University of Urbino "Carlo Bo", Via Arco d'Augusto 2, 61032 Fano, Italy

Ponnerassery S. Sudheesh
Microbiology Laboratory, Fishery Quality Control Center, Ministry of Agriculture and Fisheries Wealth, 100 Muscat, Oman

Aliya Al-Ghabshi
Microbiology Laboratory, Fishery Quality Control Center, Ministry of Agriculture and Fisheries Wealth, 100 Muscat, Oman

Nashwa Al-Mazrooei
Microbiology Laboratory, Fishery Quality Control Center, Ministry of Agriculture and Fisheries Wealth, 100 Muscat, Oman

Saoud Al-Habsi
Microbiology Laboratory, Fishery Quality Control Center, Ministry of Agriculture and Fisheries Wealth, 100 Muscat, Oman

Jie Ji
Institute of Biotechnology and Biomedicine-UAB Research Park, Autonomous University of Barcelona, 08193 Bellaterra, Spain

Debora Torrealba
Institute of Biotechnology and Biomedicine-UAB Research Park, Autonomous University of Barcelona, 08193 Bellaterra, Spain

Àngels Ruyra
Institute of Biotechnology and Biomedicine-UAB Research Park, Autonomous University of Barcelona, 08193 Bellaterra, Spain

Nerea Roher
Institute of Biotechnology and Biomedicine-UAB Research Park, Autonomous University of Barcelona, 08193 Bellaterra, Spain

Gabriella do Vale Pereira
School of Biological Sciences, Plymouth University, Plymouth, England.
Laboratory Cameroon Marine (LCM), Department of Aquaculture, Federal University of Santa Catarina (UFSC), 88061-600, Florianópolis, SC, Brazil

Gabriel Fernandes Alves Jesus
studies of core Aquaculture Pathology, UFSC, Florianópolis, SC, Brazil

Felipe do Nascimento Vieira
Laboratory Cameroon Marine (LCM), Department of Aquaculture, Federal University of Santa Catarina (UFSC), 88061-600, Florianópolis, SC, Brazil

Scheila Anelise Pereira
studies of core Aquaculture Pathology, UFSC, Florianópolis, SC, Brazil

Thiago Tetsuo Ushizima
Mar & Terra Ind. Com de pescados S/A, Itaporã, MS, Brazil

José Luiz Pedreira Mouriño
studies of core Aquaculture Pathology, UFSC, Florianópolis, SC, Brazil

Maurício Laterça Martins
studies of core Aquaculture Pathology, UFSC, Florianópolis, SC, Brazil

Esther Gómez
Area of Microbiology, Department of Functional Biology, Faculty of Medicine, IUBA, University of Oviedo, 33006 Oviedo, Asturias, Spain

Jessica Méndez
Area of Microbiology, Department of Functional Biology, Faculty of Medicine, IUBA, University of Oviedo, 33006 Oviedo, Asturias, Spain

Desirée Cascales
Area of Microbiology, Department of Functional Biology, Faculty of Medicine, IUBA, University of Oviedo, 33006 Oviedo, Asturias, Spain.

José A. Guijarro
Area of Microbiology, Department of Functional Biology, Faculty of Medicine, IUBA, University of Oviedo, 33006 Oviedo,

Asturias, Spain.

Sebastián Reyes-Cerpa
Dept. Biology, Laboratory of Virology, University of Santiago of Chile, Santiago, Chile

Kevin Maisey
Dept. Biology, Laboratory of Immunology, University of Santiago of Chile, Santiago, Chile

Felipe Reyes-López
Dept. Cell Biology, Physiology & Immunology, Unit of Animal Physiology, Autonomous University of Barcelona, Barcelona, Spain

Daniela Toro-Ascuy
Dept. Biology, Laboratory of Virology, University of Santiago of Chile, Santiago, Chile

Ana María Sandino
Dept. Biology, Laboratory of Virology, University of Santiago of Chile, Santiago, Chile; Activaq S.A., Santiago, Chile

Mónica Imarai[2]
Dept. Biology, Laboratory of Immunology, University of Santiago of Chile, Santiago, Chile

Angels Ruyra
Institute of Biotechnology and Biomedicine, the UAB Research Park, Auto`noma University of Barcelona, Barcelona, Spain

Mary Cano-Sarabia
ICN2, Catalan Institute of Nanoscience and Nanotechnology, Campus UAB, Barcelona, Spain

Simon A. MacKenzie
Institute of Biotechnology and Biomedicine, the UAB Research Park, Auto`noma University of Barcelona, Barcelona, Spain

Daniel Maspoch
ICN2, Catalan Institute of Nanoscience and Nanotechnology, Campus UAB, Barcelona, Spain
Institucio' Catalan Research and Advanced Studies (ICREA), Barcelona, Spain

Nerea Roher
Institute of Biotechnology and Biomedicine, the UAB Research Park, Auto`noma University of Barcelona, Barcelona, Spain

Michael C. Nelson
Department of Molecular and Cell Biology, University of Connecticut, Storrs, Connecticut, USA

Scott E. LaPatra
Clear Spring Foods, Inc., Research Division, Buhl, Idaho, USA

Timothy J. Welch
National Center for Cool and Cold Water Aquaculture, Agricultural Research Service, U.S. Department of Agriculture, Kearneysville, West Virginia, USAc

Joerg Graf
Department of Molecular and Cell Biology, University of Connecticut, Storrs, Connecticut, USA

Maftuch
Fisheries and Marine Science Faculty, Brawijaya University, Malang, East Java, Indonesia

E Prasetio
Fisheries and Marine Science Faculty, Brawijaya University, Malang, East Java, Indonesia

A Sudianto
Fisheries and Marine Science Faculty, Brawijaya University, Malang, East Java, Indonesia

M Rozik
Fisheries Science Faculty, Palangkaraya University, Kalimantan, Indonesia

R Nurdiyani
Fisheries and Marine Science Faculty, Brawijaya University, Malang, East Java, Indonesia

E Sanusi
Fisheries and Marine Science Faculty, Brawijaya University, Malang, East Java, Indonesia

H Nursyam
Fisheries and Marine Science Faculty, Brawijaya University, Malang, East Java, Indonesia

F Fariedah
Fisheries and Marine Science Faculty, Brawijaya University, Malang, East Java, Indonesia

Marsoedi
Fisheries and Marine Science Faculty, Brawijaya University, Malang, East Java, Indonesia

Murachman
Fisheries and Marine Science Faculty, Brawijaya University, Malang, East Java, Indonesia

Erlong Wang
Department of Basic Veterinary, College of Veterinary Medicine, Sichuan Agricultural University, Chengdu 611130, Sichuan, China

Jun Wang
Department of Basic Veterinary, College of Veterinary Medicine, Sichuan Agricultural University, Chengdu 611130, Sichuan, China

Bo Long
Department of Basic Veterinary, College of Veterinary Medicine, Sichuan Agricultural University, Chengdu 611130, Sichuan, China

Kaiyu Wang
Department of Basic Veterinary, College of Veterinary Medicine, Sichuan Agricultural University, Chengdu 611130, Sichuan, China
Key Laboratory of Animal Disease and Human Health of Sichuan Province, Sichuan Agricultural University, Chengdu 611130, Sichuan, China

Yang He
Department of Basic Veterinary, College of Veterinary Medicine, Sichuan Agricultural University, Chengdu 611130, Sichuan, China

Qian Yang
Department of Basic Veterinary, College of Veterinary Medicine, Sichuan Agricultural University, Chengdu 611130, Sichuan, China

Defang Chen
Department of Aquaculture, College of Animal Science and Technology, Sichuan Agricultural University, Chengdu 611130, Sichuan, China

Yi Geng
Department of Basic Veterinary, College of Veterinary Medicine, Sichuan Agricultural University, Chengdu 611130, Sichuan, China
Key Laboratory of Animal Disease and Human Health of Sichuan Province, Sichuan Agricultural University, Chengdu 611130, Sichuan, China

Xiaoli Huang
Department of Aquaculture, College of Animal Science and Technology, Sichuan Agricultural University, Chengdu 611130, Sichuan, China

Ping Ouyang
Department of Basic Veterinary, College of Veterinary Medicine, Sichuan Agricultural University, Chengdu 611130, Sichuan, China

Weimin Lai[1]
Department of Basic Veterinary, College of Veterinary Medicine, Sichuan Agricultural University, Chengdu 611130, Sichuan, China

Preface

Like all livestock, farmed fish can be protected from disease risks by vaccination. Vaccines have been a major factor in decreasing the use of antibiotics in aquaculture while keeping fish healthy. However, a vaccine is only effective if it is delivered properly into the fish, in the correct place and at the correct dose. Large scale Bioproduction at land or at sea comes with inherit problem of infection, disease spread and severe mortality, leading to huge economic losses and large scale outbreaks of infectious diseases. Aquaculture is not any different, a sporadic infection event in wild fish, becomes a disease outbreak in fish farming. In the event of increase in fish farming, the use of antibiotics and sulpha drugs was replaced by the concept of vaccination as a means of preventing/ controlling fish diseases and to the development of commercially available vaccines. Vaccination can contribute in not only controlling the diseases, which effects the economic and environmental sustainability of fish farming. The era of vaccine development has led to accumulation of information on immune system of fish, production of vaccines, vaccination strategies, side effects and mechanisms of vaccination. This book deals with the efforts towards vaccine development of bacterial and viral diseases.

Fisheries are based on complex resources which include hundreds of kinds of fish. Each species has its own habits, living in different kinds of water, has different market qualities but together they provide excellent protein-rich food in far greater variety than animal agriculture. Prevention and control of fish diseases in Aquaculture is high priority in aquaculture industry. Unlike treating human or other animal diseases, few drugs are available for treating diseases in fish. Therefore, control of diseases in aquaculture and fish farms relies on a combination of good management practices, use of the few approved and commercially available drugs and vaccines and prevention of infection. Vaccination is becoming an increasingly important part of aquaculture, since it is considered a cost effective method of controlling different threatening diseases.

Editor
Vikas Mishra, Ph.D.

INTRODUCTION

A vaccine is a biological preparation that provides active acquired immunity to a particular disease. A vaccine typically contains an agent that resembles a disease-causing micro-organism and is often made from weakened or killed forms of the microbe, its toxins or one of its surface proteins. The agent stimulates the body's immune system to recognize the agent as a threat, destroy it, and keep a record of it, so that the immune system can more easily recognize and destroy any of these micro-organisms that it later encounters. Vaccines can be prophylactic (example: to prevent or ameliorate the effects of a future infection by any natural or "wild" pathogen), or therapeutic (e.g., vaccines against cancer are also being investigated).

The administration of vaccines is called vaccination. The effectiveness of vaccination has been widely studied and verified; for example, the influenza vaccine, the HPV vaccine, and the chicken pox vaccine. Vaccination is the most effective method of preventing infectious diseases; widespread immunity due to vaccination is largely responsible for the worldwide eradication of smallpox and the restriction of diseases such as polio, measles, and tetanus from much of the world. The World Health Organization (WHO) reports that licensed vaccines are currently available to prevent or contribute to the prevention and control of twenty-five infections.

The terms vaccine and vaccination are derived from Variolae vaccinae (smallpox of the cow), the term devised by Edward Jenner to denote cowpox. He used it in 1798 in the long title of his Inquiry into the...Variolae vaccinae...known.[as]...the Cow Pox, in which he described the protective effect of cowpox against smallpox. In 1881, to honour Jenner, Louis Pasteur proposed that the terms should be extended to cover the new protective inoculations then being developed

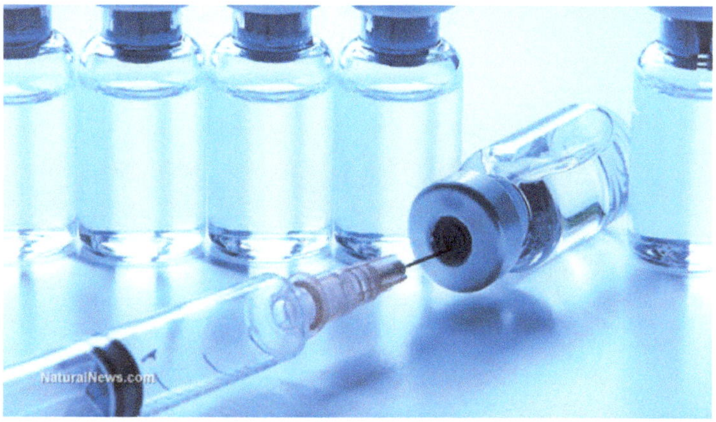

Overview of Fish Vaccination

The culture environment, be it in tanks, cages or ponds, seawater or freshwater, can stress the fish due to high stocking densities, reduced oxygen levels, sub-optimal water quality and high pathogen loads. Therefore, farmed fish, similar to farmed animals, have an increased susceptibility to infectious diseases. Such diseases may be bacterial, viral, fungal or parasitic in origin. Prophylactic treatments and good management practices can usually prevent or reduce the susceptibility to diseases. Although, antibiotics can overcome bacterial diseases, consumer health and food safety issues prevent their use in aquaculture. Moreover, established viral diseases cannot be treated. Therefore, vaccination is the best alternative to combat bacterial and viral diseases. It has been a key tool in the success of the livestock, pig, poultry and salmon industries in the fight against infectious diseases. In salmon farming, the use of vaccines is now so routine that all fish stocked in sea cages are vaccinated. It has been proven to be cost effective. It is also the key reason for the success of that industry. However, in Asia, fish vaccination is still a new concept that is not very well known and understood among farmers and professionals working in the area.

What are vaccines?

Vaccines are various preparations of antigens derived from specific pathogenic organisms that are rendered non-pathogenic. They stimulate the immune system and increase the resistance to disease from subsequent infection by the specific pathogen. Vaccination can be compared with an insurance policy where it is worth paying a basic fee for a policy that would later cover the costs of a more expensive disease or accident that may occur. Similarly, vaccination is a preventive measure that protects fish against a future disease and the associated costs due to morbidity, mortality and therapeutic treatment. However, just as an insurance policy will cover the costs of an accident only if this fits the clauses of the insurance contract, a vaccine only protects against specific diseases. For example, a vaccine against Streptococcus iniae infection will protect the vaccinated fish against Streptococcus iniae but it will not protect the fish against Vibrio anguillarum. A vaccine can be either water or oil based. Typically, injection vaccines are oil based as the oil provides adjuvant qualities. This means that the oil increases the effectiveness of the vaccine as well as the duration of the desired protection. The choice of vaccine depends on the particular case. It will depend on whether protection can be obtained, the duration of the protection possible versus the required duration, the final cost of the vaccine in relation to the benefit to the farmer and the registration limitations imposed by authorities in the countries where the vaccine is marketed. There are three major modes for the application of vaccines: oral, immersion and injection. However in all cases, only healthy fish should be vaccinated as it is a preventive measure and not a cure.

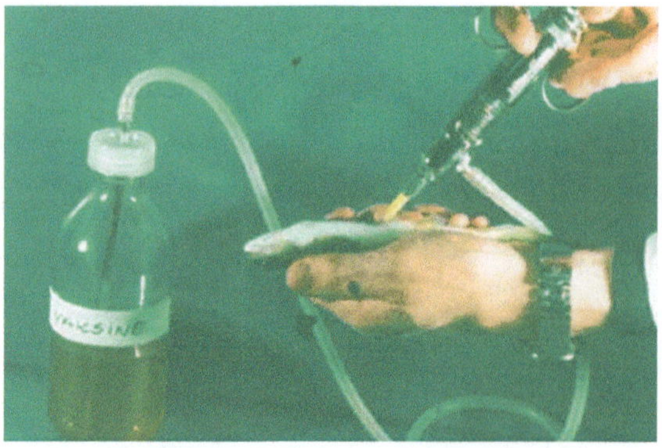

Oral vaccination

With oral vaccination, the vaccine is either mixed with the feed, coated on top of the feed (topdressed) or bio-encapsulated. When antigens are to be incorporated in feed, the heat sensitivity of the antigen has to be considered. When vaccines are used as top dressing in feed, a coating agent is usually applied, either to prevent leaching of the antigen from the pellets or to prevent breakdown of the antigen in the acidic environment of the fish stomach. For sensitive antigens, various microencapsulation methods are being evaluated and tested.

Bio-encapsulation is used where fish fry are to be vaccinated. In this case, live feed, such as Artemia nauplii, copepods or rotifers, are incubated in a vaccine suspension after which they are fed to the fry. Since these live organisms are non-selective filter feeders, they will accumulate the antigen in their digestive tract and as such, transform themselves into living microcapsules.

Oral vaccination has the advantage that it is a very easy vaccine administration method with no stress to the fish. However, oral vaccines have a very short term stability once mixed with the feed. In most cases, only limited protection can be obtained and the duration of protection can be rather short. Moreover, although oral vaccination is the preferred method from a fish farmer's perspective, at present there are few, if any, effective oral vaccines in the market.

Effectiveness

Vaccines have historically been the most effective means to fight and eradicate infectious diseases. Limitations to their effectiveness, nevertheless, exist. Sometimes, protection fails because the host's immune system simply does not respond adequately or at all. Lack of response commonly results from clinical factors such as diabetes, steroid use, HIV infection or age. It also might fail for genetic reasons if the host's immune system includes no strains of B cells that can generate antibodies suited to reacting effectively and binding to the antigens associated with the pathogen.

Even if the host does develop antibodies, protection might not be adequate; immunity might develop too slowly to be effective in time, the antibodies might not disable the pathogen completely, or there might be multiple strains of the pathogen, not all of which are equally susceptible to the immune reaction. However, even a partial, late, or weak immunity, such as a one resulting from cross-immunity to a strain other than the target strain, may mitigate an infection, resulting in a lower mortality rate, lower morbidity and faster recovery.

Adjuvants commonly are used to boost immune response, particularly for older people (50–75 years and up), whose immune response to a simple vaccine may have weakened.

Maurice Hilleman's measles vaccine is estimated to prevent 1 million deaths every year.

The efficacy or performance of the vaccine is dependent on a number of factors:

- the disease itself (for some diseases vaccination performs better than for others)
- the strain of vaccine (some vaccines are specific to, or at least most effective against, particular strains of the disease)
- whether the vaccination schedule has been properly observed.
- idiosyncratic response to vaccination; some individuals are "non-responders" to certain vaccines, meaning that they do not generate antibodies even after being vaccinated correctly.
- assorted factors such as ethnicity, age, or genetic predisposition.

Immersion vaccination

Skin epithelium and gills have mechanisms to protect fish in a broad as well as specific way. Immersion vaccination works on the ability of mucosal surfaces to recognize pathogens they had been in contact with. When fish are immersed in water containing the diluted vaccine, the suspended antigens from the vaccine may be adsorbed by the skin and gills. Then, specialized cells, such as antibody-secreting cells, present in the skin and gill epithelium will be activated and will protect the fish when fish are exposed to the live pathogen at a later stage. Other cells located in the epithelium of skin and gills, such as antigen presenting cells (macrophages), also absorb vaccine antigens and transport them to specialized tissues where the systemic immune response builds up.

In immersion vaccination, there are two application methods: dip and bath. In dip vaccination, fish are immersed for a very short duration, usually 30 seconds, in a highly concentrated vaccine solution, usually 1 part vaccine product to 9 parts water. With bath vaccination, fish are exposed for a longer period, usually one to several hours, in a lower concentration of vaccine.

Injection vaccination

Initially, fish farmers may not favour injection vaccination as they fear that the stress resulting from the handling and injection of the fish will cause high mortality. In contrast, many studies and farmers' experience in the salmon industry have shown that there is no mortality associated with the vaccination process per se, although some weak fish may die due to the handling process.

Light anaesthesia of fish is needed for injection vaccination. This decreases the stress due to vaccination, prevents mechanical injuries and helps the fish to recover faster from the handling. When injection vaccination is performed properly, mortality immediately after vaccination should not exceed 0.25%. Higher mortalities indicate incorrect use of anaesthetic, excessive handling/stress of the fish, incorrect needle insertion (e.g., rupturing the spleen), vaccination of diseased or weakened fish, lack of oxygen in the anaesthesia vessel, etc.

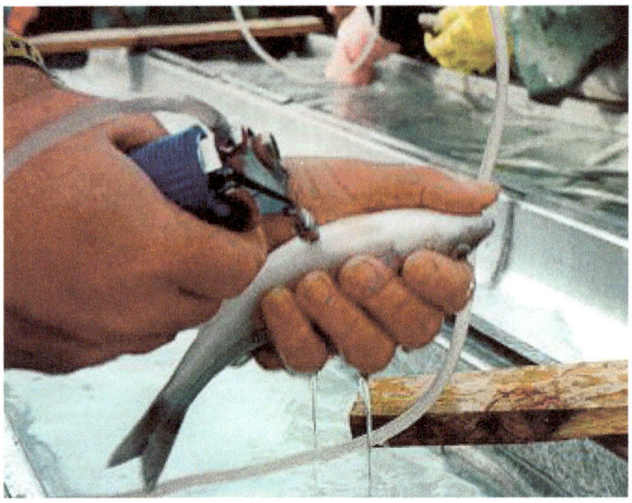

Injection vaccines can be administered by intramuscular or intraperitoneal (in the abdominal cavity) injection, but the latter is by far the most common. As intraperitoneal injection vaccination involves depositing the vaccine in the abdominal cavity, it is important that the needle should penetrate the targeted abdominal wall of fish by 1 to 2 mm. Short needles might deposit the vaccine in the musculature and cause inflammation and a bad immune response.

The right thickness of needle is also important. If the vaccine is viscous, it might pass through the needle with difficulty. In contrast, a needle that is too large can induce sores and secondary infections at the injection site, and allow the vaccine to flow back. The rapidity of the immune response is temperature dependant. It usually takes several weeks before good immune protection is developed as a result of vaccination. Therefore, it is important not to stress the fish in the weeks following vaccination as stress is known to suppress the immune system.

Injection vaccination has a number of major advantages that makes it a preferred vaccination method. Injection vaccination provides for a long duration of protection, i.e., for over a year, and it allows for multiple antigens to be combined in a single vaccine and, therefore, in a single administration. In addition, the fish farmer is assured that every fish in the population has received the vaccine and at the correct dose. At present, the most complex products provide protection against 6 antigens (5 bacterial and 1 viral) and, in the near future, heptavalent vaccines will become available. The injection volumes per fish are usually 0.1 or 0.2 ml and give protection throughout the production cycle of most farmed species. Injections are in general superior to any other vaccine application method; however, from a practical point of view, they can only be applied to fish of 10g or more.

Different types of vaccines exist:

- Inactivated vaccines are the most common in aquaculture. They are produced by inactivating the disease-causing micro-organism with chemicals or heat.
- Live, attenuated vaccines. To make a live, attenuated vaccine, the disease-causing microorganism is grown under special laboratory conditions that cause it to lose its virulence or disease-causing properties. Intervet's Aquavac ESC®, a vaccine used against Edwardsiella ictaluri in Channel catfish (Ictalurus punctatus) in the USA is an example.
- Subunit vaccines are a more modern type of vaccine, developed from antigenic fragments that are able to evoke an immune response. Subunit vaccines can be made by purification of parts of the actual micro-organism or they can be made in the laboratory using genetic engineering techniques. An example is Intervet's Compact® IPN used against infectious pancreatic necrosis virus infections of salmon in Chile.
- Recombinant vector vaccines. A vaccine vector or carrier is a weakened bacterium into which harmless parts of genetic material from another disease-causing micro-organism have been inserted.
- DNA vaccines are developed using plasmids containing information on the expression (of part) of the antigen of interest.
- Synthetic/peptide vaccines are synthetically prepared antigenic epitopes in a suitable carrier or adjuvant.
- Anti-idiotypic vaccines contain antibodies developed so as to mimic an antigen.

CHAPTER 1

COLUMNARIS DISEASE IN FISH: EMPHASIS ON BACTERIUM-HOST INTERACTIONS

Annelies Maria Declercq, Freddy Haesebrouck, Wim Van den Broeck, Peter Bossier, and Annemie Decostere

THE AGENT

History and Taxonomy

Flavobacterium columnare (*F. columnare*), the causative agent of columnaris disease, belongs to the family *Flavobacteriaceae* [1, 2, 3]. Columnaris disease was first described by Davis among warm water fishes from the Mississippi River [4]. Although unsuccessful in cultivating the etiological agent, Davis described the disease and reported large numbers of slender, motile bacteria present in the lesions [4]. Upon examining a wet mount preparation of these lesions, column-like structures formed by these bacteria were evident. The organism was hence named *Bacillus columnaris* and the disease elicited columnaris disease. Over the decades the taxonomic status of the pathogen has changed several times since the pioneering work of Davis [4]. In 1944, Ordal and Rucker were the first to isolate the bacterium from a natural outbreak of columnaris disease among sockeye salmon (*Onchorhynchus nerka*) [5]. A diluted culture medium was used to grow the bacterium. Based on cellular morphology, they identified the bacterium as a myxobacterium. Organisms classified in the order *Myxobacteria* are long, thin Gram-negative rods that are motile on agar media by a creeping or flexing motion. They have a life cycle composed of vegetative cells, microcysts (resting cells), and fruiting bodies, or only vegetative cells and microcysts [6]. Ordal and Rucker reported that the myxobacterium from columnaris disease produced both fruiting bodies and microcysts and named the organism *Chondrococcus columnaris* [5]. Garnjobst assigned the bacterium to the family *Cytophagaceae* as *Cytophaga columnaris* after isolating a pathogenic bacterium resembling *Chondrococcus columnaris* morphologically, but not producing microcysts [7]. Bernardet and Grimont reclassified the organism and placed it in the family *Cytophagaceae* and the genus *Flexibacter*, assigning it as *Flexibacter columnaris* [8]. Finally, in 1996, the bacterium received its current name, *Flavobacterium columnare*, based on DNA-rRNA hybridization data and protein and fatty acid profiles [3]. In 1999, the *F. columnare* cluster was subdivided in three genomovars based on differences in 16S rRNA sequences, restriction fragment length polymorphism (RFLP) and DNA-DNA-hybridization [9].

Citation: Annelies Maria Declercq, Freddy Haesebrouck, Wim Van den Broeck, Peter Bossier, and Annemie Decostere. Columnaris disease in fish: emphasis on bacterium-host interactions, DOI:10.1186/1297-9716-44-27.

Morphology and Biochemical Characteristics

The morphological and biochemical characteristics of *F. columnare* are summarized in Table 1. For a full biochemical profile of *F. columnare*, the reader is referred to Bernardet and Bowman [1].

Table 1. Morphological and biochemical characteristics of *F. columnare* (adapted from [1, 3, 10, 11, 12])

Characteristic	Description
Growth condition	Strictly aerobic
Gram-stain	Gram-negative
Morphology	Long, slender gliding rods of 4 to 10 µm and 0.3 to 0.5 µm wide. In aging cultures spheroplasts may occur

Capsule	Described to be absent [10] or present [11] depending on the adopted strain
Congo red absorption	Present due to an extracellular galactosamine glycan in the mucus and the production of flexirubin-type pigments
H_2S-production	Present
Degradation of crystalline cellulose	Absent
Degradation of complex acidic polysaccharides of connective tissue	Present

Epidemiology

F. columnare is distributed worldwide in fresh water sources and may infect many different wild and cultured freshwater fish species, such as (but not limited to) carp, channel catfish, goldfish, eel, perch, salmonids and tilapia [1, 12, 13, 14, 15, 16, 17]. This disease also assails many tropical freshwater aquarium fish [12, 14]. In the channel catfish (*Ictalurus punctatus*) industry in the United States, *F. columnare* is the second most prevalent bacterium, after *Edwardsiella ictaluri*, to cause disease and mortality [18, 19], with yearly losses estimated at 30 million dollars [20]. This organism can also be part of the bacterial microbiota of freshwater fish, eggs and the rearing waters the fish live in [21]. Fish may reside in a clinically healthy carrier status harbouring an isolate remaining from a previous outbreak of columnaris disease and in this way act as an infection source for other fish [22, 23]. Fujihara and Nakatani reported that rainbow trout surviving a *F. columnare* infection can release up to 5×10^3 colony forming units/mL/h of viable bacteria into tankwater [22]. The gills were shown to be the major release site of this pathogen. Dead fish would be able to spread the disease at a higher transmission rate compared to living fish [24].

Several studies have indicated the potential for *F. columnare* to survive for extended periods in water. Survival was demonstrated to be influenced by physical and chemical characteristics of the surrounding water. Fijan indicated that *F. columnare* can survive up to 16 days at 25°C in hard, alkaline water with a high organic load [25]. Soft water with 10 mg/L $CaCO_3$, especially when acid or with a low organic content, does not provide a favorable environment for the organism [25]. Chowdhury and Wakabayashi determined that calcium, magnesium, potassium and sodium ions all are important for long-term survival of *F. columnare* in water [26]. Ross and Smith found that survival of *F. columnare* in static, sterile river water was directly related to temperature, with a higher survival percentage at lower temperatures [27]. The bacterium can keep its infectivity in lake water in laboratory conditions for at least five months [28]. *F. columnare* is also capable of surviving in sterile river mud [6]. Apparently, mud slurry often contains sufficient nutrients to maintain viability of *F. columnare* longer than sterile river water. In this case, however, the percentage survival of *F. columnare*seeded into mud seems to be higher at 25°C than at 5°C. Temperatures below 5°C are even detrimental to *F. columnare* cells in mud. *F. columnare* also grows well on particulate fish feed [29]. When surviving outside the host, *F. columnare* can change from a virulent to a less virulent form with an altered colony morphology, probably to save energy [28]. It has been suggested that *F. columnare* strains at fish farms originate from environmental waters and that the farm environment and practices may select for virulent strains that cause outbreaks in the farm [28, 30].

THE DISEASE

Clinical Signs, Histopathology, Ultrastructural Features and Haematology

F. columnare causes acute to chronic infections and typically affects the gills, the skin and fins. The clinical manifestation of columnaris disease amongst others is dependent on the virulence of the eliciting strain. In a study of Rucker et al., the strains of low virulence induced slow progressive infections at water temperatures above 21°C and caused massive tissue damage before death occurred [31]. Strains of high virulence caused fulminating infections and killed young salmon (*Salmo salar*) in 12 to 24 h at 20°C. Ordinarily, these fish did not show gross tissue damage at the time of death [31]. The same was found in a study of Pacha and Ordal [32] and Foscarini [33]. The gross pathology observed in the fish experimentally infected with strains of *F. columnare* of high virulence was usually very limited. Apparently, death occurred before gross external manifestations of the disease appeared. However, some of the last fish to die did show macroscopically visible signs [34]. Besides the virulence of the strain being a determinant factor, in coldwater and temperate fish, age also seems to have an important impact on the severity of the clinical signs. In young fish, the disease develops acutely and mostly damages the gills (Figure 1). In adults, the disease may adopt an acute, subacute or chronic course. When the disease course is acute or subacute in adult fish, yellowish areas of necrotic tissue can appear in the gills ultimately resulting in complete gill destruction (Figure 2) [1, 32, 35].

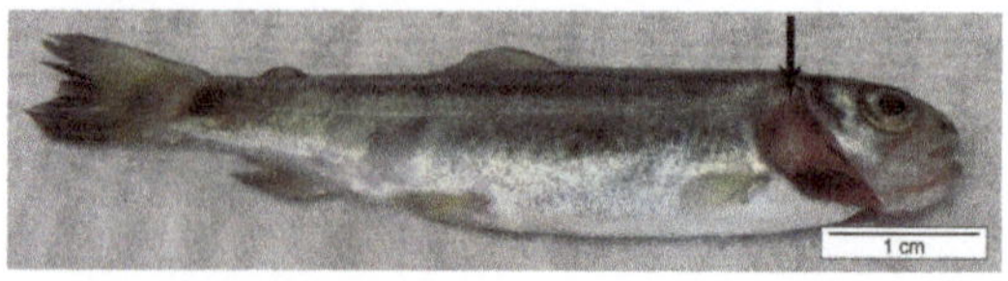

Figure 1. *F. columnare* elicited gill lesions in rainbow trout (*Oncorhynchus mykiss*) fry (operculum removed). In young fish, the disease is mostly acute and the gill is the major site of damage. The lesions are exhibited by pale necrotic areas (arrow, bar = 1 cm).

Figure 2. Gill lesions in a shubunkin (*Carassius auratus*) (operculum removed) caused by *F. columnare*. Yellowish-white areas of de-

generation are visible in the ventral part of the first gill arch (arrow). When the *F. columnare* infection spreads rapidly throughout the gill lamellae, the fish may die in a short period of time without any other apparent lesions. Respiratory distress, caused by damage to the gills, appears to be the cause of death (bar = 1 cm).

In chronic cases, it takes longer before gill damage appears and skin lesions may develop as well [1, 32, 35]. On the body, small lesions start as areas of pale discolorations of the skin, which usually are surrounded by a zone with a distinct reddish tinge. This mostly begins at the base of the dorsal fin. Fin deterioration then occurs, starting from the lesion at the base of the fin and progressing to the outer edge, the opposite to normal finrot. The lesions then begin extending laterally from their common location at the base of the dorsal fin to encircle the fish resembling a "saddle-back" (Figure 3). The disease therefore is referred to as "saddle-back disease" [1, 32, 36]. Finrot is also often present [1, 37].

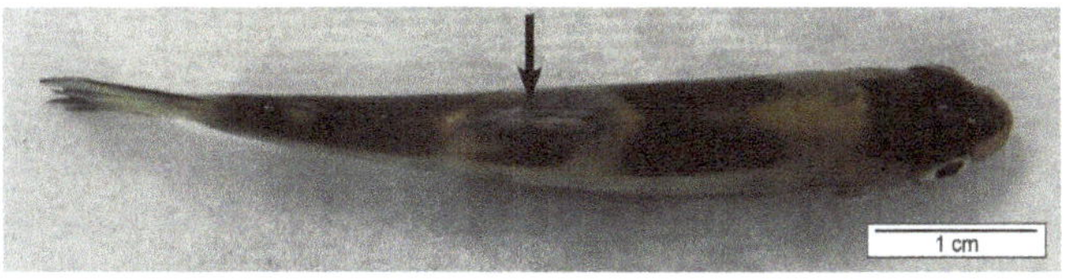

Figure 3. *F. columnare* induced a saddleback lesion (arrow) in rainbow trout fry. The lesion is visible as a discoloration starting around its common location at the base of the dorsal fin and extending laterally to encircle the fish resembling a saddle. Hence, the descriptive term "saddle-back" is often used and the disease is denoted as "saddle-back disease" (bar = 1 cm).

In rainbow trout, the area around the adipose fin may become dark and show erosions. These lesions expand to the peduncle, hence the name "peduncle disease" [1]. The lesions may progress cranially and caudally and even into the deeper skin layers, exposing the musculature leading to deep ulcers (Figures 4 and 5) [1, 35]. The lesions typically are covered with yellowish-white mucus [1].

Figure 4. Skin ulceration (arrow) in a minnow (*Phoxinus phoxinus*) caused by *F. columnare.* The lesion has progressed into deeper skin layers, exposing the musculature. The edge of the ulceration displays a distinct reddish tinge and its centre is covered with yellowish-white mucus (bar = 1 cm).

Figure 5. Tail fin erosion in a minnow caused by *F. columnare.* Columnaris disease led to complete disappearance of the upper half of the fin and exposing the underlying musculature (arrow, bar = 500 µm).

Ulceration of the oral mucosa also occurs, resulting in mouthrot. These mouth lesions are more lethal than are the skin lesions, since the painful oral lesions render the fish anorectic and lead to death due to starvation. Moreover, the disease spreads easily to the mandible and the maxilla. Secondary infections with fungi or other bacteria may deteriorate the situation and can be seen together with the filamentous bacteria [38]. In tropical fish, this clinical sign led to the disease being termed "cotton wool disease" or "mouth fungus" [1]. Finrot can also be present in tropical fish [37]. Lesions can be restricted to local skin discoloration, with or without ulceration, and degeneration of underlying muscle fibers [1, 12, 39].

The skin or gills need to be abraded, for bacteria to enter the bloodstream and cause systemic infections [40]. However, Hawke and Thune have isolated *F. columnare* from internal organs without any external lesions appearing [18]. Foscarini described that the pathological changes to the gill structure caused by columnaris disease go hand in hand with cardiac alterations [33]. The first day after infection, bradycardia was noticed with the formation of hyperplastic gill lesions. Degenerative processes of the lamellae in the following days resulted in a compensatory tachycardia. This research suggests that the interaction between the impaired gill vascular blood circulation and the cardiac changes could result in the death of the fish [33].

Light microscopic examination of the affected gill tissue reveals the disappearance of the normal structure of primary and secondary filaments (Figure 6) [32, 33, 38, 41]. In the initial phase, proliferation of epithelial cells of the gill filaments can be accompanied by an increase of mucous cells [33]. The proliferating tissue can occlude the space between adjacent gill lamellae. In more advanced stages, the occlusion can be total causing the gill lamellae to be completely surrounded by the propagating tissue. Congestion of gill lamellae occurs due to accumulation of blood masses and inflammatory cell infiltration can be noticed. Edema causes lifting of the surface epithelium of gill lamellae from the underlying capillary bed. In more advanced stages of the disease, fusion of gill lamellae and/or gill filaments appears [32, 33, 38, 41]. Complete clubbing of gill filaments can finally result in circulatory failure and extensive internal hemorrhage [33]. Moreover, huge clusters of *F. columnare* can be found on the cell surface and/or in between necrotic sites (Figure 6, detail).

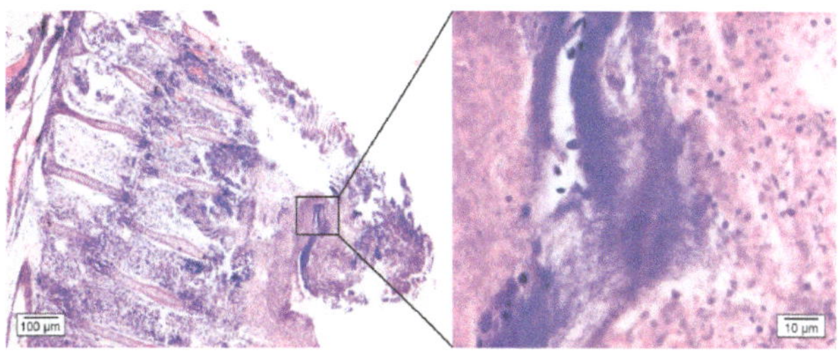

Figure 6. Gill section of a koi carp infested with *F. columnare.* In the left figure, extensive loss of branchial structures is visible. This is an advanced stage of the disease in which the filaments and the lamellae have fused and the gill epithelium is destroyed (H&E, bar = 100 µm). Complete clubbing of gill filaments may finally result in circulatory failure and extensive internal hemorrhage. A detail of this is depicted in the right figure, where *F. columnare*bacteria are visible

as long, slender, purplish structures in between remnants of the gill tissue (H&E, bar = 10 µm).

Columnaris disease may cause acute ulcerative dermatitis extending into the hypodermis and the muscle. Waterlogging can be present. The latter appears when the osmotic barrier is broken and thus forces water into the tissues, leading to severe dermal edema. Rupture of pigment cells with the loss of melanocytes can also be seen. Columns or hay-stack-like aggregates of bacteria can be gathered between dermal collagen fibers. The typical long and slender bacterial cells can easily be noted upon inspecting hematoxylin and eosin (H&E) or Giemsa stained sections from affected tissue where they appear bluish-purple and blue, respectively [38]. When all these changes occur rapidly, they may proceed to severe necrosis and sloughing of the epidermis [32, 38].

Scanning electron microscopic (SEM) pictures of affected gill arches reveal the presence of rod-shaped bacterial cells, approximately 0.3-0.5 µm wide and 3-10 µm long (Figure 7). These long, thin bacteria adhere on the surface of the gills and appear to be aggregated rather than evenly distributed across the gill epithelium [42, 43]. Transmission electron microscopic (TEM) examination of gill tissue shows numerous long, slender bacteria in close contact with the gill tissue (Figure 8).

Figure 7. Gill of a rainbow trout infested with *F. columnare*. This scanning electron microscopic (SEM) picture of an affected gill arch reveals the presence of long, thin, rod-shaped bacterial cells, approximately 0.3-0.5 µm wide and 3-10 µm long (SEM, bar = 5 µm).

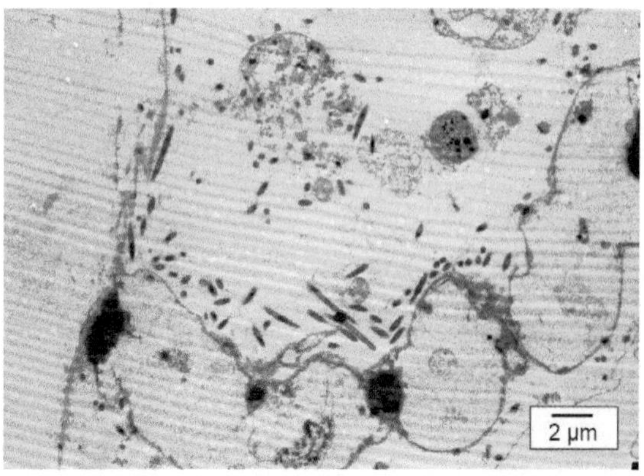

Figure 8. Gill lamellae of a koi carp following infection by *F. columnare*. This transmission electron microscopic picture clearly demonstrates the loss of structure of the gill lamellae. The *F. columnare* bacteria are lining up around the remnants of these lamellae, and are closely associated with the gill epithelium (TEM, bar = 2 µm).

Bullard and McElwain were the first to describe the ultrastructural features of saddleback lesions associated with experimental infections of *F. columnare* in channel catfish and zebrafish (*Danio rerio*) using SEM [44]. Channel catfish skin lesion samples had margins typified by shed epidermal cells and lesion centers that exhibited a multitude of rod-shaped bacterial cells, approximately 3-10 µm long and 0.3-0.5 µm wide, intermixed with cellular debris. Zebrafish skin lesion samples displayed a multitude of rod-shaped bacterial cells and exhibited similar ultrastructural changes. Scales were missing or, when present, denuded of epidermis.

In a cutaneous columnaris challenge model in koi carp, significant changes in blood parameters were observed in the infected fish [45]. For the hematologic parameters, a significant decrease was noted in Packed Cell Volume (PCV), haemoglobin concentration, red blood cell count, mean corpuscular volume and absolute lymphocyte counts. As for the biochemical parameters, marked hyponatremia, hypochloridemia and hyperglycemia were observed. Calcium and magnesium levels dropped only slightly and total serum protein and albumin-like protein concentrations decreased mildly. Alkaline phosphatase (ALP), aspartate aminotransferase (AST), lactate dehydrogenase (LDH) and creatine kinase (CK) secretions were significantly increased [45]. In a study by Řehulka and Minařík blood parameters from brook trout suffering from an acute, natural outbreak of *F. columnare* also revealed anemia, but levels of mean corpuscular volume and mean corpuscular haemoglobin were higher [15]. Total protein levels fell much below physiological parameters in infected fish. Calcium concentrations were reduced significantly. Blood Urea Nitrogen (BUN) measurements were much higher compared to normal levels. The catalytic activity of AST, alanine aminotransferases and LDH reached multiples of normal values. In contrast, ALP concentrations decreased. Hypoglycemia was noted. No data were shown on sodium or chlorine [15].

Diagnosis

Timely detection of this pathogenic agent is important to prevent its spreading and to reduce the economic loss to fish farmers.

The isolation of *F. columnare* is possible from external lesions, provided that the samples are taken from the edge of recent lesions [1]. *F. columnare* requires low nutrient media. No growth of *F. columnare* is reported on trypticase soy agar (TSA), nutrient agar or Marine 2216 agar [1]. In 1944, *F. columnare* was first cultured on Cytophaga agar, a nutrient poor medium [5]. Since then, several media, including Shieh [46] and TYES [47] have been developed in an attempt to improve growth of the bacterium. Growth does not occur in media that contain NaCl concentrations of over 0.5% or that have a pH lower than six [1]. Depending on the strain, the bacterium grows between 15 and 37°C [9, 12], with an optimal growth occurring between 25°C and 30°C. Colonies appear after 24 to 48 h of incubation [12]. Glucose does not improve growth. Considering that *F. columnare* is easily overgrown by contaminating bacteria, selective media have been developed based on inherent resistance of *F. columnare* to different antimicrobial agents, including polymyxin and neomycin [6, 8,25]. Decostere et al. added one µg/mL tobramycin to modified Shieh medium, demonstrating this to be an effective selective supplement for isolating *F. columnare*[41]. *F. columnare* largely displays two colony types on solid media; smooth and rhizoid [48]. Kunttu et al. characterized a third, rough colony morphology variant [49]. The rhizoid colony variants were assigned virulent and moderately adherent, the non-rhizoid rough colony variants non-virulent and highly adherent, and the smooth colony variants non-virulent and poorly adherent [49]. Colonies of *F. columnare* are notorious for their sometimes strong adherence to the agar. This trait is also exhibited in broth cultures where yellow, filamentous clumps of bacterial cells can form a thick ring at the surface of a glass recipient [12, 50]. Adherence may be lost after several subcultures. Colonies can be recognized by their distinctive yellow pigmentation. The yellow color is due to the production of flexirubin pigments [1]. A method to identify the typical protein profile of *F. columnare* is whole-cell protein analysis [3]. Genomovar ascription has been performed using 16S-restriction fragment length polymorphism to divide *F. columnare* into three genomovars I-III [9]. Compared to RFLP, single strand conformation polymorphism (SSCP) would have an improved resolution power in the study of intraspecies diversity in *F. columnare*[51]. Polymerase chain reaction (PCR) has gained interest for definitive identification of *F. columnare* based on the selective amplification of the 16S ribosomal RNA gene using species-specific primers [52, 53]. This technique gives a conclusive identification of the organism within a few hours and eliminates the need for biochemical testing which is laborious and sometimes inconclusive [52].

Besides isolation, other methods may be used for detecting this pathogen. These include serological methods such as the enzyme-linked immunosorbent assay [54] and the fluorescent antibody test [55, 56]. Both have proven to be efficient and rapid for diagnosing columnaris disease. A loop-mediated isothermal amplification method (LAMP) for rapid detection of *Flavobacterium columnare* from infected fish organs (gills, skin and head kidneys) was established in channel catfish [57]. PCR may also be adopted, having the advantage of being able to detect very low levels of *F. columnare*[52, 53]. In addition, Panangala et al. have developed a TaqMan-based real-time PCR targeting a 113 bp nucleotide region of the chondroitin AC lyase gene of *F. columnare*[58]. This PCR is specific, sensitive and reproducible for the detection and quantitation of *F. columnare* in tissues (blood, gills and kidney) of infected fish. Moreover, real-time PCR-based methods are distinctively more advantageous than conventional PCR as they eliminate the need for detection of amplified products by gel electrophoresis, thus reducing costs, time and labour [58]. Length heterogeneity PCR (LH-PCR) is based on comparing naturally varying lengths of 16S rDNA PCR products between bacterial groups [59]. LH-PCR has also proven effective in detecting *F. columnare* in fish tissue [60].

PATHOGENESIS

Various reports have noted the difference in virulence among *F. columnare* strains [11, 12, 40, 42, 61, 62, 63,64, 65]. In the last decade, various studies have attempted to elucidate the pathogenesis of columnaris disease. Although significant progress has been made, still many questions remain unanswered on how this pathogenic organism elicits disease, and information concerning the bacteriological events preceding disease and death is scarce. Li et al. related this lack of knowledge to the absence of an efficient molecular manipulation system for *F. columnare*, especially a plasmid-based inframe knockout system [66], despite two recent reports on the establishment of genetic manipulation system for the bacterium [67, 68]. Li et al. identified the type I restriction-modification system (R-M system) in *F. columnare* to improve electroporation efficiency and suggested that it would be of significant interest to examine the composition and diversity of R-M systems in strains of *F. columnare* in order to set up a suitable genetic manipulation system for the bacterium [66]. Only very recently, the full genome of *F. columnare* ATCC 49512 was sequenced [69].

Colonization

The colonization of the fish tissue is to be regarded as a complex multistep process which can be subdivided into the stages of attraction, adhesion and aggregation, requiring a step-by-step analysis. The exact factors mediating colonization have however not yet received the full attention they merit and to date remain largely unidentified.

The research group of Klesius et al. demonstrated by means of the traditional capillary tube method that the mucus from the skin and gills of catfish promotes chemotaxis of *F. columnare*[63]. The gliding motility of this bacterium is well known [1, 12, 41]. Indeed, the observation of a drop of bacteria grown in broth under a phase contrast microscope shows the slowly forward and backward gliding of *F. columnare*. Pate and Ordal described that the location of fibrillar structures spanning the gap between the outer membrane and the mucopeptide layer might play a role in the gliding motility of the *F. columnare* bacterial cells [70]. Although Klesius et al. acknowledged that they were not able to fully define the role that chemotaxis plays in the virulence of *F. columnare*, the chemotactic response of the more virulent genomovar II isolates suggested that chemotaxis could be associated with virulence [63]. LaFrentz and Klesius developed a culture independent method to quantify the chemotactic response of *F. columnare* to skin mucus using blind-well chemotaxis chambers [71] generating similar results as stated above [63]. At least three carbohydrate-binding receptors (D-mannose, D-glucose and *N*-acetyl-D-glucosamine) associated with the capsule of *F. columnare* might be involved in chemotactic responses [62]. The gliding motility gene *gld* H was found to be significantly ($P < 0.001$) upregulated in *F. columnare* as soon as five minutes post-exposure to catfish mucus. When pretreated with D-mannose, there was no upregulation of gliding motility genes [62].

The ability to adhere is a prerequisite for the successful colonization of the host tissue. Decostere et al. performed a series of both in vivo and in vitro experiments and found that a highly virulent strain adhered more readily to the gill tissue of black mollies (*Poecilia reticulata*) than did the low virulence strain [11, 12, 42]. This research group underscored that the adhesion of *F. columnare* to the gill tissue constitutes an important step in the pathogenesis of columnaris disease. Bader et al. adopted an adhesion defective mutant of *F. columnare* in immersion challenge trials and found that the mortality was reduced with 75% and occurred 24 h later compared to the strains that still possessed the adhesion capacities, confirming the findings of the former research group [48]. In channel catfish genomovar II is considered to be more virulent than genomovar I [51, 72]. Challenge of rainbow trout with genomovar I and II isolates of *F. columnare* demonstrated a difference in the cumulative percent mortalities (CPM), with the genomovar II isolates inducing significantly higher CPM [73]. Olivares-Fuster et al. compared adhesion of *F. columnare* genomovar I and II strains to the skin and gill of channel catfish and the gill of zebrafish (*Danio rerio*) [43]. At 0.5 h post-challenge, both strains adhered to the gill of channel catfish at comparable levels, but significant differences in adhesion were found in later datapoints over time. They concluded that particular strains of *F. columnare* exhibit different levels of adhesion to their fish hosts and that adhesion to fish tissues is not sufficient to cause columnaris disease. The same statement was previously made by Klesius et al. [63].

Kunttu et al. have shown that colony morphology affects the adhesion capacity of *F. columnare* to polystyrene and put forward the hypothesis that there is a link between virulence and rhizoid colony morphology [49, 74]. The formation of different colony morphologies could be caused by changes in the cell surface components of the bacteria. It was also found that *F. columnare* changes colony morphology during experimental long-term storage in lake water, indicating the importance of different colony morphologies to bacterial survival [64, 74,75]. Arias et al. did note that long-term starvation appears to decrease cell fitness and resulted in loss of virulence [75]. However, whether adhesion impairment was the cause of the observed virulence loss was not investigated.

F. columnare produces two types of mucus. The first one is an acidic polysaccharide and is made visible by ruthenium red staining. Another type of mucus is a basic, partially acetylated polygalactosamine, which cannot be stained with ruthenium red. Pate and Ordal described a capsular material which coated the surface of the bacterial cell that could be stained with ruthenium red [70]. They stated that the ruthenium red-positive material was probably an acid mucopolysaccharide that might be involved in the adhesive properties of the cells. The exact role of the mucopolysaccharides was however not further illuminated in this study. Decostere et al. demonstrated that the adherence capabilities of a highly virulent *F. columnare* strain to the gill were significantly reduced following treatment of the bacteria with sodium metaperiodate or incubating them with D-glucose, *N*-acetyl-D-glucosamine, D-galactose and D-sucrose [11]. Treatment with pronase or trypsin did not cause any significant inhibition of adhesion [11]. The same research group noted that the highly virulent strain had a thick capsule with a regular and dense appearance, whereas the capsule of the low virulent strain was much thinner. This made them to speculate that a lectin-like carbohydrate substance incorporated in the capsule might be partially responsible for the adhesion to the gill tissue.

Sun et al. conducted the first transcriptomic profiling of host responses to columnaris disease following an experimental immersion challenge by using illumina-based RNA-sequencing expression profiling [76]. A rhamnose-binding lectin (RBL) was detected as by far the most highly upregulated gene observed in their differentially expressed set, with expression increasing 105-fold by four hours following infection. This upregulation dramatically decreased at the later verified timepoints (24 h and 48 h), suggesting the importance of this gene during early infection events [76]. Immunohistochemical staining with antisera against an RBL in rainbow trout revealed the presence of these RBLs in mucous cells of the gill and in various cells related to innate immunity [77]. This expression pattern is consistent with the finding of *F. columnare* bacterial cell aggregates around goblet cells in common carp [11] and catfish [43] following experimental challenge with a highly virulent strain [76]. Beck et al. identified two distinct catfish families with differential susceptibilities to columnaris disease, with the one family completely resistant and the other susceptible [78]. In the susceptible family, an acute and robust upregulation in catfish RBL was observed following challenge that persisted for at least 24 h. After exposure of the catfish to different doses of the putative RBL ligands L-rhamnose and D-galactose, these sugars were found to protect the fish against columnaris disease, likely through competition with *F. columnare* binding of host RBL. Moreover, RBL expression was upregulated in fish fasted for 7 d (as compared to fish fed to satiation daily), yet expression levels returned to those of satiated fish within 4 h after re-feeding. These findings highlight putative roles for RBL in the context of columnaris disease and reveal new aspects linking RBL regulation to feed availability [79]. Further studies are needed to further pinpoint the (various) factor(s) responsible for mediating adhesion to the fish tissue.

The adhesion to the fish tissue was shown to be impacted by various environmental parameters. Using a gill perfusion model [61, 80], Decostere et al. noted that the adhesion of a highly virulent strain to the gill tissue was enhanced by a number of factors, including immersion of the gill in divalent ion water, the presence of nitrite or organic matter, and high temperatures [61]. The positive effect of high temperatures on the adhesion has also been demonstrated by Kunttu et al. [49]. Adhesion seems to decrease in vitro as salinity goes up [79]. Expression of adhesins by bacteria is regulated on two different levels. One is directed by environmental sensing and transcription of specific regulatory elements, and the other is a random switching on and off of adhesion genes by submission of the bacterial population to unpredictable environmental conditions [81]. If and how both phenomena are established in *F. columnare* remains to be elucidated.

An aggregative adhesion pattern of a highly virulent *F. columnare* strain onto gill tissue is a distinct feature in both in vivo and organ culture experiments [35, 42, 53, 61]. This results in an irregular gill surface covered by a thick mat consisting of numerous clumps of *F. columnare* bacterial cells, most likely impeding oxygen uptake and causing death of the fish. These microcolonies are not observed when a low virulence strain is used. Biofilm formation capacity was demonstrated in vitro for one *F. columnare* strain when exposed to mucus [67]. These features point towards biofilm formation potentially being an important stage in the pathogenicity of *F. columnare*, warranting its further investigation.

The developmental switch to the biofilm state is commonly regulated by quorum sensing. Quorum sensing (QS) is a mechanism of gene regulation in which bacteria coordinate the expression of certain genes in response to the presence or absence of small signal molecules. In many Gram-negative bacteria, the signal molecule is an *N*-acylhomoserine lactone (AHL) [82]. Additionally, a signaling molecule known as autoinducer-2 (AI-2) may also be employed [83]. Wagner-Döbler et al. found short-chain AHL-type activity in *Flavobacterium*sp., but no AHL-presence could be confirmed using gas chromatography-mass spectrometry (GS-MS) [84]. Using light chromatography-mass spectrometry (LC-MS),

Romero et al. described the presence of short-type AHL activity in the culture media of nine *Tenacibaculum maritimum* strains, biofilm-forming members of the phylum *Bacteroidetes*, formerly referred to as "*Cytophaga-Flavobacterium-Bacteroides*" group [82]. To our knowledge, so far, QS by AHL or AI-2 has not yet been demonstrated in *F. columnare*.

Exotoxins, Bacteriocins and Endotoxins

It is known that polysaccharide degradation in combination with the secretion of various extracellular enzymes participate in the destruction of skin, muscle and gill tissue [1], enhancing pathogenicity. In culture, *F. columnare* produces an enzyme that degrades chondroitin sulfates A and C and hyaluronic acid, the complex polysaccharides of connective tissue. This so-called chondroitin AC lyase acts specifically on a group of acidic mucopolysaccharides found primarily in animal connective tissue [40]. No correlations were found between host origin, geographic distribution, and amount of enzyme produced by different isolates [85]. AC lyase is alleged to play a role in the virulence of *F. columnare*[49, 65]. Though high AC lyase activity solely would not be enough to induce virulence in *F. columnare* strains, both high AC lyase activity and gliding motility of the bacteria would be needed for *F. columnare* to be virulent [49]. Proteases also contribute to damaging the tissue or enhancing invasive processes [40]. Newton et al. isolated and partially characterized proteases of 23 isolates of *Flavobacterium columnare* derived primarily from channel catfish raised in the southeastern United States [86]. The bacterial isolates were divided into two groups according to the apparent molecular masses of proteases after zymographic resolution by non-reducing, non-denaturing sodium dodecyl sulfate–polyacrylamide gel electrophoresis (SDS–PAGE) with gelatin as the protease substrate. Isolates of group one, produced two proteases with apparent molecular masses of 53.5 and 58 kilodaltons. The isolates of group two, revealed three proteases with apparent molecular masses of 44.5, 48 and 59.5 kilodaltons. All isolates degraded gelatin and casein. Seven out of 23 isolates degraded elastin. More protease was produced in a medium with low nutrients and salts than in media with higher concentrations of nutrients. Moreover, a sharp increase was seen in protease production during the first 24 h of incubation and these levels dropped only slightly in the remaining days of the experiment [86].

Bacteria not only need to enter the host tissue, they also need to eliminate competitive bacteria. Different strains of *F. columnare* release specific, non-transmissible, bactericidal substances equivalent to colicins of*Escherichia coli* into the environment to reduce competition from other bacterial strains [87]. It was postulated that cells of *F. columnare* also possess multiple specific receptors for the bacteriocins, and, consequently, the cells are sensitive only to those bacteriocins for which the cell possesses receptors [87]. Antagonism of*Pseudomonas sp.* MT5 against *F. columnare* bacteria was found to be very strong in agar assays [88]. However, antagonistic baths of the *Pseudomonas* bacterial strain could not yet prevent nor treat a *F. columnare* infection following experimental challenge [23].

Differences in LPS composition between highly and low virulent strains of *F. columnare* retrieved from channel catfish have been detected [89]. Analysis of LPS by immunoblotting revealed that an avirulent mutant of a *F. columnare* isolate lacked the high molecular weight components of LPS present in virulent isolates. Based on these differences of LPS and total protein profiles, the research group was able to discriminate the attenuated mutant from other *F. columnare* strains [89]. Kunttu et al. determined LPS-profiles of different colony morphology variants of *F. columnare* in rainbow trout [49]. Colony morphology variants of the same strain produced a similar single LPS band. However, there were size differences between different strains. Both research groups used different LPS extraction and detection methods, rendering comparison of the obtained results impossible [49].

Interaction with the Fish Immune System

When encountering an infection, the first system activated is the innate immune system. It is assumed that the surface mucus layer, as a first physical-immunological barrier, plays an important adhesive role and that it is part of the innate host resistance of fish to disease [43]. Antibacterial characteristics of the fish mucus against*F. columnare* have been demonstrated. In an experimental challenge trial of cutaneous columnaris disease in koi carp [45], lesions were noted on locations where mucus had been removed. After incubating *F. columnare*inoculated agar plates, a lower number of colonies was counted on plates to which mucus was added. Fluorescent microscopy of a stain-based bacterial viability assay also revealed a higher number of dead bacteria in *F. columnare* cultured with mucus [45]. Staroscik and Nelson compared growth, biofilm formation, extracellular protease production and changes in protein expression of a highly virulent *F. columnare* strain cultured in media supplemented with juvenile Atlantic salmon (*Salmo salar* L.) skin mucus with the same media without mucus [67]. Interestingly, and in contrast to the reasoning resulting from the above mentioned studies, salmon surface mucus promoted the growth of *F. columnare*, induced the bacteria to grow as a biofilm and increased extracellular protease activity [67]. This might indicate that skin mucus of different fish species responds differently to the bacteria, or that it is the *F. columnare* strain which is critical in determining the antibacterial capacities of the mucus.

The fish skin in itself also forms a barrier against pathogens. Aranishi et al. immersed Japanese eel (*Anguilla japonica*) in a *F. columnare* suspension to monitor dermal nonspecific stress responses [90]. They found that cathepsins B and L activities in the infected fishes increased more than 1.5-fold over their initial values over a 48 h period, along with a 4.5-fold increase in bacteriolytic activity. These cathepsins likely participate in bacteriolysis associated with Japanese eel skin and their activities may represent an important nonspecific response of eels [90].

Some studies reveal that *F. columnare* might be able to avoid parts of the immune system. In a study by Ourth and Bachinski, the catfish alternative complement pathway (ACP) was inhibited by large amounts of sialic acid contained by Gram-negative bacterial pathogens, including *F. columnare*[10]. Sialic acid seemed to be the determining factor for the pathogenicity of *F. columnare*, as very little or no bactericidal activity was produced against this bacterium by the catfish ACP. Furthermore, the latter greatly increased after removal of sialic acid with neuraminidase [10]. Another recurrent feature in *F. columnare* infections, is the lack of an inflammatory response as observed upon inspecting affected tissues microscopically. This resulted in the hypothesis that *F. columnare* triggers the endogenous programmed cell death machinery of immune cells to evade the immune system. Do Vale et al. have already proven that apoptosis can be a very powerful pathogenic strategy by inducing lysis of phagocytic cells [91]. Sun et al. speculated that negative regulation of one of the central innate immune signaling pathways NF-κB, may be the result of immune evasion or manipulation by *F. columnare* via secreted toxins [76]. Furthermore, high levels of inducible nitric oxide synthases (iNOS), apoptotic-promoting interferon (IFN) and other members of oxidative stress responses and apoptotic pathways, such as caspase 8 and G3BP1 (Rasputin), were observed following a *F. columnare* infection. Rasputin plays an inhibitory role in negative regulation of apoptosis and was highly downregulated at all examined time-points [76]. This could explain why hardly any inflammatory cells appear in early infections with *F. columnare* as was described by Morrison et al. [36].

Ourth and Wilson demonstrated that *F. columnare* is resistant to the bactericidal action of serum of non-immunized catfish via the ACP [92]. Ourth and Bachinski confirmed this finding [93], but also demonstrated that the classical, antibody-mediated complement pathway is highly effective in killing *F. columnare*[10]. The virulence of *F. columnare* can also be influenced by transferrin [94]. The survival time of fish experimentally challenged with *F. columnare* by intraperitoneal injection with iron-free human transferrin (Sigma) was reduced when iron was injected prior to exposure. The effect of iron was only evident in one of the two strains examined when the challenge was delivered via immersion. These results indicate that iron depletion may limit the virulence of *F. columnare* more in systemic infections than in external infections. This hypothesis is supported by data indicating that administration of transferrin prior to challenge increased survival after challenge by injection but had little or no effect on bath-challenged fish [94]. However, the results found were not consistent for all strains tested and no statistical data were presented.

Several immunization experiments adopting different administration routes, have proven that fish can be protected from subsequent *F. columnare* infections by activating the adaptive immune system [20, 22, 95, 96,97, 98]. High agglutinin titers and good protection were obtained in trout following subcutaneous or intraperitoneal injection with heat-killed *F. columnare* cells [22]. Schachte and Mora obtained a high agglutinin titer in channel catfish by intramuscular injection of heat-inactivated cells of the pathogen, but the actual level of protection was not examined [99]. Becker and Fujihara reported that rainbow trout injected with heat-killed cells produced an agglutinating titer of 1:5120, and that 60-70% of the trout later survived an injection of 10^6live *F. columnare* cells [96]. Tilapia (*Oreochromis niloticus* (L.)) could mount a significant humoral response in plasma and cutaneous mucus to *F. columnare* after intraperitoneal immunization with formalin-killed sonicated cells in Freund's complete adjuvant [97]. Protection levels were not investigated in the latter study. Protection was obtained after oral immunization with heat-killed or formalin-killed cells of *F. columnare* in the fish feed of three-month-old coho salmon (*Oncorhynchus kisutch*) [22, 100]. However, the protection as observed in these studies did not go hand-in-hand with high agglutinin titers. Bath-immunization with a bacterin was shown to protect carp against experimental challenge but antibodies against *F. columnare* were not detected in sera from immunized fish [101]. Song's work with bacterins demonstrated unequivocally that immersion vaccination could result in high levels of protection, but field test results were inconsistent [102]. Song was able to demonstrate that there was cross-protection between isolates and that there may be a common protective antigen among the strains tested [102]. Polyvalent vaccines have also been tested. Using intraperitoneal injections of a combination of formalin-killed *F. columnare*, *Aeromonas salmonicida* and*Yersinia ruckeri* antigens, interference from *A. salmonicida* antigen was shown to suppress responses to two other antigens [100]. Commercially available oral and bathing vaccines have been successfully tested in largemouth bass (*Micropterus salmoides*) fry and salmon, respectively [95, 103]. Vaccination trials are further elaborated on below.

IMPORTANCE OF ENVIRONMENTAL FACTORS

Karvonen et al. described the effect of global warming on the prevalence of different fish parasites and bacteria [104]. *F. columnare* could be one of the many taking advantage of this phenomenon. Indeed, transmission of columnaris disease is more efficient in higher temperatures [30, 105]. Holt et al. found that when steelhead trout (*Salmo gairdneri*) or coho salmon (*0. kisutch*) experimentally infected with *F. columnare*were held in water at 12 to 20°C, mortality increased with temperature [106]. As stated above, adhesion to gill tissue of a highly virulent *F. columnare* strain is enhanced at increased temperature [61] and chondroitin AC lyase activity of this pathogen increases along with the temperature [107]. The influence of rearing density and water temperature in rainbow trout was studied by Suomalaien et al. [105]. Normal rearing densities with high temperatures (23°C) proved to increase both transmission rate of columnaris disease and mortality in the fish. Normal densities at low temperatures (18°C) did not affect mortality, but increased the transmission rate of columnaris disease [105].

Columnaris disease is furthermore influenced by water quality. Decostere et al. observed significantly higher bacterial titers on the gills when organic matter or nitrite were added to an organ bath when performing ex vivo trials with *F. columnare*[61]. They discussed that organic matter could concentrate nutrients to feed the bacteria and that degrading enzymes could be kept in close contact with the host tissue. Bacterial titers were furthermore markedly lower in gills placed in an organ bath with distilled water with or without 0.03% NaCl compared to the titers of gills suspended in Ringer solution or in formulated water containing divalent ions (magnesium and calcium) [61]. Morris et al. noted that the survival of the fish exposed to *F. columnare*significantly increased as unionized ammonia concentrations increased [108]. These results suggest that complex interactions can complicate prediction of the responses of fish to concurrent chemical stressors and bacterial pathogens [108]. Bandilla et al. described that co-infections of ectoparasites with *F. columnare*increased the susceptibility of rainbow trout to the bacterial pathogen [109]. Compared with single infections, the mortality was significantly higher and the onset of disease condition occurred earlier in fish which were concomitantly infected by the parasite *Argulus coregoni* and *F. columnare*[109].

EXPERIMENTAL TRIALS/PATHOGENICITY TESTS

A reliable and reproducible experimental infection model is crucial for studying bacterium-host interactions and evaluating the efficacy of both curative and preventive measures. Most researchers recognize, in experimental challenges with *F. columnare,* the fine and delicate balance and complex interplay between the bacterial cells, fish and environment in the successful reproduction of columnaris disease.

Hitherto, columnaris disease has been reproduced experimentally in black mollies (*Poecilia sphenops*) [12], channel catfish [43, 44, 110, 111], eel [94], golden shiner (*Notemigonus crysoleucas*) [112], koi carp [45], rainbow trout [49, 65, 105, 113], tilapia [94] and zebrafish [43, 44].

In the various experimental infection trials, several ways to both cultivate and harvest the bacterial cells for inoculation were adopted. Different media were used to cultivate the bacteria, including Hsu-Shotts or modified Hsu-Shotts medium [79, 114], Ordal's medium [115], Anacker and Ordal's medium also referred to as "Cytophaga medium" [94], *F. columnare* growth medium broth [115] and Shieh or modified Shieh medium [12,23, 43, 44, 45, 105]. The bacteria were mostly grown on a shaker [43, 44, 45, 75, 109, 110, 111]. Temperatures and incubation times of the bacteria varied from 21°C to up to 30°C and from 24 to 48 h, respectively.

To reproduce columnaris disease experimentally, largely two inoculation routes were adopted, viz. bath (immersion) and injection. With regard to immersion challenges, mostly full-grown broth cultures were added to the water with water temperatures of the immersion water varying from 18°C to 30°C [12, 23, 42, 43, 44,111]. Only in a minority of cases were the bacterial cells first harvested through centrifugation and consequently resuspended before being added to the inoculation bath [33, 45, 114]. Times during which the fish were bath exposed varied from 15 min [13, 72, 105] over 30 min [12, 23, 43, 44, 45, 109] to 45 min [42], one hour [79, 94] and longer [110, 111]. When the bacterial cells were administered through injection, then they were retrieved from centrifuged broth cultures following discard of the supernatant [12] or immediately taken from a full-grown broth culture without primary centrifugation [13, 58, 94]. A third inoculation route consisted of adding dead fish that were first injected intraperitoneally with a virulent *F. columnare* strain, to the aquarium water in which fish were immersed [20]. In some researches, the successful reproduction of columnaris disease depended on or the severity of the elicited disease increased by abrading the skin or gills of the fish [45, 94, 105] whereas in another study, clipping the fin had no effect on success after immersion challenge with *F. columnare*[12].

The infection route, which determines the way the bacterial cells are to be grown/collected, also defines the disease producing capacity of the adopted strains and the disease picture they elicit. Indeed, especially for low virulent strains, higher morbidity and mortality rates were noted after injection of fish as compared to inoculation through immersion [12]. Pacha and Ordal reported a similar finding [34]. They stated that contact with highly virulent strains induced infection and disease contraction more than intramuscular injection, whereas injection of low virulent bacteria more readily induced infection and disease contraction than did contact [34].

Environmental conditions can also highly influence morbidity and mortality rates during a challenge trial. To illustrate the dramatic effects of water temperature on the level of mortalities, the investigation of Holt et al. is particularly relevant [106]. This team challenged steelhead trout (*Oncorhynchus mykiss*), Chinook salmon (*O. tshawytscha*) and coho salmon with *F. columnare*, via the water-borne route. At a water temperature of 9.4°C, there were no mortalities attributable to *F. columnare*. By increasing the temperature to 12.2°C, mortality became as high as 4-20%; at 20.5°C, all the steelhead trout and coho salmon and 70% of the Chinook salmon died [106]. High (23°C) rearing temperatures also increased mortality significantly in rainbow trout compared to lower (18°C) temperatures [105]. Fish density at fish farms is also a key player in influencing mortality in an outbreak of columnaris disease as mortality rates started earlier and remained higher when fish were stocked at high densities [105]. Water flow is another important factor since it acts as a determining factor with regard to the contact time between the possibly present bacteria and the host tissue. High mortality rates were observed in elvers kept in standing water, while in aquaria with running water, mortality in elvers was reduced by half [94].

CONTROL

Preventive Measures

Management plays a key role in the prevention of the disease. Cunningham et al. showed that some commonly recorded production variables (feed consumption, pond depth, ammonia levels and stocking events) were associated with columnaris disease outbreaks and, if monitored, could help identify "at risk" ponds prior to disease outbreaks [116]. Suomalainen et al. pointed out that reduction of fish density could be used in the prevention of columnaris disease especially if water temperature is high [105]. As lower rearing density can also decrease the transmission of ectoparasites and penetrating endoparasites, it could be an efficient tool in ecological disease management as a whole [105]. High nitrite levels and organic load can stimulate the adherence capacity of *F. columnare*[61], and therefore it is important to control these parameters as well. Furthermore, water treatment could aid in averting a bacterial outbreak. Conrad et al. reported that ozone treatment of water significantly reduced the numbers of added *F. columnare*, which could be a practical method of prevention [117]. Salt and acidic bath treatments could be used to disinfect water contaminated by *F. columnare*[23]. An in vivo immersion challenge of *F. columnare* in channel catfish and goldfish (*Carassius auratus* L.) revealed decreasing mortality as salinity goes up, with significantly lower and no mortalities when salinity reaches values of 1.0‰ and between 3 and 9‰, respectively [79]. If the fish can be adapted to salt levels of at least 1.0‰, this method could be used as a possible preventive measure in columnaris disease. Shoemaker et al. suggested that in the absence of natural food, juvenile channel catfish should be fed at least once every other day to apparent satiation in order to maintain normal physiological function and improve resistance to *F. columnare*, since deprivation reduced innate resistance of catfish to columnaris disease [118].

Besides optimizing and adjusting management practices, chemical agents can also be adopted as a preventive approach. Davis concluded that the development or intensification of columnaris disease could be prevented by treating the fish for 20 min in a copper sulfate ($CuSO_4$) bath at 37 mg/L (1:30 000) or by adding copper sulfate to pond water at 0.5 mg/L [4]. Dipping the fish one at a time in a 1:2000 copper-sulfate for one to two minutes was also proven to be effective in the prevention of the disease. Rogers suggested the addition of potassium permanganate ($KMnO_4$) to the water at 2 mg/L [119]. Darwish et al. also confirmed the prophylactic value of $KMnO_4$ at doses around 2 mg/L [120]. Prophylactic treatment of channel catfish with 15 mg/L chloramine-T reduced fish mortality from a *F. columnare* infection from 84–100% to 6–14% [121]. Thomas-Jinu and Goodwin demonstrated the efficacy of prophylactically given oxytetracycline against mortality in channel catfish and also reported zero mortality for the combination of sulphadimethoxine and ormetoprim in feed prior to bacterial challenge with four highly virulent strains of *F. columnare*[111].

Another method to prevent columnaris disease is through vaccination. Although vaccination trials have not always been successful, success rates have increased as knowledge on fish immunity and its role in the defense against bacterial diseases continues to expand. Bath immunization with a bacterin was shown to protect carp against experimental challenge, but no agglutinin could be found in sera from immunized fish [101]. Immersion of channel catfish in a bacterin, when performed each year, induced a significant decrease in mortality compared to unvaccinated fish [98]. Fujihara and Nakatani obtained protection against columnaris disease in 3-month-old coho salmon by oral immunization with heat-killed cells of *F. columnare* incorporated into fish feed [22]. Ransom proved that prolonged feeding (over three months) of formalin-killed bacteria provided high levels of protection [100]. Ourth and Bachinski proposed that strains containing sialic acid could serve as potential vaccine strains for columnaris disease [10]. As stated above, immunization with formalin-killed sonicated cells in Freund's complete adjuvant injected intraperitoneally in tilapia resulted in a significant systemic humoral response within two weeks and antibody levels almost tripled following secondary immunization [97]. At 10 weeks postimmunization, the mean antibody titer remained significantly elevated. Antibodies were also observed in cutaneous mucus of these fish at six and eight weeks postimmunization [97]. An attenuated immersion vaccine currently is registered for the use in channel catfish in the USA [20]. Fry between 10 to 48 days post hatch that were vaccinated through immersion achieved a relative percent survival (RPS) between 57 and 94% following *F. columnare* challenge. This vaccine was also proven to be efficient in largemouth bass fry resulting in RPS values between 74 and 94%, depending on the vaccine dose [20]. Bebak et al. tested a commercial oral vaccine in largemouth bass fry [95]. Vaccinated fish had a 43% lower risk of death by *F. columnare* during the field trial [95]. An immersion vaccine consisting of a bacterin of *F. columnare*was also brought to the market in the USA as an aid in the prevention of columnaris disease in healthy salmonids of over three grams [103].

Probiotics appear to be a promising way in the prevention of different bacterial diseases in aquaculture [122]. Boutin et al. isolated different strains of commensal bacteria from the skin mucus of unstressed brook charr (*Salvelinus fontinalis*) which in vitro revealed antagonistic effects against *F. columnare*[123]. The strains were mixed and used to treat columnaris disease in an in vivo experiment. This resulted in a significant decrease of mortality indicating the potential use of these probiotic candidates in the efficient and durable management of columnaris disease. The immunostimulants β-glucan and β-hydroxy-β-methylbutyrate raised the levels of immune function parameters, but did not improve survival in challenge trials with *F. columnare* at any concentration of the stimulants used [113]. The research group of Sink et al. demonstrated that mortality rates in golden shiners fed high-fat diets with a dairy-yeast prebiotic were significantly lower after a challenge trial with *F. columnare*[112].

Recent studies have demonstrated genetic variation in resistance towards *F. columnare*[73, 78, 124]. Arias et al. presented experimental data on the susceptibility to columnaris disease of hybrid catfish (female channel catfish × male blue catfish (*I. furcatus*)) (C×B) [124]. Under experimental conditions, C×B hybrids were significantly more resistant to columnaris disease caused by a highly virulent strain of *F. columnare* belonging to genomovar II compared to channel catfish and blue catfish. Beck et al. also found one of the two investigated catfish families to be completely resistant towards *F. columnare* resulting in no mortality after inoculation with the bacteria [78]. These *F. columnare* resistant families could be of great financial importance in the catfish industry. Interestingly, LaFrentz et al. found that the two families that exhibited the highest CPM after *F. columnare* challenges, had the lowest CPM following *E. ictaluri* challenge, the latter being the most prevalent bacterium to cause disease and mortality in the catfish industry [73]. Further research on larger numbers of families is needed to determine whether there is any genetic correlation between resistance to *E. ictaluri* and resistance to *F. columnare*, the two leading bacterial diseases in the catfish industry [125].

Curative Approach

Treatment of columnaris disease using antimicrobial agents has known different success rates. External treatments are possible only in early stages of the disease, when the infection is still superficial [6]. Drugs which have been used effectively in bath therapies are chloramphenicol [126], nifurpirinol [127, 128], nifurprazine [129, 130] and oxolinic acid [131, 132]. If the disease is in an advanced stage and/or signs of septicaemia are observed, it is necessary to administer antimicrobials in the feed. Oxytetracycline given orally for up to 10 days proved effective in early as well as advanced outbreaks of columnaris disease in Pacific salmon (*Salmo salar*) [6, 133]. Lack of success of orally administered oxytetracycline has also been reported [134]. Sulfonamides, such as sulfamerazine and sulfamethazine, can be used orally but would be less effective than other drugs [6, 135]. Nitrofuran can also be administered orally for 3 to 5 days [6, 96, 130]. Gaunt et al. demonstrated the efficacy of florfenicol in the feed against columnaris disease in channel catfish [136]. Darwish et al. also illustrated the clear benefit of florfenicol against a mixed infection of *A. hydrophila* and *F. columnare* in Sunshine bass (hybrid striped bass, *Morone chrysops* female × *Morone saxatilis* male) [115]. The excessive use of antimicrobial agents to withstand *F. columnare* has its negative attributes though [20]. These include possible allergic reactions elicited in the user after food contact [137]. Potential impacts on human health resulting from the emergence of drug-resistant bacteria and the associated risk of transfer of these resistant traits to the environment and human-associated bacteria are also a major concern [137]. Declercq et al. demonstrated the in vitro multiple resistance of *F. columnare* strains originating from ornamental fish toward several clinically important antibiotics, such as quinolones and tetracyclines [138]. The results obtained in this study appeal for less prudent use of antimicrobials especially in the ornamental fish industry and therefore urges to limit their use and to focus on the development of alternative curative and preventive measures against columnaris disease.

Besides resorting to antimicrobial agents, chemicals have also been adopted in the curative treatment of columnaris disease. In a study by Thomas-Jinu and Goodwin, the herbicide Diquat® (Zeneca Agricultural Products, Wilmington, DE, USA) was shown to significantly reduce channel catfish mortalities to zero percent after challenge with *F. columnare*[111]. The herbicide has also proven to be effective in the treatment of columnaris disease in salmonids [6, 139]. Copper sulfate [4] and potassium permanganate [119] are among the older chemicals used for treatment and prevention of columnaris disease in pond fishes. The organic load in water affects the efficacy of potassium permanganate, but methods are available to estimate that organic load and compensate by adjusting the level of the chemical [6]. Darwish et al. suggested that copper sulfate has clear therapeutic value against *F. columnare* infections in channel catfish when treated in an ultralow flow-through system during 4 h [110]. Thomas-Jinu and Goodwin on the other hand proved the inefficacy of this same chemical against columnaris disease, which might be due to the advanced stage of the experimental infection at the time of treatment [111]. An in vitro assay of commercial products containing peracetic acid was proven to be effective against *F. columnare* infection [140]. Laanto et al. reported some *Flavobacterium* sp. phage lysates to inhibit growth or lyse the bacterial cultures [141]. The authors recommended that the causative agent of this strong inhibition or lysis should be studied further for the possibility of developing antimicrobial agents. Prasad et al. also described the successful use of *F. columnare* phage FCP1 to combat columnaris disease in walking catfish (*Claries batrachus*) [142]. Phage treatment led to disappearance of gross symptoms, resulted in a negative bacteriological test, a detectable phage, and 100% survival in experimentally infected *C. batrachus*. The result of this study opens new perspectives for the treatment of columnaris disease elicited by antimicrobial resistant *F. columnare* strains [142].

CONCLUSION

Despite the worldwide importance of columnaris disease and the multitude of studies focusing on its causative agent, only snippets of the bacterium-host interactions have hitherto been exposed. This results in major knowledge gaps still prevailing on amongst others how the pathogen is able to establish and maintain a grip on the skin and gill tissue and elicit disease and mortality. Methods for rapid identification of the bacterium in the environment, host tissue or in culture have indeed evolved rapidly and some techniques permit to trace even a few bacterial cells. However, the rationale for the difference in virulence among *F. columnare* strains remains to be unraveled. Full genome sequencing of the reference strain has been an important step towards gaining insight into the pathogenesis of columnaris disease. Efficient molecular manipulation systems could be of great importance to further explore this domain. Genetic knockouts could in effect illuminate virulence factors involved in the pathogenesis of this disease. Although various models to experimentally induce columnaris disease are available, there is a lack of a standardized experimental model hampering the comparison of retrieved data in between research groups and the drawing of consistent conclusions. Consistency in the cultivation protocol of the bacteria in terms of incubation temperature, infection dose, exposure time and infection route is desirable. Extrapolation of a standardized model from one to another fish species is not possible rendering it necessary to set-up an experimental model for each fish sensitive species. Besides further elucidating the pathogenesis of columnaris disease, the development of efficient curative and effective preventive measures also needs to continue. With the upcoming antimicrobial resistance, focus should be laid on the search for preventive measures. Management plans trying to optimize or adjust fish densities and control water quality parameters are a first critical step in controlling columnaris disease. Vaccines, chemotherapeutics and probiotics could possibly have a promising future in the prevention and control of the disease as some studies have already demonstrated. Another interesting envisaged method to mitigate columnaris disease is the creation of a fish breeding program introducing *F. columnare* resistant fish. Data generated through pathogenesis studies will allow for the further development and optimization of cost-effective and environmentally friendly prevention methods of columnaris disease in the various susceptible fish species.

ACKNOWLEDGEMENTS

The authors thank Prof. Paul Simoens of the Department of Morphology of the Faculty of Veterinary Medicine at Ghent University for critically reviewing this manuscript. Prof. Koen Chiers of the Department Bacteriology, Pathology and Avian Diseases of the Faculty of Veterinary Medicine at Ghent University is gratefully acknowledged for his expertise and help with the interpretation of the fish pathology. Prof. Maria Cornelissen and Leen Pieters of the Department of Anatomy, Embryology, Histology and Medical Physics of the Faculty of Medicine and Health Sciences at Ghent University are accredited for their cooperation. Furthermore, we also give thanks to Lobke De Bels, Jurgen De Craene, Bart De Pauw and Patrick Vervaet for their technical assistance. The financial support by the Special Research Grant (Bijzonder Onderzoeksfonds, BOF, grant number 01Z06210) of Ghent University, Belgium is gratefully acknowledged.

AUTHORS' CONTRIBUTIONS

AM Declercq: Gathering information and papers, writing the outlines, writing the paper; A Decostere: Proposing the subject, reviewing and editing the paper; F Haesebrouck: Critically reviewing the manuscript; W Van den Broeck: Interpretation of histological, SEM- and TEM- pictures, critically reviewing the manuscript, P Bossier: Critically reviewing the manuscript. All authors agreed on outlines and the final version of the paper.

REFERENCES

1. Bernardet JF, Bowman JP: The genus *Flavobacterium*. The Prokaryotes: A Handbook on the Biology of Bacteria: Volume 7: Proteobacteria: Delta and Epsilon Subclasses. Deeply Rooting Bacteria. Edited by: Dworkin M, Falkow S. 2006, New York: Springer Science+Business Media, LLC, 481-531.
2. Bernardet JF, Nakagawa Y, Holmes B: Proposed minimal standards for describing new taxa of the family*Flavobacteriaceae* and emended description of the family. Int J Syst Evol Microbiol. 2002, 52: 1049-1070. 10.1099/ijs.0.02136-0.
3. Bernardet JF, Segers P, Vancanneyt M, Berthe M, Kersters K, Vandamme P: Cutting a Gordian knot: emended classification and description of the genus *Flavobacterium*, emended description of the family *Flavobacteriaceae*, and proposal of *Flavobacterium hydatis* nom. nov. (basonym, *Cytophaga aquatilis* Strohl and Tait 1978). Int J Syst Evol Bacteriol. 1996, 46: 128-148. 10.1099/00207713-46-1-128.
4. Davis HS: A new bacterial disease in freshwater fishes. United States Bureau of Fisheries Bulletin. 1922, 38: 37-63.
5. Ordal EJ, Rucker RR: Pathogenic myxobacteria. PSEBM. 1944, 56: 15-18.
6. Bullock GL, Hsu TC, Shotts EB: Columnaris disease of fishes. USFWS Fish Disease Leaflet. 1986, 1986: 1-9.
7. Garnjobst L: *Cytophaga columnaris* (Davis) in pure culture: a myxobacterium pathogenic to fish. J Bacteriol. 1945, 49: 113-128.
8. Bernardet J-F, Grimont PAD: Deoxyribonucleic acid relatedness and phenotypic characterization of *Flexibacter columnaris* sp. nov., norn. rev., *Flexibacter psychrophilus* sp. nov., nom. rev., and *Flexibacter maritimus* Wakabayashi, Hikida, and Masumura 1986. IJSB. 1989, 39: 346-354.
9. Triyanto , Wakabayashi H: Genotypic diversity of strains of *Flavobacterium columnare* from diseased fishes. Fish Pathol. 1999, 34: 65-71. 10.3147/jsfp.34.65.
10. Ourth DD, Bachinski LM: Bacterial sialic acid modulates activation of the alternative complement pathway of channel catfish (*Ictalurus punctatus*). Dev Comp Immunol. 1987, 11: 551-564. 10.1016/0145-305X(87)90044-9.
11. Decostere A, Haesebrouck F, Van Driessche E, Charlier G, Ducatelle R: Characterization of the adhesion of*Flavobacterium columnare (Flexibacter columnaris)* to gill tissue. J Fish Dis. 1999, 22: 465-474. 10.1046/j.1365-2761.1999.00198.x.
12. Decostere A, Haesebrouck F, Devriese LA: Characterization of four *Flavobacterium columnare* (*Flexibacter columnaris*) strains isolated from tropical fish. Vet Microbiol. 1998, 62: 35-45. 10.1016/S0378-1135(98)00196-5.
13. Figueiredo HCP, Klesius PH, Arias CR, Evans J, Shoemaker CA, Pereira DJ, Peixoto MTD: Isolation and characterization of strains of *Flavobacterium columnare* from Brazil. J Fish Dis. 2005, 28: 199-204. 10.1111/j.1365-2761.2005.00616.x.
14. Morley NJ, Lewis JW: Consequences of an outbreak of columnaris disease (*Flavobacterium columnare*) to the helminth fauna of perch (*Perca fluviatilis*) in the Queen Mary reservoir, south-east England. J Helminthol. 2010, 84: 186-192. 10.1017/S0022149X09990459.
15. Řehulka J, Minařík B: Blood parameters in brook trout *Salvelinus fontinalis* (Mitchill, 1815), affected by columnaris disease. Aquac Res. 2007, 38: 1182-1197. 10.1111/j.1365-2109.2007.01786.x.
16. Soto E, Mauel MJ, Karsi A, Lawrence ML: Genetic and virulence characterization of *Flavobacterium columnare* from channel catfish (*Ictalurus punctatus*). J Appl Microbiol. 2008, 104: 1302-1310. 10.1111/j.1365-2672.2007.03632.x.
17. Suomalainen L-R, Bandilla M, Valtonen ET: Immunostimulants in prevention of columnaris disease of rainbow trout,*Oncorhynchus mykiss* (Walbaum). J Fish Dis. 2009, 32: 723-726. 10.1111/j.1365-2761.2009.01026.x.
18. Hawke JP, Thune RL: Systemic isolation and antimicrobial susceptibility of *Cytophaga columnaris* from commercially reared channel catfish. J Aquat Anim Health. 1992, 4: 109-113. 10.1577/1548-8667(1992)004<0109:SIAASO>2.3.CO;2.
19. Wagner BA, Wise DJ, Khoo LH, Terhune JS: The epidemiology of bacterial diseases in food-size channel catfish. J Aquat Anim Health. 2002, 14: 263-272. 10.1577/1548-8667(2002)014<0263:TEOBDI>2.0.CO;2.
20. Shoemaker CA, Klesius PH, Drennan JD, Evans J: Efficacy of a modified live *Flavobacterium columnare* vaccine in fish. Fish Shellfish Immun. 2011, 30: 304-308. 10.1016/j.fsi.2010.11.001.
21. Barker GA, Smith SN, Bromage NR: Effect of oxolinic acid on bacterial flora and hatching success rate of rainbow trout,*Oncorhynchus mykiss*, eggs. Aquaculture. 1990, 9: 205-222.
22. Fujihara MP, Nakatani RE: Antibody protection and immune responses of rainbow trout coho salmon to *Chondrococcus columnaris*. J Fish Res Bd Canada. 1971, 28: 1253-1258. 10.1139/f71-191.
23. Suomalainen L-R, Tiirola M, Valtonen ET: Treatment of columnaris disease of rainbow trout: low pH and salt as possible tools?. Dis Aquat Organ. 2005, 65: 115-120.
24. Kunttu HMT, Valtonen ET, Jokinen EI, Suomalainen L-R: Saprophytism of a fish pathogen as a transmission strategy. Epidemics. 2009, 1: 96-100. 10.1016/j.epidem.2009.04.003.
25. Fijan FJ: Antibiotic additives for isolation of *Chondrococcus columnaris* from fish. Appl Microbiol. 1968, 17: 333-334.
26. Chowdhury MBR, Wakabayashi H: Survival of four major bacterial fish pathogens in different types of

experimental water. Bangladesh J Microbiol. 1990, 7: 47-54.

27. Ross AJ, Smith CA: Effect of temperature on survival of *Aeromonas liquefaciens*, *Aeromonas salmonicida*,*Chondrococcus columnaris*, and *Pseudomonas fluorescens*. Progr Fish Cult. 1974, 36: 51-52. 10.1577/1548-8659(1974)36[51:EOTOSO]2.0.CO;2.

28. Kunttu HMT, Sundberg L-R, Pulkkinen K, Valtonen ET: Environment may be the source of *Flavobacterium columnare*outbreaks at fish farms. Environ Microbiol Rep. 2012, 4: 398-402. 10.1111/j.1758-2229.2012.00342.x.

29. Wakabayashi H: Columnaris disease. Bacterial diseases of fish. Edited by: Inglis V, Roberts RJ, Bromage NR. 1993, Oxford, England: Blackwell Scientific Publications, 23-39.

30. Pulkkinen K, Suomalainen L-R, Read AF, Ebert D, Rintamäki P, Valtonen ET: Intensive fish farming and the evolution of pathogen virulence: the case of columnaris disease in Finland. Proc Biol Sci. 2010, 277: 593-600. 10.1098/rspb.2009.1659.

31. Rucker RR, Earp BJ, Ordal EJ: Infectious diseases of Pacific salmon. Trans Am Fish Soc. 1953, 83: 297-312.

32. Pacha RE, Ordal EJ: Histopathology of experimental columnaris disease in young salmon. J Comp Path. 1967, 77: 419-423. 10.1016/0021-9975(67)90027-8.

33. Foscarini R: Induction and development of bacterial gill disease in the eel (*Anguilla japonica*) experimentally infected with *Flexibacter columnaris:* pathological changes in the gill vascular structure and in cardiac performance. Aquaculture. 1989, 78: 1-20. 10.1016/0044-8486(89)90002-1.

34. Pacha RE, Ordal EJ: Myxobacterial diseases of salmonids. A symposium on diseases of fishes and shellfishes. Edited by: American Fisheries Society special publication, Snieszko SF. 1970, 243-257. 5

35. Decostere A: *Flavobacterium columnare* infections in fish: the agent and its adhesion to the gill tissue. Verhandelingen - KAGB. 2002, 64: 421-430.

36. Morrison C, Cornick J, Shum G, Zwicker B: Microbiology and histopathology of "saddle-back" disease of underyearling Atlantic salmon, *Salmo salar*. J Bacteriol. 1981, 78: 225-230.

37. Decostere A, Haesebrouck F: Outbreak of columnaris disease in tropical aquarium fish. Vet Rec. 1999, 144: 23-24. 10.1136/vr.144.1.23.

38. Ferguson HW: Systemic Pathology of Fish. 2006, London: Scottian Press, 2

39. Michel C, Messiaen S, Bernardet J-F: Muscle infections in imported neon tetra, *Paracheirodon innesi* Myers: limited occurrence of microsporidia and predominance of severe forms of columnaris disease caused by an Asian genomovar of *Flavobacterium columnare*. J Fish Dis. 2002, 25: 253-263. 10.1046/j.1365-2761.2002.00364.x.

40. Dalsgaard I: Virulence mechanisms in *Cytophaga psychrophila* and other *Cytophaga*-like bacteria pathogenic for fish. Ann Rev Fish Dis. 1993, 3: 127-144.

41. Decostere A, Haesebrouck F, Devriese L: Development of a medium for the selective isolation of *Flavobacterium columnare* from diseased fish. J Clin Microbiol. 1997, 35: 322-324.

42. Decostere A, Haesebrouck F, Charlier G, Ducatelle R: The association of *Flavobacterium columnare* strains of high and low virulence with gill tissue of black mollies (*Poecilia sphenops*). Vet Microbiol. 1999, 67: 287-298. 10.1016/S0378-1135(99)00050-4.

43. Olivares-Fuster O, Bullard SA, McElwain A, Llosa MJ, Arias CA: Adhesion dynamics of *Flavobacterium columnare* to channel catfish *Ictalurus punctatus* and zebrafish *Danio rerio* after immersion challenge. Dis Aquat Organ. 2011, 96: 221-227. 10.3354/dao02371.

44. Bullard SA, McElwain A, Arias CR: Scanning electron microscopy of "saddleback" lesions associated with experimental infections of *Flavobacterium columnare* in channel catfish, *Ictalurus punctatus* (*Siluriformes: Ictaluridae*), and zebrafish,*Danio rerio* (*Cypriniformes: Cyprinidae*). J World Aquacult Soc. 2011, 42: 906-913. 10.1111/j.1749-7345.2011.00527.x.

45. Tripathi NK, Latimer KS, Gregory CR, Ritchie BW, Wooley RE, Walker RL: Development and evaluation of an experimental model of cutaneous columnaris disease in koi *Cyprinus carpio*. J Vet Diagn Invest. 2005, 17: 45-54. 10.1177/104063870501700109.

46. Shieh HS: Studies on the nutrition of a fish pathogen, *Flexibacter columnaris*. Microbios Letters. 1980, 13: 129-133.

47. Holt RA: Cytophaga psychrophila, the causative agent of bacterial cold-water disease in salmonid fish. 1988, Corvallis, OR: Oregon State University

48. Bader JA, Shoemaker CA, Klesius PH: Production, characterization and evaluation of virulence of an adhesion defective mutant of *Flavobacterium columnare* produced by β-lactam selection. Lett Appl Microbiol. 2005, 40: 123-127. 10.1111/j.1472-765X.2004.01641.x.

49. Kunttu HMT, Jokinen EI, Valtonen ET, Sundberg L-R: Virulent and nonvirulent *Flavobacterium columnare* colony morphologies: characterization of chondroitin AC lyase activity and adhesion to polystyrene. J Appl Microbiol. 2011, 111: 1319-1326. 10.1111/j.1365-2672.2011.05149.x.

50. Bernardet JF: "*Flexibacter columnaris*": first description in France and comparison with bacterial strains from other origins. Dis Aquat Organ. 1989, 6: 37-44.

51. Olivares-Fuster O, Baker JL, Terhune JS, Shoemaker CA, Klesius PH, Arias CR: Host-specific association between*Flavobacterium columnare* genomovars and fish species. Syst Appl Microbiol. 2007, 30: 624-633. 10.1016/j.syapm.2007.07.003.

52. Darwish AM, Ismaiel AA, Newton JC, Tang J: Identification of *Flavobacterium columnare* by a species-specific polymerase chain reaction and renaming of ATCC43622 strain to *Flavobacterium johnsoniae*. Mol Cell Probes. 2004, 18: 421-427. 10.1016/j.mcp.2004.07.002.

53. Welker TL, Shoemaker CA, Arias CR, Klesius PH: Transmission and detection of *Flavobacterium columnare* in channel catfish *Ictalurus punctatus*. Dis

Aquat Organ. 2005, 63: 129-138.

54. Shoemaker CA, Shelby RA, Klesius PH: Development of an indirect ELISA to detect humoral response to *Flavobacterium columnare* infection of channel catfish, *Ictalurus punctatus*. J Appl Aquaculture. 2003, 14: 43-52. 10.1300/J028v14n01_03.

55. Panangala VS, Shelby RA, Shoemaker CA, Klesius PH, Mitra A, Morrison EE: Immunofluorescent test for simultaneous detection of *Edwardsiella ictaluri* and *Flavobacterium columnare*. Dis Aquat Organ. 2006, 68: 197-207.

56. Speare DJ, Markham RJ, Despres B, Whitman K, MacNair N: Examination of gill from salmonids with bacterial gill disease using monoclonal antibody probes for *Flavobacterium branchiophilum* and *Cytophaga columnaris*. J Vet Diagn Invest. 1995, 7: 500-505. 10.1177/104063879500700413.

57. Yeh H-Y, Shoemaker CA, Klesius PH: Sensitive and rapid detection of *Flavobacterium columnare* in channel catfish*Ictalurus punctatus* by a loop-mediated isothermal amplification method. J Appl Microbiol. 2006, 100: 919-925. 10.1111/j.1365-2672.2006.02853.x.

58. Panangala VS, Shoemaker CA, Klesius PH: TaqMan real-time polymerase chain reaction assay for rapid detection of*Flavobacterium columnare*. Aquac Res. 2007, 38: 508-517. 10.1111/j.1365-2109.2007.01695.x.

59. Suzuki M, Rappe MS, Giovanni SJ: Kinetic bias in estimates of coastal picoplankton community structure obtained by measurements of small-subunit rRNA gene PCR amplicon length heterogeneity. Appl Environ Microbiol. 1998, 64: 4522-4529.

60. Suomalainen L-R, Tiirola MA, Valtonen ET: Effect of *Pseudomonas sp.* MT5 baths on *Flavobacterium columnare* infection in rainbow trout and on microbial diversity on fish skin and gills. Dis Aquat Organ. 2005, 63: 61-68.

61. Decostere A, Haesebrouck F, Turnbull JF, Charlier G: Influence of water quality and temperature on adhesion of high and low virulence *Flavobacterium columnare* strains to isolated gill arches. J Fish Dis. 1999, 22: 1-11. 10.1046/j.1365-2761.1999.00132.x.

62. Klesius PH, Pridgeon JW, Aksoy M: Chemotactic factors of *Flavobacterium columnare* to skin mucus of healthy channel catfish (*Ictalurus punctatus*). FEMS Microbiol Lett. 2010, 310: 145-151. 10.1111/j.1574-6968.2010.02060.x.

63. Klesius PH, Shoemaker CA, Evans JJ: *Flavobacterium columnare* chemotaxis to channel catfish mucus. FEMS Microbiol Lett. 2008, 288: 216-220. 10.1111/j.1574-6968.2008.01348.x.

64. Kunttu H: Characterizing the bacterial fish pathogen *Flavobacterium columnare* and some factors affecting its pathogenicity. Jyväskyla Stud Biol Environ Sci. 2010, 206: 1-69.

65. Suomalainen L-R, Tiirola M, Valtonen ET: Chondroitin AC lyase activity is related to virulence of fish pathogenic*Flavobacterium columnare*. J Fish Dis. 2006, 29: 757-763. 10.1111/j.1365-2761.2006.00771.x.

66. Li N, Zhang LQ, Zhang J, Liu ZX, Huang B, Zhang SH, Nie P: Type I restriction-modification system and its resistance in electroporation efficiency in *Flavobacterium columnare*. Vet Microbiol. 2012, 160: 61-68. 10.1016/j.vetmic.2012.04.045.

67. Staroscik AM, Nelson DR: The influence of salmon surface mucus on the growth of *Flavobacterium columnare*. J Fish Dis. 2008, 31: 59-69.

68. Zhang J, Zou H, Wang LF, Li N, Wang GT, Huang B, Nie P: Construction of two selective markers for integrative/conjugative plasmids in *Flavobacterium columnare*. Chin J Oceanol Limnol. 2012, 30: 269-278. 10.1007/s00343-012-1077-z.

69. Tekedar HC, Karsi A, Gillaspy AF, Dyer DW, Benton NR, Zaitshik J, Vamenta S, Banes MM, Gülsoy N, Aboko-Cole M, Waldbieser GC, Lawrence ML: Genome sequence of the fish pathogen *Flavobacterium columnare* ATCC 49512. J Bacteriol. 2012, 194: 2763-2764. 10.1128/JB.00281-12.

70. Pate JL, Ordal EJ: The fine structure of *Chondrococcus columnaris*. III. The surface layers of *Chondrococcus columnaris*. J Cell Biol. 1967, 35: 37-51. 10.1083/jcb.35.1.37.

71. LaFrentz BR, Klesius PH: Development of a culture independent method to characterize the chemotactic response of*Flavobacterium columnare* to fish mucus. J Microbiol Meth. 2009, 77: 37-40. 10.1016/j.mimet.2008.12.011.

72. Shoemaker CA, Olivares-Fuster O, Arias CR, Klesius PH: *Flavobacterium columnare* genomovar influences mortality in channel catfish (*Ictalurus punctatus*). Vet Microbiol. 2008, 127: 353-359. 10.1016/j.vetmic.2007.09.003.

73. LaFrentz BR, LaPatra SE, Shoemaker CA, Klesius PH: Reproducible challenge model to investigate the virulence of*Flavobacterium columnare* genomovars in rainbow trout *Oncorhynchus mykiss*. Dis Aquat Organ. 2012, 101: 115-122. 10.3354/dao02522.

74. Kunttu HMT, Suomalainen L-R, Jokinen EI, Valtonen ET: *Flavobacterium columnare* colony types: connection to adhesion and virulence?. Microb Pathog. 2009, 46: 21-27. 10.1016/j.micpath.2008.10.001.

75. Arias CR, LaFrentz S, Cai W, Olivares-Fuster O: Adaptive response to starvation in the fish pathogen *Flavobacterium columnare:* cell viability and ultrastructural changes. BMC Microbiol. 2012, 12: 266-276. 10.1186/1471-2180-12-266.

76. Sun F, Peatman E, Li C, Liu S, Jiang Y, Zhou Z, Liu Z: Transcriptomic signatures of attachment, NF-κB suppression and IFN stimulation in the catfish gill following columnaris bacterial infection. Dev Comp Immunol. 2012, 38: 169-180. 10.1016/j.dci.2012.05.006.

77. Tateno H, Yamaguchi T, Ogawa T, Muramoto K, Watanabe T, Kamiya H, Saneyoshi M: Immunohistochemical localization of rhamnose-binding lectins in the steelhead trout (*Oncorhynchus mykiss*). Dev Comp Immunol. 2002, 26: 543-550. 10.1016/S0145-305X(02)00007-1.

78. Beck BH, Farmer BD, Straus DL, Li C, Peatman E: Putative roles for a rhamnose binding lectin in

Flavobacterium columnare pathogenesis in channel catfish *Ictalurus punctatus*. Fish Shellfish Immun. 2012, 33: 1008-1015.

79. Altinok I, Grizzle JM: Effects of low salinities on *Flavobacterium columnare* infection of euryhaline and freshwater stenohaline fish. J Fish Dis. 2001, 24: 361-367. 10.1046/j.1365-2761.2001.00306.x.
80. Decostere A, Henckaerts K, Ducatelle R, Haesebrouck F: An alternative model to study the association of rainbow trout (*Oncorhynchus mykiss* L.) pathogens with the gill tissue. Lab Anim. 2002, 36: 396-402. 10.1258/002367702320389053.
81. Ofek I, Hasty DL, Doyle RJ: Adhesins as Bacterial Cell Surface Structures: General Concepts of Structure, Biogenesis, and Regulation. Bacterial adhesion to animal cells and tissues. Edited by: ASM Press. 2003, Washington D.C.: ASM Press, 77-81.
82. Romero M, Avendaño-Herrera R, Magariños B, Cámara M, Otero A: Acylhomoserine lactone production and degradation by the fish pathogen *Tenacibaculum maritinum*, a member of the *Cytophaga-Flavobacterium_ Bacteroides* (CFB) group. FEMS Microbiol Lett. 2009, 304: 131-139.
83. Camilli A, Bassler BL: Bacterial small-molecule signaling pathways. Science. 2006, 311: 1113-1116. 10.1126/science.1121357.
84. Wagner-Döbler I, Thiel V, Eberl L, Allgaier M, Bodor A, Meyer S, Ebner S, Henning A, Pukall R, Schulz S: Discovery of complex mixtures of novel long-chain quorum sensing signals in free-living and host-associated marine*Alphaproteobacteria*. Chem Bio Chem. 2005, 6: 2195-2206. 10.1002/cbic.200500189.
85. Griffin BR: Characteristics of a chondroitin AC lyase produced by *Cytophaga columnaris*. T Am Fish Soc. 1991, 120: 391-395. 10.1577/1548-8659(1991)120<0391:CO ACAL>2.3.CO;2.
86. Newton JC, Wood TM, Hartley MM: Isolation and partial characterization of extracellular proteases produced by isolates of *Flavobacterium columnare* derived from channel catfish. J Aquat Anim Health. 1997, 9: 75-85. 10.1577/1548-8667(1997)009<0075:IAPCOE>2.3. CO;2.
87. Anacker RL, Ordal EJ: Studies on the myxobacterium *Chondrococcus columnaris*. II. Bacteriocins. J Bacteriol. 1959, 78: 33-40.
88. Tiirola M, Valtonen ET, Rintamaki-Kinnunen P, Kulomaa MS: Diagnosis of flavobacteriosis by direct amplification of rRNA genes. Dis Aquat Organ. 2002, 51: 93-100.
89. Zhang Y, Arias CR, Shoemaker CA, Klesius PH: Comparison of lipopolysaccharide and protein profiles between*Flavobacterium columnare* strains from different genomovars. J Fish Dis. 2006, 29: 657-663. 10.1111/j.1365-2761.2006.00760.x.
90. Aranishi F, Mano N, Nakane M, Hirose H: Epidermal response of the Japanese eel to environmental stress. Fish Physiol Biochem. 1998, 19: 197-203. 10.1023/A:1007746514851.
91. Do Vale A, Marques F, Silva MT: Apoptosis of sea bass (*Dicentrarchus labrax* L.) neutrophils and macrophages induced by experimental infection with *Photobacterium damselae* subsp. *piscicida*. Fish Shellfish Immun. 2003, 15: 129-144. 10.1016/S1050-4648(02)00144-4.
92. Ourth DD, Wilson EA: Bactericidal serum response of the channel catfish against Gram-negative bacteria. Dev Comp Immunol. 1982, 6: 579-583. 10.1016/S0145-305X(82)80044-X.
93. Ourth DD, Bachinski LM: Bactericidal response of channel catfish *Ictalurus punctatus* by the classical and alternative complement pathways against bacterial pathogens. J Appl Ichtyol. 1987, 3: 42-45. 10.1111/j.1439-0426.1987.tb00449.x.
94. Kuo S-C, Chung HY, Kou GH: Studies on artificial infection of the gliding bacteria in cultured fishes. Fish Pathol. 1981, 15: 309-314. 10.3147/jsfp.15.309.
95. Bebak J, Matthews M, Shoemaker C: Survival of vaccinated, feed-trained largemouth bass fry (*Micropterus salmoides floridanus*) during natural exposure to *Flavobacterium columnare*. Vaccine. 2009, 27: 4297-4301. 10.1016/j.vaccine.2009.05.026.
96. Becker CD, Fujihara MP: The bacterial pathogen, Flexibacter columnaris and its epizootiology among Columbia river fish. 1978, Washington, D.C.: American Fisheries Society
97. Grabowski LD, LaPatra SE, Cain KD: Systemic and mucosal antibody response in tilapia, *Oreochromis niloticus* (L.), following immunization with *Flavobacterium columnare*. J Fish Dis. 2004, 27: 573-581. 10.1111/j.1365-2761.2004.00576.x.
98. Moore AA, Eimers ME, Cardella MA: Attempts to control *Flexibacter columnaris* epizootics in pond-reared channel catfish by vaccination. J Aquat Anim Health. 1990, 2: 109-111. 10.1577/1548-8667(1990)002<0109:ATCCEI >2.3.CO;2.
99. Schachte JHJ, Mora EC: Production of agglutinating antibodies in the channel catfish (*Ictalurus punctatus*) against*Chondrococcus columnaris*. J Fish Res Board Canada. 1973, 30: 116-118. 10.1139/f73-017.
100. Ransom DP: Immune responses of salmonids: (a) oral immunization against Flexibacter columnaris. (b) effects of combining antigens in parenterally administered polyvalent vaccines. 1975, Corvallis, OR: MS Thesis. Oregon State University
101. Liewes EW, Van Dam RH: Procedures and application of the in vitro fish leukocyte stimulation assay. Dev Comp Immunol. 1982, 2: 223-232.
102. Song Y-L: Comparison of Flexibacter columnaris strains isolated from fish in North America and other areas of the Pacific Rim. PhD Thesis. 1986, Corvallis, OR: Oregon State University
103. AFS-FCS (2011):*Guide to using drugs, biologics, and other chemicals in aquaculture. American fisheries society fish culture section.*http://www.fws.gov/fisheries/aadap/AFS-FCS%20documents/GUIDE_FEB_2011.pdf,
104. Karvonen A, Rintamäki P, Jokela J, Valtonen ET:

Increasing water temperature and disease risks in aquatic systems: climate change increases the risk of some, but not all, diseases. Int J Parasitol. 2010, 40: 1483-1488. 10.1016/j.ijpara.2010.04.015.

105. Suomalainen L-R, Tiirola M, Valtonen ET: The influence of rearing conditions on *Flavobacterium columnare* infection of rainbow trout. J Fish Dis. 2005, 28: 271-277. 10.1111/j.1365-2761.2005.00631.x.

106. Holt RA, Sanders JE, Zinn JL, Fryer JL, Pilcher KS: Relation of water temperature to *Flexibacter columnaris* infection in steelhead trout (*Salmo gairdneri*) coho (*Oncorhynchus kisutch*), and chinook (*0. tshawytscha*) salmon. J Fish Res Board Can. 1975, 32: 1553-1559. 10.1139/f75-182.

107. Suomalainen L-R, Reunanen H, Ijäs R, Valtonen ET, Tiirola M: Freezing induces biased results in the molecular detection of *Flavobacterium columnare*. Appl Environ Microbiol. 2006, 72: 1702-1704. 10.1128/AEM.72.2.1702-1704.2006.

108. Morris JM, Snyder-Conn E, Foott JS, Holt RA, Suedkamp MJ, Lease HM, Clearwater SJ, Meyer JS: Survival of lost river suckers (*Deltistes luxatus*) challenged with *Flavobacterium columnare* during exposure to sublethal ammonia concentrations at pH9.5. Arch Environ Contam Toxicol. 2006, 50: 256-263. 10.1007/s00244-004-0194-x.

109. Bandilla M, Valtonen ET, Suomalainen L-R, Aphalo PJ, Hakalahti T: A link between ectoparasite infection and susceptibility to bacterial disease in rainbow trout. Int J Parasitol. 2006, 36: 987-991. 10.1016/j.ijpara.2006.05.001.

110. Darwish AM, Mitchell A, Straus DL: Evaluation of a 4-h static copper sulphate treatment against experimental infection of*Flavobacterium columnare* in channel catfish (*Ictalurus punctatus*). Aquaculture Res. 2011, 43: 688-695.

111. Thomas-Jinu S, Goodwin AE: Acute columnaris infection in channel catfish, *Ictalurus punctatus* (Rafinesque): efficacy of practical treatments for warmwater aquaculture ponds. J Fish Dis. 2004, 27: 23-28. 10.1046/j.1365-2761.2003.00504.x.

112. Sink TD, Lochmann RT, Goodwin AE, Mareceaux E: Mortality rates in golden shiners fed high-fat diets with or without a dairy-yeast prebiotic before challenge with *Flavobacterium columnare*. N Am J Aquaculture. 2007, 69: 305-308. 10.1577/A06-078.1.

113. Kunttu HMT, Valtonen ET, Suomalainen L-R, Vielma J, Jokinen IE: The efficacy of two immunostimulants against*Flavobacterium columnare* infection in juvenile rainbow trout (*Oncorhynchus mykiss*). Fish Shellfish Immun. 2009, 26: 850-857. 10.1016/j.fsi.2009.03.013.

114. Bader JA, Nusbaum KE, Shoemaker CA: Comparative challenge model of *Flavobacterium columnare* using abraded and unabraded channel catfish, *Ictalurus punctatus* (Rafinesque). J Fish Dis. 2003, 26: 461-467. 10.1046/j.1365-2761.2003.00479.x.

115. Darwish AM, Bebak JA, Schrader KK: Assessment of Aquaflor®, copper sulphate and potassium permanganate for control of *Aeromonas hydrophila* and *Flavobacterium columnare* infection in sunshine bass, *Morone chrysops* female x*Morone saxatilis* male. J Fish Dis. 2012, 35: 637-647. 10.1111/j.1365-2761.2012.01393.x.

116. Cunningham FL, Jack SW, Hardin D, Wills RW: Pond-Level risk factors associated with columnaris disease on Mississippi commercial catfish farms. J Aquat Anim Health. 2012, 24: 178-184. 10.1080/08997659.2012.675932.

117. Conrad JF, Holt RA, Kreps TD: Ozone disinfection of flowing water. Prog Fish-Cult. 1975, 37: 134-136. 10.1577/1548-8659(1975)37[134:ODOFW]2.0.CO;2.

118. Shoemaker CA, Klesius PH, Lim C, Yildirim M: Feed deprivation of channel catfish, *Ictalurus punctatus* (Rafinesque), influences organosomatic indices, chemical composition and susceptibility to *Flavobacterium columnare*. J Fish Dis. 2003, 26: 553-561. 10.1046/j.1365-2761.2003.00489.x.

119. Rogers WA: Principal diseases of catfish: how to identify and fight them. Fish Farming Ind. 1971, 2: 20-26.

120. Darwish AM, Mitchell AJ, Straus DL: Evaluation of potassium permanganate against an experimental subacute infection of *Flavobacterium columnare* in channel catfish, *Ictalurus punctatus* (Rafinesque). J Fish Dis. 2009, 32: 193-199. 10.1111/j.1365-2761.2008.01015.x.

121. Riley TA: Treatment of Flavobacterium columnare and toxicity of chloramine-T in channel catfish. 2000, Ictalurus punctatus. MS thesis: Auburn University, Auburn, Alabama

122. Merrifield DL, Dimitroglou A, Foey A, Davies SJ, Baker RTM, Bøgwald J, Castex M, Ringø E: The current status and future focus of probiotic and prebiotic applications for salmonids. Aquaculture. 2010, 302: 1-18. 10.1016/j.aquaculture.2010.02.007.

123. Boutin S, Bernatchez L, Audet C, Derôme N: Antagonistic effect of indigenous skin bacteria of brook charr (*Salvelinus fontinalis*) against *Flavobacterium columnare* and *F. psychrophilum*. Vet Microbiol. 2012, 155: 355-361. 10.1016/j.vetmic.2011.09.002.

124. Arias CR, Cai W, Peatman E, Bullard SA: Catfish hybrid *Ictalurus punctatus* × *I. furcatus* exhibits higher resistance to columnaris disease than the parental species. Dis Aquat Organ. 2012, 100: 77-81. 10.3354/dao02492.

125. LaFrentz BR, Shoemaker CA, Booth NJ, Peterson BC, Ourth DD: Spleen index and mannose-binding lectin levels in four channel catfish families exhibiting different susceptibilities to *Flavobacterium columnare* and *Edwardsiella ictaluri*. J Aquatic Anim Health. 2012, 24: 141-147. 10.1080/08997659.2012.675936.

126. Snieszko SF: Columnaris disease of fishes. USFWS Fish Leafl. 1958, 46: 1-3.

127. Amend DF, Ross AJ: Experimental control of columnaris disease with a new nitrofuran drug, P-7138. Prog Fish-Cult. 1970, 32: 19-25. 10.1577/1548-8640(1970)32[19:ECOCDW]2.0.CO;2.

128. Ross AJ: *In vitro* studies with nifurpirinol (P-7138) and bacterial fish pathogens. Prog Fish-Cult. 1972, 34: 18-20. 10.1577/1548-8640(1972)34[18:IVSWNP]2.0.CO;2.

129. Shiraki K, Miyamoto F, Sato T, Sonezaki I, Yano K: Studies on a new chemotherapeutic agent nifurprazine (HB-115) against fish infectious diseases. Fish Pathol. 1970, 4: 130-137. 10.3147/jsfp.4.130.

130. Deufel J: Prophylactic measures against bacterial diseases of salmonid fry (Prophylaktische Massnahmen gegen bakterielle Erkrankungen der Salmonidenbrut). Osterr Fisch. 1974, 27: 1-5.

131. Endo T, Ogishima K, Hayasaka H, Kaneko S, Ohshima S: Application of oxolinic acid as a chemotherapeutic agent against infectious diseases in fishes-I. Antibacterial activity, chemotherapeutic effects and pharmacokinetics of oxolinic acid in fishes. Bull Jpn Soc Sci Fish. 1973, 39: 165-171. 10.2331/suisan.39.165.

132. Soltani M, Shanker S, Munday BL: Chemotherapy of *Cytophaga/Flexibacter*-like bacteria (CFLB) infections in fish: studies validating clinical efficacies of selected antimicrobials. J Fish Dis. 1995, 18: 555-565. 10.1111/j.1365-2761.1995.tb00360.x.

133. Wood JW: Treatment and drugs. Diseases of Pacific Salmon: their prevention and treatment. 1974, State of Washington: Department of Fisheries, Hatchery Division, 7-16. 2

134. Koski P, Hirvelä-Koski V, Bernardet JF: *Flexibacter columnaris* infection in arctic char (*Salvelinus alpinus* (L.)). First isolation in Finland. Bull Eur Ass Fish Pathol. 1993, 31: 66-69.

135. Wolf K, Snieszko SF: Uses of antibiotics and other antimicrobials in therapy of diseases of fishes. Antimicrob Agents Chemother. 1963, 161: 597-603.

136. Gaunt PS, Gao D, Sun F, Endris R: Efficacy of florfenicol for control of mortality caused by *Flavobacterium columnare*infection in channel catfish. J Aquat Anim Health. 2010, 22: 115-122. 10.1577/H09-057.1.

137. Serrano PH: Responsible use of antibiotics in aquaculture. FAO Fish Tech Pap. 2005, 469: 1-97.

138. Declercq AM, Boyen F, Van den Broeck W, Bossier P, Karsi A, Haesebrouck F, Decostere A: Antimicrobial susceptibility pattern of *Flavobacterium columnare* isolates collected worldwide from 17 fish species. J Fish Dis. 2013, 36: 45-55. 10.1111/j.1365-2761.2012.01410.x.

139. Darwish AM, Mitchell AJ: Evaluation of diquat against an acute experimental infection of *Flavobacterium columnare* in channel catfish, *Ictalurus punctatus* (Rafinesque). J Fish Dis. 2009, 32: 401-408. 10.1111/j.1365-2761.2009.01024.x.

140. Marchand P-A, Phan T-M, Straus DL, Farmer BD, Stüber A, Meinelt T: Reduction of *in vitro* growth in *Flavobacterium columnare* and *Saprolegnia parasitica* by products containing peracetic acid. Aquac Res. 2012, 43: 1861-1866. 10.1111/j.1365-2109.2011.02995.x.

141. Laanto E, Sundberg L-R, Bamford JKH: Phage specificity of the freshwater fish pathogen *Flavobacterium columnare*. Appl Environ Microbiol. 2011, 77: 7868-7872. 10.1128/AEM.05574-11.

142. Prasad Y, Arpana , Kumar D, Sharma AK: Lytic bacteriophages specific to *Flavobacterium columnare* rescue catfish, *Clarias batrachus* (Linn.) from columnaris disease. J Environ Biol. 2011, 32: 161-168.

CHAPTER 2 STRATEGIES AND HURDLES USING DNA VACCINES TO FISH

Linn B Hølvold, Anne I Myhr, and Roy A Dalmo

INTRODUCTION

There are quite a limited number of DNA vaccine strategies that have been successful in giving significant protection against fish diseases. The excellent exceptions are DNA vaccination against infectious hematopoietic necrosis virus (IHNV) at commercial level and against viral hemorrhagic septicemia virus (VHSV) at experimental/small scale level [1]. A promising strategy to increase the DNA vaccine efficacy, against other viral diseases, is to target the construct and/ transgene to antigen presenting cells. Wang et al. [2] have presented an overview on how DNA vaccines can be targeted to antigen presenting cells (APC) and dendritic cells (DC) using molecular assemblies with the resulting immune responses. Another strategy is to encapsulate the DNA vaccine in certain carriers with the intention to increase transgene and MHC expression. There are many particulate carriers that have been explored to present and target DNA vaccines to desired cells and tissues, one of them being poly (D,L-lactic-co-glycolic)-acid (PLGA) particles.

DNA VACCINES AND VACCINATION

The definition of DNA vaccination as provided by The Norwegian Biotechnology Advisory Board [3] is "the intentional transfer of genetic material (DNA or RNA) to somatic cells for the purpose of influencing the immune system". This sets it slightly apart from gene therapy, which in the same report is referred to as an introduction of novel gene(s) for purposes other than influencing the immune system. The mechanism of a DNA vaccine can in many ways be likened to that of a virus, as it requires the same cellular machinery in order to replicate and since it also triggers immune responses normally seen with viral infections [4]. Unlike conventional viral vaccines based on subunits or killed virus, a DNA vaccine may conserve the structure and hence also antigenicity of a transgenic antigen/protein [5].

Citation: Linn B Hølvold, Anne I Myhr, and Roy A Dalmo. Strategies and hurdles using DNA vaccines to fish. Veterinary Research, 2014; 45: 21. DOI:10.1186/1297-9716-45-21.

IMMUNE RESPONSES TO DNA VACCINATION – MAINLY IN MICE AND BIRDS

A vital attribute of DNA vaccines is the ability to induce all three arms of adaptive immunity, namely; helper T-cells, cytotoxic T cells (CTLs) and antibodies, although they were initially investigated in the attempt to find ways of delivering antigen to major histocompatibility complex (MHC) class I and induce T helper 1 (Th1) responses [6, 7]. Professional APCs are the cells that primarily contribute to the immune responses to DNA vaccination. Direct transfection of DCs provides the most efficient priming of naïve CTLs, and is perhaps the major mechanism for priming of these cells [8]. APCs are also able to take up exogenous antigens and process them for presentation by either MHC class II, or MHC class I following transfer of the antigen to the cytosol (cross-priming) [9]. These responses are vital in the cellular immune response following DNA vaccination. The expression of reporter genes is not only higher in fish, but also seems to have a longer duration. Transgene expression has been detected at the injection site as long as two years after injection of glass catfish (*Kryptopterus bicirrhus*) [10], and Tonheim et al. detected both supercoiled (sc) DNA and luciferase expression at the injection site 535 days after intramuscular injection of Atlantic salmon (*Salmo salar* L.) with pDNA [11].

The immunogenicity of DNA vaccines stems not only from the expression of the GOI, but also from properties of the plasmid vectors themselves. The ability of foreign nucleic acids to induce interferons (IFNs) in mouse fibroblasts was discovered by two independent research groups as early as in 1963 [12]. The most studied pDNA property in terms of possible adjuvant effects is CpG motifs, regions in the DNA where a cytosine nucleotide occurs next to a guanine nucleotide and the two are linked by a phosphodiester bond. In vertebrates these sequences are highly methylated, whereas they show a much lower methylation frequency in viral and bacterial DNA (such as pDNA). These CpG motifs can act as pathogen-associated molecular pattern molecules (PAMPs) and are recognized as danger signals by the vertebrate immune system, resulting in a release of cytokines, macrophage activation, a differentiation of Th1 effector cells as well as B-cell proliferation and secretion of antibodies [13]. Plasmid DNA containing optimized CpG content may be likely used in DNA vaccine strategies.

DNA VACCINATION AGAINST NOVIRHABDOVIRUSES

The first DNA vaccination of fish took place in 1996, when Anderson et al. immunized rainbow trout (*Onchorhyncus mykiss*) against IHNV [14]. Since then several trials have been performed for a wide variety of fish species and pathogens and in 2005 a vaccine against IHNV infection in salmonids (Apex-IHN®, Novartis Animal Health) was also one of the first DNA vaccine ever to be cleared for marketing (by the Canadian Food Inspection Agency). In 1999 the injection of Atlantic salmon with pCMV4-G (plasmid-encoded glycoprotein) from a rainbow trout IHNV isolate induced significant protection against challenges with IHNV, even though the salmon were much larger than the rainbow trout in previous challenge studies [15]. DNA vaccination of fish has been shown to induce both innate and adaptive immune responses similar to what is seen in mammalian species, and seems especially efficient against novirhabdoviruses (like VHSV and IHNV). These are simple RNA viruses with either five or six genes and a single viral surface protein (glycoprotein, or G protein) that acts as the protective antigen [16].

In rainbow trout an immunization against VHSV enables the induction of cell-mediated immune responses encompassing both CTLs and natural killer (NK) cells and has also been shown to significantly reduce the replication of virus during challenge [17]. Interestingly, when Cuesta et al. looked at the innate and adaptive responses in vaccinated and control fish after challenge, they found the highest increase in vaccinated fish to be that of innate immune responses [18]. Lorenzen et al. [19] have also demonstrated the importance of innate responses in early antiviral defense, wherein rainbow trout were subjected to VHSV challenge following vaccination with pIHN-G (plasmid-encoded IHNV glycoprotein). Whereas protection at late stages of the challenge could only be conferred by previous immunization with pVHS-G, the two vaccines induced similar levels of immune responses and protection during the first week following challenge.

In conclusion, anti novirhabdovirus response governed by DNA vaccines relies much on the immunogenicity of the G-protein resulting in long-lasting protection mediated by cellular and humoral responses. In addition, there is an innate and specific response following DNA vaccination that may be protective in a short-term perspective.

Other RNA viruses, or the larger DNA viruses, often offer more difficulty in identifying a protective antigen, although viral surface protein genes are almost always chosen for DNA vaccines [2]. Some pathogens, such as infectious pancreatic necrosis virus (IPNV), primarily cause disease in fry. Whereas vaccination by injection is highly impractical at this stage [20], the vaccination of post-smolts has been shown to induce protection upon challenge [21]. Good protection did, however, require the use of plasmids encoding all the large poly-proteins of the IPNV.

RECENT DNA VACCINATION LABORATORY TRIALS

DNA vaccines and their effects against several viral and bacterial diseases in fish have been reviewed by Tonheim et al.[22], Kurath [23], Redding and Weiner [24], and Gomez-Casado et al. [25]. Since these reviews were published, several new results using DNA vaccine strategies have been reported (Table 1).

Table 1. Experimental DNA vaccines and their protection in fish following experimental infection

Pathogen	Gene inserted	Host	Administration route/adjuvant	Protection	References
IHNV	IHNV-G plus suicidal gene	Rainbow trout	I.m/none	Yes	[26]
IHNV	IHNV-G; different genogroups	Rainbow trout	I.m/none	Yes; cross-protection	[27]
IHNV	IHNV-G	Rainbow trout	Oral/PLGA	No	[28]
VHSV	*E. tarda* as delivery vehicle of the vaccine	Olive/Japanese flounder (*Paralichthys olivaceus*)	I.m	Yes	[29]
IPNV	VP2		Oral/alginate	Yes	[30]
IPNV	VP2; Segment A of TA strain	Atlantic salmon	I.m	No	[31]
SAV	E1 and E2	Atlantic salmon	I.m	No	[32]
Megalocytivirus	86-residue VP	Turbot (*Scophthalmus maximus*)	I.m	Yes	[33]
Edwardsiella tarda	Eta6-FliC chimeric protein	Japanese flounder (*Paralichthys olivaceus*)	I.m	Yes	[34]

E. tarda	D15-like surface antigen	Japanese flounder	I.m	Yes	[35]
E. tarda	Eta2	Japanese flounder	i.m	Yes	[36]
Streptococcus iniae	sagF, sagG and sagI	Japanese flounder	I.m	Yes	[37]
S. iniae	Sia10 delivered by *E. tarda*	Japanese flounder	Oral/alginate, immersion- boosted	Yes	[38]
Vibrio harveyi	FlaA	Yellow grouper*(Epinephelus awoara)*	I.m	Yes	[39]
V. harveyi	DegQ or/and Vhp1	Japanese flounder	I.m	Yes	[40]
V. alginolyticus	FlaA	Red snapper (*Lutjanus campechanus*)	I.m	Yes	[41]
S. iniae and *V. anguillarum*	Sia10 and/or OmpU	Turbot	I.m	Yes, cross-protection	[37]
Flavobacterium psychrophilum	Hsp60, hsp70	Rainbow trout	I.m	No	[42]
Cryptocaryon irritans	iAg	Orange spotted grouper (*E. coioides)*	I.m	Yes	[43]
Cryptobia salmocitica	Metalloprotease	Atlantic salmon and rainbow trout	I.m	Partly	[44]
Ichthyophthrius multifiliis	Immunobilization antigens and cystein protease	Rainbow trout	I.m, gene gun and air pressure	No	[45]

Apparently, DNA vaccination may also confer protection against bacteria and parasites – but not against all. Bacteria and parasites may express and harbor numerous different antigens dependent on their life cycle, this make a DNA vaccination strategy more complicated. There are developed "traditional" oil adjuvanted vaccines against a number of bacterial diseases, where the strategies may be quite efficient preventing infection and disease. This is in contrast to some parasitic infections/ attachment such as sea lice - where no immune prophylaxis exists at present at an industrial scale.

ADMINISTRATION AND DISTRIBUTION OF DNA VACCINES

Intramuscular Injection

Intramuscular injection is widely applied for pDNA delivery in fish and generally results in strong expressions of transgene at the injection site [22]. Studies in mice have found the dispersion of pDNA immediately following intramuscular injection to take place primarily between the muscle body and epimysium (connective tissue that enclosing the entire muscle). Myocytes and mononuclear cells take up pDNA after administration [22], but despite a rapid initiation of uptake the subsequent uptake is slow and cells along the muscle fibers have been shown to be transfected over a period of hours following injection. With very small fish this initial dispersion of a vaccine might be enough to ensure the perfusion of intact pDNA to more distant tissues, while in large fish the injected volume will mainly rest along the needle trajectory [22]. The transportation of pDNA to and from blood to other tissues has been reported for various fish species [22]. Plasmids have been recovered from sites such as liver, spleen, head-kidney, heart and intestine for some time after injection, but mainly persist at the site of injection. Degradation of the pDNA starts within five minutes following injection of mice, with as much as 95-99% of the initial pDNA amount degraded within 90 min [22]. The rate of degradation in the tissue of cold-water fish remains to be determined. The extent of histopathological changes at the injection site following intramuscular DNA delivery in fish appear to increase with an increase in vaccination dose [46], but vaccination will generally induce only moderate local tissue damage in form of degeneration of myocytes, hemorrhages and a transient influx of inflammatory cells [20].

Other Routes of Delivery

Other routes of pDNA administration that have been investigated in fish are intravenous, intraperitoneal, oral delivery, and particle bombardment [1, 21]. Accumulation of naked pDNA took place primarily in the heart, kidney and liver following intravenous administration [22], whereas oral delivery resulted in a recovery of DNA fragments from the pyloric region, kidney, spleen and liver – assessed by revers-transcription polymerase chain reaction (RT-PCR) [28].

UPTAKE OF PLASMID DNA IN FISH AND MAMMALIAN SPECIES

A wide variety of mammalian cell types has been shown to take up pDNA – reviewed by Budker et al. [47], but so far uptake in fish has only been reported for myocytes, head kidney macrophages and endocardial endothelial cells (EECs) [22]. The exact mechanisms by which myocytes take up pDNA remains to be determined, but several suggestions have been made [48]. It was previously theorized that direct injections caused temporary membrane disruptions and/or pores that allowed for the entry of pDNA [5], but studies have shown that such disruptions in fact work to abolish transfection [12].

Several PPRs have been shown to bind nucleic acids including plasmid DNA, reviewed by Desmet et al. [49]. Of the number of cell-surface receptors investigated in terms of DNA binding and uptake, scavenger receptors (SRs) in particular have been a subject of interest. These receptors comprise a broad family of membrane proteins capable of binding a wide range of anionic ligands and are present on several different cell types [50]. The uptake of pDNA by SRs has been demonstrated in Atlantic cod (*Gadus morhua*) atrial EECs [51], but although the SRs may bind DNA they appear not to be essential for the immunostimulatory activity of CpG DNA [52]. This indicates that SRs activity may be essential for the transgene expression – as they may bind and facilitate degradation of high amount of plasmid DNA before expression occurs.

FACTORS INFLUENCING TRANSFECTION AND TRANSGENE EXPRESSION

Nucleic acids have poor intrinsic transfection efficiency due to their large size and negative charge [53]. The processes of gene transcription and translation are carried out by the cell's own machinery, and may be influenced by a variety of factors such as pDNA vector design, pDNA concentrations, administration volumes, age and size of the fish, water temperatures and method as well as route of administration.

It has been well documented that the DNA topoform has a strong influence on the efficiency of transfection [54]. Supercoiled DNA is frequently reported as the most potent topoform followed by open circular (oc) forms, whereas linearization of the DNA has been shown to nearly abolish the expression of transgene and linearized pDNA. The choice of promoter also has a profound influence on the level of transgene expression [55] and the cytomegalovirus (CMV) promoter has often been reported as the most effective. Because of this, the CMV promoter is the most commonly applied in gene transfection studies and its potency has been demonstrated also in fish [5]. The use of an intron and poly-adenylation (termination) signals further improves expression [55].

Despite the inherent adjuvant effects of DNA vaccines, it has been shown that immune responses triggered by DNA vaccination may in fact limit the duration of transgene expression [55]. The lack of anti-Luc antibodies means that both levels and duration of expression are commonly higher in studies applying luciferase compared to more immunogenic antigens, and immune responses have only been observed with the application of large pDNA doses and potent adjuvants such as Freund´s complete adjuvant (FCA) [56]. A co-administration of vectors encoding Luc and immunogenic proteins (β-gal, G protein) decreases both the strength and duration of luciferase expression, as the initiated antigen specific cytotoxic responses work to eliminate transfected cells expressing the transgenic protein [56]. The stimulation of PRRs may also induce responses that can be detrimental to transgene expression. The hallmark cytokines of the inflammatory response, TNF-α and IL-1β, have both been shown to inhibit transgene expression in vitro and in vivo [57], as have the Th1 associated cytokines IFN-γ and IFN-α [58]. Levels of inhibition have in fact been shown to correlate with the levels of CpG-induced cytokines, and synergistic effects were also observed. The inhibitory effect takes place at the mRNA (post-transcriptional) level, hence not causing vector degradation, inhibition of total cellular protein synthesis or elimination of infected/transfected cells [58].

Whereas intramuscular injection is widely acknowledged as the superior method of administration to achieve high levels of transgene expression [1], the result still depends on factors such as dose, volume and fish size and age [59]. Transgene expression has been found to be higher in young and growing fish [5], and small fish sizes appear to favor not only the distribution of pDNA throughout tissue but the distribution of transgene expression as well [5]. For example expression of luciferase in thymus, gills, spleen and kidney has been reported for small fish (< 5 g), although the highest expression is consistently detected at the site of injection in myocytes, infiltrating cells and epithelial cells lining small capillaries [20, 60].

The first reporter gene study in fish indicated the existence of a maximum above where there would be no further increase in expression [60]. This observation is supported by later findings in zebra fish (*Danio rerio*) and rainbow trout [5], and there are indications that excessive DNA concentrations may actually reduce transgene expression [21]. The injection of large volumes might contribute to a spatial distribution by creating temporary gaps between fibers [61], which appears to induce higher expression as well as reduce the variations commonly observed in in vivo transfection studies [5]. A pre-injection of an isotonic solution created the same effect, and lead to both higher and more equal transgene expression among individuals after DNA injection [61].

ADVANTAGES, DISADVANTAGES AND CHALLENGES OF DNA VACCINES

The potency of DNA vaccines for inducing the different branches of both innate and adaptive immunity has already been described. DNA vaccines also show high efficiency when given at early life stages [62], and provide the benefit of inducing protective immunity over a wide span of temperatures [63].

The advantages of DNA vaccination still stretch beyond merely the immunological capacities. Looking at the concept from a manufacturer or/and investor's standpoint, DNA vaccines are relatively inexpensive and easy to produce. The processes required for production are identical for all DNA vaccines, and the ease of cloning also enables rapid modifications in a way that is generally not obtainable with conventional vaccine preparations [60].

Potential side effects include e.g. risk of autoimmunity, immune tolerance against the expressed antigen, too high CTL response resulting in myositis, chromosomal integration, and injection site inflammation and tissue destruction [1, 64].

APPLICATION OF MOLECULAR ADJUVANTS TO INCREASE TRANSGENE IMMUNOGENICITY

To increase immunogenicity of a given DNA vaccine one may consider optimization by including co-injected adjuvant, either being traditional aluminum salts, polysaccharides (e.g. zymosan, glucans, chitosan), different liposomes, synthetic polymers and TLR agonists. Moreover, plasmid encoded cytokine adjuvants may also be one approach to increase the immunogenicity. IL-2, IFN-γ, IL-12, GM-CSF and IL-15 have been shown to modulate immune responses when co-encoded by the DNA vaccine [65]. One may also apply the concept of immune modulating effects chemokines, transcription factors and/or co-stimulatory factors assembled into the plasmid vectors. Using fish models, only one report on the use of cytokine adjuvants encoded by the same plasmid as the expression plasmid has been published. Caipang et al. used Japanese flounder interferon regulatory factor-1 (IRF-1) cloned into a plasmid DNA vector containing the major capsid protein gene of sea bream iridovirus. The antibody levels of fish injected with this vaccine were not significant higher with the control plasmid without IRF-1 gene [66]. Unfortunately, no experimental virus challenge was carried out to monitor vaccine efficacy. To increase vaccine potency and efficacy of poor performing DNA vaccines one should explore strategies such as inclusion of molecular adjuvants perhaps in combination with targeting carrier systems such as nano- and microparticles. Although transfection levels are often low, studies have demonstrated a persistence of transgene expression at the injection site that might coincide with a time when the fish would normally be ready for slaughter [67]. There have been experiments with suicidal DNA vaccines for fish, where the plasmid vector includes a protein to induce apoptosis after an immune response has been triggered [26].

PLGA PARTICLES AS CARRIER SYSTEMS FOR DNA VACCINES – FOCUS ON FISH

Besides the proven efficacy of G-protein expressing plasmids against novirhabdoviruses, there is an urgent need to develop high performing vaccines against other viral diseases. In this aspect, DNA vaccines delivered by several kinds of particles may show promise. There is considerable research being done on the application of carrier systems and/or DNA-complexes to facilitate higher levels of entry and reduce degradation, reviewed by Saade et al. [65]. The potential of PLGA particles as adjuvants and carriers for DNA vaccine delivery has received considerable attention in mammalian studies [68]. In spite of this, reports on the use of PLGA particles for DNA delivery to fish are nearly non-existent. In 2008, Tian et al. were the first to report on the use of PLGA microcapsules containing a plasmid vaccine for the oral immunization of Japanese flounder against lymphocystis disease virus (LCDV) [69]. Following immunization they detected transgene expression in gills, intestine, spleen and kidney from fish vaccinated with encapsulated pDNA. The encapsulated pDNA also induced higher levels of antibodies compared to control fish injected with naked pDNA. Tian and Yu later demonstrated a significant increase in resistance to LCDV infection after oral administration of a pDNA vaccine encapsulated into PLGA nanoparticles [70]. Finally, the last study so far reported was by Adomako et al., who used PLGA nanoparticles incorporated into feed for oral delivery of a DNA vaccine against IHNV to rainbow trout [28]. They found that particles were mostly taken up in the posterior gut, but that a significant number of fish that showed uptake in gastrointestinal epithelial cells did not demonstrate detectable levels of transgene expression. Although antibody responses could be detected in fish given high doses of pDNA, the RPS at a six weeks post-vaccination challenge was still only 22%. No VHSV/IHNV DNA vaccine trials using nano- or micro-particle vehicles have been carried out so far. It would have been very interesting to compare the vaccine efficacy of naked plasmid encoding for VHSV and IHNV G-protein to corresponding particle delivered vaccines.

The exploitation of PLGA encapsulated DNA vaccines in fish is new, currently there are no information whether this strategy is a way to go to increase efficacy of DNA vaccines for fish. There are, however, indications that pDNA encapsulated in PLGA nanoparticles induce a antiviral immune response in salmon - at a higher level than what is achieved with only pDNA [71].

ISSUES OF MAKING DNA ENCAPSULATED PLGA PARTICLES

The w1/o/w2 method described previously is frequently used for the encapsulation of pDNA into PLGA particles, but results are highly variable with regard to encapsulation efficiency and loading as well as DNA degradation and release. The hydrophilic nature of DNA complicates the process of encapsulation as it increases the risk of plasmid diffusion into the w2 phase during solvent evaporation [72], and the encapsulation process also offers multiple challenges in terms of preservation of bioactivity – i.e. keeping the supercoiled DNA topoform intact.

Among the factors known to affect the integrity of the pDNA to be encapsulated are the polymer composition and molecular weight (Mw), shear force, preparation temperatures, solvents and the concentration and Mw of the applied stabilizer [73]. In general, polymers with a high Mw also result in the highest encapsulation efficiencies and lower the burst release of plasmid DNA [73].

TRANSGENE EXPRESSION AND IMMUNE RESPONSES BY PLGA-ENCAPSULATED PDNA

Encapsulated DNA has been shown to be more potent than naked DNA at mediating transgene expression in vitro in a variety of mammalian cells types [74]. However, in vivo studies report on a superiority of naked DNA in eliciting transgene expression not only compared to PLGA particles, but also to other formulated DNA vaccines as well [74].

Small particles (< 100 nm) are often shown to be internalized more rapidly, and also show the highest transfection efficiencies [75].

The adsorption of plasmid DNA onto the surface of PLGA particles, rather than encapsulation, has been reported to result in a higher transgene expression, but the expression declined more rapidly [48].

In addition to the pro-inflammatory cytokines induced by empty PLGA nano- and micro-particles (IL-1β and TNF-α) [76], the use of different particle sizes might influence the resulting cytokine profile after administration of encapsulated or particle-bound CpG DNA [77]. In addition to pro-inflammatory cytokines, nanoparticles have been shown to also enable an induction of antiviral cytokines such as type I IFNs in addition [77]. The encapsulation of pDNA encoding antigens has also been shown to elicit CTL-responses, even with pDNA-amounts that elicited no such responses after naked administration [77]. Moreover, encapsulated pDNA also enhanced the total antibody response at high doses, while inducing a more rapid and complete sero-conversion when lower doses were applied [78].

The injection of PLG microspheres into muscle has been shown to result in a foreign body response, with a large influx of different inflammatory cells that appear largely related to microspheres especially at later time-points [79]. These infiltrating cells were also the ones that were primarily transfected, an observation similar to that found in another study [80].

OTHER PARTICLES IN VACCINE DELIVERY TO FISH

The extensive use of PLGA as vaccine delivery vehicles can largely be attributed to their high biocompatibility as well as the ease with which the particles can be prepared. There are other means to increase the level of transgene expression that may increase vaccine efficacy, given that the transgenes produced display fair immunogenic properties, by using tailored particle systems – as outlined by Nguyen et al. [73]. Other particles have also been investigated as carrier systems for vaccination of fish (Table 2), either on their own or in combination.

Table 2. Compounds tested as adjuvants and gene vaccine delivery systems to fish species, published from 2007 to date

Particles	Carrier molecule(s)	Assembly	Species
Alginate	ß(1,4)-D-mannuronic and α(1,4)-L-guluronic acid residues	Encapsulation	Japanese flounder [81]
Chitosan	ß(1,4)-D-glucosamine and N-acetyl-D-glucosamine copolymer	Encapsulation	Nile tilapia (*Oreochromis niloticus)*[82]
			Japanese flounder [83]
			Asian sea bass (*Lates calcifer*) [84]
			Turbot [85]
Liposomes	Artificial lipid bilayer vesicles	Encapsulation	Kelp grouper (*E. bruneus*) [86]
Polycaprolactone	Biodegradable, synthetic polymer	Coating	Indian major carp (*Labeo rohito*) [87]
Calcium phosphate	Inorganic, biodegradable and biocompatible material	Coating	Indian major carp [88]

CURRENT CHALLENGES IN THE USE OF PLGA PARTICLES AS CARRIERS/ADJUVANTS

Despite a growing number of optimistic reports on the adjuvant/carrier properties of PLGA particles, there are still many hindrances to be overcome. One of these is the preparation method, where different drugs/vaccines require different conditions. While some compounds are easily encapsulated, others – like DNA vaccines – are more difficult to encapsulate in an efficient manner. Low encapsulation efficiencies not only result in low antigen loading, but also mean that a large amount of drug/vaccine goes to waste during particle preparation. The detrimental effects that encapsulation may have on certain antigens such as pDNA are also major limiting factors at the moment, and need to be resolved. Particle preparation is also low-scale work as of yet, with considerable effort to be made before the process is optimized on a large scale suitable for mass-production.

CONCERNS REGARDING PLGA NANO- AND MICRO-PARTICLES

Any construct/compound, when brought down to sub-micron sizes, will exhibit new and potentially harmful characteristics [89]. The small size means they can interact with biological membranes in an entirely new way, thus inducing responses not seen with larger constructs. As with all new applications it takes time to fully survey the potential side effects. In vitro studies using PLGA nano- and micro-particles have so far not revealed any toxic effects, even at large doses [90]. The various concerns regarding the use of PLGA particles in aquaculture vaccines have recently been reviewed by Nielsen et al. [91].

SAFETY AND REGULATORY ASPECTS BY DNA VACCINES

Safety aspects include potential effects on the vaccinated animals, the environment and the consumer [1, 64,92]. Other safety issues include potential shed of the vaccine to the environment from the vaccinated animal and by predatory animals [93]. Human safety does also, although with low probability, include potential effects by self-injection by vaccinators. These safety aspects need to be taken into consideration by relevant authorities when safety aspects are to be documented.

The Canadian Food Inspection Agency (CFIA) that gives authority to the Veterinary Biologics Section (VBS) of the Animal Health and Production Division (AHPD) approved the IHNV DNA vaccine for commercial use in Canada in 2005 [94]. Five aspects were taken into consideration before the vaccine was approved: (i) public perception and acceptance, (ii) regulatory and environmental concerns, (iii) risk-benefit, (iv) feasibility of producing the vaccine at a scale and cost appropriate for the fish industry, and (v) intellectual property issues [91]. At present there has not been approved any DNA vaccine to be used in aquaculture in Europe. To obtain marketing authorization within Europe, a new veterinary medicine has to meet the criteria and requirements of the EU pharmaceutical legislation for both medical and veterinary applications. To achieve a marketing application a MAA (marketing authorization application) has to be submitted to the European Medicines Agency (EMA). The EMA has drafted guidelines for the veterinary use of DNA vaccines [95]. The European guidelines include several aspects to be considered in order to conduct a risk assessment of DNA vaccines. This includes i) the possibility of pDNA integrating into the chromosome, ii) concerns about possible adverse effects on the immune system, iii) risks posed by the additional use of genes encoding cytokines or co-stimulatory molecules or iv) undesirable biological activity by the expressed antigen itself. At present there are work undertaken within EMA with the intention to revise and update the guidance for DNA vaccines and the goal is to provide a specific guidance document [96]. This work is especially concerned to the use of DNA vaccines to humans and if the outcome will have influence on the veterinary use of DNA vaccines is therefore uncertain. The guidelines prepared by the FDA (US) points to some of the same areas as in EU, and recommends that safety testing should include tests on vaccine immunogenicity, effects from cytokines and other immunomodulatory genes, autoimmunity, local reactogenicity and systemic toxicity and studies of bio-distribution, persistence and integration [97].

The issue of plasmid persistence and chromosomal integration of DNA vaccine is of relevance for both safety and policy [92, 93]. Norwegian authorities may for example, due to uncertainties with regard to whether DNA vaccines persist degradation in tissues and organs label DNA vaccinated fish as a GMO [98]. If a DNA vaccinated animal is considered to be a GMO the producers has also to meet the requirements of the EU environmental legislation on the deliberate release of GMOs (Directive 2001/18/EC). The objective of an environmental risk assessment (ERA) in accordance with Directive 2001/18/EC is to identify and assess on a case-by-case basis the potential harmful effects of a GMO for humans, animals (domestic and wildlife), plants, microorganisms and the environment. The EMA has developed two specific guidelines for the preparation of ERA to facilitate adoptions of the requirements and the methodology of the Directive to GMO-containing medical products. Although these guidelines have mainly been used for MAA of GM vaccines, they may provide necessary information and procedures that can also be used to perform ERA for DNA vaccines and vaccinated animals.

The process of obtaining a market authorization for a new fish vaccine is a both time consuming and expensive. To overcome some of the legal uncertainties there is a need to assess potential consequences on a case-by-case basis and to be aware of that a number of factors such as the inserted gene, other gene inserts of the pDNA, injection site and amount of vaccine injected are of importance. Safety and regulatory uncertainties are related to distribution and degradation of the DNA after injection and includes that there is a need of research on the (i) the stability of the DNA vaccine, (ii) plasmid persistence, (iii) unintended immunological impacts, and (iv) potential for integration of the pDNA into the chromosome of the recipient organism [93]. These uncertainties are also of relevance for consumer acceptance and the markets confidence in safety. The GMO issues may only be of concern for countries that have specific GMO legislation and that demands that GMOs and products thereof need to be labeled. In USA and Canada there are no requirements for labeling of food containing GMOs, and they do not have specific GMO legislation. In Europe due to the uncertainties with regard to the persistence of a DNA vaccine, the vaccinated fish may need to be labeled as a GMO. For example European countries that employ restrictions on GMO may not import the DNA vaccinated fish, and if the fish need to be labeled as a GMO this will have an influence on consumer willingness to buy it. Ultimately may these uncertainties prevent the use of DNA vaccines [92].

CONCLUSIONS

There is a crucial need to increase efficacy of DNA vaccines against persistent and hard-to-combat viral infections, this can be met by: (i) Application of vaccine carriers to increase the uptake in antigen presenting cells – followed by enhanced presentation of transgene peptides/antigens, (ii) Use of nano-scale particles to increase the level of cross presentation by such cells – this may also be beneficial to produce antibody response as well as cell mediated immunity, and (iii) Employing additional adjuvants such as TLR ligands, other than RNA and/or DNA, ligands to boost the response considerably. Safety and regulatory uncertainties are related to distribution and degradation of the DNA after injection and more effort needs to be put into gaining understanding of the mechanisms of pDNA uptake, from the moment of administration until the stage of transcription and translation in the nucleus.

ACKNOWLEDGEMENTS

This project was funded by the Research Council of Norway (project nos. 182035 and 183204/S40) and Tromsø Research Foundation ("Induction and assessment of T cell immunity to virus antigens in salmonids).

AUTHORS' CONTRIBUTIONS

LBH has made the most substantial contribution to this review, while RAD and AIM has revised the manuscript critically. All authors read and approved the final manuscript.

REFERENCES

1. Evensen Ø, Leong JA: DNA vaccines against viral diseases of farmed fish. Fish Shellfish Immun. 2013, 35: 1751-1758. 10.1016/j.fsi.2013.10.021.
2. Wang G, Pan L, Zhang Y: Approaches to improved targeting of DNA vaccines. Hum Vaccin. 2011, 7: 1271-1281. 10.4161/hv.7.12.17983.
3. Board TNBA: Regulation of DNA vaccines and gene therapy on animals. 2003, [http://www.bion.no/filarkiv/2010/07/2006_03_regulation_of_DNA_vaccines.pdf]
4. Ballesteros NA, Saint-Jean SS, Perez-Prieto SI, Coll JM: Trout oral VP2 DNA vaccination mimics transcriptional responses occurring after infection with infectious pancreatic necrosis virus (IPNV). Fish Shellfish Immunol. 2012, 33: 1249-1257. 10.1016/j.fsi.2012.09.004.
5. Heppell J, Lorenzen N, Armstrong NK, Wu T, Lorenzen

E, Einer-Jensen K, Schorr J, Davis HL: Development of DNA vaccines for fish: vector design, intramuscular injection and antigen expression using viral haemorrhagic septicaemia virus genes as model. Fish Shellfish Immunol. 1998, 8: 271-286. 10.1006/fsim.1997.0133.

6. Liu MA: DNA vaccines: an historical perspective and view to the future. Immunol Rev. 2011, 239: 62-84. 10.1111/j.1600-065X.2010.00980.x.

7. Raz ETH, Sato Y, Corr M, Dudler JA, Swain SL, Spiegelberg HL, Varson DA: Preferential induction of a Th1 immune response and inhibition of specific IgE antibody formation by plasmid DNA immunization. Proc Natl Acad Sci U S A. 1996, 93: 5141-5145. 10.1073/pnas.93.10.5141.

8. Sbai H, Schneider J, Hill AV, Whalen RG: Role of transfection in the priming of cytotoxic T-cells by DNA-mediated immunization. Vaccine. 2002, 20: 3137-3147. 10.1016/S0264-410X(02)00251-7.

9. Dixon B, Stet RJM: The relationship between major histocompatibility receptors and innate immunity in teleost fish. Dev Comp Immunol. 2001, 25: 683-699. 10.1016/S0145-305X(01)00030-1.

10. Dijkstra JM, Okamoto H, Ototake M, Nakanishi T: Luciferase expression 2 years after DNA injection in glass catfish (*Kryptopterus bicirrhus*). Fish Shellfish Immunol. 2001, 11: 199-202. 10.1006/fsim.2000.0303.

11. Tonheim TC, Leirvik J, Løvoll M, Myhr AI, Bøgwald J, Dalmo RA: Detection of supercoiled plasmid DNA and luciferase expression in Atlantic salmon (*Salmo salar* L.) 535 days after injection. Fish Shellfish Immunol. 2007, 23: 867-876. 10.1016/j.fsi.2007.03.015.

12. Rotem Z, Cox RA, Isaacs A: Inhibition of virus multiplication by foreign nucleic acid. Nature. 1963, 197: 564-566. 10.1038/197564a0.

13. Klinman DM: Immunotherapeutic uses of CpG oligodeoxynucleotides. Nat Rev Immunol. 2004, 4: 248-257.

14. Anderson ED, Mourich DV, Fahrenkrug SC, LaPatra S, Shepherd J, Leong JAC: Genetic immunization of rainbow trout (*Onchorhynchus mykiss*) against infectious hematopoietic necrosis virus. Mol Mar Biol Biotechnol. 1996, 5: 114-122.

15. Traxler GS, Anderson E, LaPatra SE, Richard J, Shewmaker B, Kurath G: Naked DNA vaccination of Atlantic salmon *Salmo salar* against IHNV. Dis Aquat Organ. 1999, 38: 183-190.

16. Purcell MK, Laing KJ, Winton JR: Immunity to fish rhabdoviruses. Viruses. 2012, 4: 140-166. 10.3390/v4010140.

17. Purcell MK, Nichols KM, Winton JR, Kurath G, Thorgaard GH, Wheeler P, Hansen JD, Herwig RP, Park LK: Comprehensive gene expression profiling following DNA vaccination of rainbow trout against infectious hematopoietic necrosis virus. Mol Immunol. 2006, 43: 2089-2106. 10.1016/j.molimm.2005.12.005.

18. Cuesta A, Tafalla C: Transcription of immune genes upon challenge with viral hemorrhagic septicemia virus (VHSV) in DNA vaccinated rainbow trout (*Oncorhynchus mykiss*). Vaccine. 2009, 27: 280-289. 10.1016/j.vaccine.2008.10.029.

19. Lorenzen N, Lorenzen E, Einer-Jensen K, LaPatra SE: Immunity induced shortly after DNA vaccination of rainbow trout against rhabdoviruses protects against heterologous virus but not against bacterial pathogens. Dev Comp Immunol. 2002, 26: 173-179. 10.1016/S0145-305X(01)00059-3.

20. Lorenzen E, Lorenzen N, Einer-Jensen K, Brudeseth B, Evensen Ø: Time course study of in situ expression of antigens following DNA-vaccination against VHS in rainbow trout (*Oncorhynchus mykiss* Walbaum) fry. Fish Shellfish Immunol. 2005, 19: 27-41. 10.1016/j.fsi.2004.10.009.

21. Gomez-Chiarri M, Livingston SK, Muro-Cacho C, Sanders S, Levine RP: Introduction of foreign genes into the tissue of live fish by direct injection and particle bombardment. Dis Aquat Organ. 1996, 27: 5-12.

22. Tonheim TC, Bøgwald J, Dalmo RA: What happens to the DNA vaccine in fish? A review of current knowledge. Fish Shellfish Immunol. 2008, 25: 1-18. 10.1016/j.fsi.2008.03.007.

23. Kurath G: Biotechnology and DNA vaccines for aquatic animals. Rev Sci Tech. 2008, 27: 175-196.

24. Redding L, Weiner DB: DNA vaccines in veterinary use. Expert Rev Vaccines. 2009, 8: 1251-1276. 10.1586/erv.09.77.

25. Gomez-Casado E, Estepa A, Coll JM: A comparative review on European-farmed finfish RNA viruses and their vaccines. Vaccine. 2011, 29: 2657-2671. 10.1016/j.vaccine.2011.01.097.

26. Alonso M, Chiou PP, Leong JA: Development of a suicidal DNA vaccine for infectious hematopoietic necrosis virus (IHNV). Fish Shellfish Immunol. 2011, 30: 815-823. 10.1016/j.fsi.2011.01.001.

27. Penaranda MMD, LaPatra SE, Kurath G: Specificity of DNA vaccines against the U and M genogroups of infectious hematopoietic necrosis virus (IHNV) in rainbow trout (*Oncorhynchus mykiss*). Fish Shellfish Immunol. 2011, 31: 43-51. 10.1016/j.fsi.2011.03.003.

28. Adomako M, St-Hilaire S, Zheng Y, Eley J, Marcum RD, Sealey W, Donahower BC, LaPatra S, Sheridan PP: Oral DNA vaccination of rainbow trout, *Oncorhynchus mykiss* (Walbaum), against infectious haematopoietic necrosis virus using PLGA [Poly(D, L-Lactic-Co-Glycolic Acid)] nanoparticles. J Fish Dis. 2012, 35: 203-214. 10.1111/j.1365-2761.2011.01338.x.

29. Choi SH, Kim MS, Kim KH: Immunization of olive flounder (*Paralichthys olivaceus*) with an auxotrophic *Edwardsiella tarda* mutant harboring the VHSV DNA vaccine. Fish Shellfish Immunol. 2012, 33: 569-574. 10.1016/j.fsi.2012.06.013.

30. Heras AI D l, Rodríguez Saint-Jean S, Perez-Prieto SI: Immunogenic and protective effects of an oral DNA vaccine against infectious pancreatic necrosis virus in fish. Fish Shellfish Immunol. 2010, 28: 562-570. 10.1016/j.fsi.2009.12.006.

31. Munang'andu HM, Fredriksen BN, Mutoloki S,

Brudeseth B, Kuo T-Y, Marjara IS, Dalmo RA, Evensen Ø: Comparison of vaccine efficacy for different antigen delivery systems for infectious pancreatic necrosis virus vaccines in Atlantic salmon (*Salmo salar* L.) in a cohabitation challenge model. Vaccine. 2012, 30: 4007-4016. 10.1016/j.vaccine.2012.04.039.

32. Xu C, Mutoloki S, Evensen Ø: Superior protection conferred by inactivated whole virus vaccine over subunit and DNA vaccines against salmonid alphavirus infection in Atlantic salmon (*Salmo salar* L.). Vaccine. 2012, 30: 3918-3928. 10.1016/j.vaccine.2012.03.081.
33. Zhang M, Hu YH, Xiao ZZ, Sun Y, Sun L: Construction and analysis of experimental DNA vaccines against megalocytivirus. Fish Shellfish Immunol. 2012, 33: 1192-1198. 10.1016/j.fsi.2012.09.010.
34. Jiao XD, Zhang M, Hu YH, Sun L: Construction and evaluation of DNA vaccines encoding *Edwardsiella tarda* antigens. Vaccine. 2009, 27: 5195-5202. 10.1016/j.vaccine.2009.06.071.
35. Sun Y, Liu CS, Sun L: Construction and analysis of the immune effect of an *Edwardsiella tarda* DNA vaccine encoding a D15-like surface antigen. Fish Shellfish Immunol. 2011, 30: 273-279. 10.1016/j.fsi.2010.10.020.
36. Sun Y, Liu CS, Sun L: Comparative study of the immune effect of an *Edwardsiella tarda* antigen in two forms: Subunit vaccine vs DNA vaccine. Vaccine. 2011, 29: 2051-2057. 10.1016/j.vaccine.2011.01.013.
37. Sun Y, Hu YH, Liu CS, Sun L: Construction and comparative study of monovalent and multivalent DNA vaccines against*Streptococcus iniae*. Fish Shellfish Immunol. 2012, 33: 1303-1310. 10.1016/j.fsi.2012.10.004.
38. Sun Y, Hu YH, Liu CS, Sun L: A *Streptococcus iniae* DNA vaccine delivered by a live attenuated *Edwardsiella tarda* via natural infection induces cross-genus protection. Lett Appl Microbiol. 2012, 55: 420-426.
39. Qin YX, Su YQ, Wang SF, Yan QP: Immunogenicity and protective efficacy of *Vibrio harveyi* pcFlaA DNA vaccine in*Epinephelus awoara*. Chin J Oceanol Limn. 2009, 27: 769-774.
40. Hu YH, Sun L: A bivalent *Vibrio harveyi* DNA vaccine induces strong protection in Japanese flounder (*Paralichthys olivaceus*). Vaccine. 2011, 29: 4328-4333. 10.1016/j.vaccine.2011.04.021.
41. Liang HY, Wu ZH, Jian JC, Huang YC: Protection of red snapper (*Lutjanus sanguineus*) against *Vibrio alginolyticus* with a DNA vaccine containing flagellin flaA gene. Lett Appl Microbiol. 2011, 52: 156-161. 10.1111/j.1472-765X.2010.02981.x.
42. Plant KP, LaPatra SE, Cain KD: Vaccination of rainbow trout, *Oncorhynchus mykiss* (Walbaum), with recombinant and DNA vaccines produced to *Flavobacterium psychrophilum* heat shock proteins 60 and 70. J Fish Dis. 2009, 32: 521-534. 10.1111/j.1365-2761.2009.01008.x.
43. Priya TAJ, Lin YH, Wang YC, Yang CS, Chang PS, Song YL: Codon changed immobilization antigen (iAg), a potent DNA vaccine in fish against *Cryptocaryon irritans* infection. Vaccine. 2012, 30: 893-903. 10.1016/j.vaccine.2011.11.102.
44. Tan CW, Jesudhasan P, Woo PTK: Towards a metalloprotease-DNA vaccine against piscine cryptobiosis caused by*Cryptobia salmositica*. Parasitol Res. 2008, 102: 265-275.
45. Jorgensen LV, Sigh J, Kania PW, Holten-Andersen L, Buchmann K, Clark T, Rasmussen JS, Einer-Jensen K, Lorenzen N: Approaches towards DNA vaccination against a skin ciliate parasite in fish. PloS One. 2012, 7: e48129-10.1371/journal.pone.0048129.
46. Garver KA, Conway CM, Elliott DG, Kurath G: Analysis of DNA-vaccinated fish reveals viral antigen in muscle, kidney and thymus, and transient histopathologic changes. Mar Biotechnol. 2005, 7: 540-553. 10.1007/s10126-004-5129-z.
47. Budker V, Budker T, Zhang GF, Subbotin V, Loomis A, Wolff JA: Hypothesis: naked plasmid DNA is taken up by cells *in vivo* by a receptor-mediated process. J Gene Med. 2000, 2: 76-88.
48. Zhong Q, Chinta DM, Pamujula S, Wang H, Yao X, Mandal TK, Luftig RB: Optimization of DNA delivery by three classes of hybrid nanoparticle/DNA complexes. J Nanobiotechnol. 2010, 8: 6-10.1186/1477-3155-8-6.
49. Desmet CJ, Ishii KJ: Nucleic acid sensing at the interface between innate and adaptive immunity in vaccination. Nat Rev Immunol. 2012, 12: 479-491. 10.1038/nri3247.
50. Mukhopadhyay S, Gordon S: The role of scavenger receptors in pathogen recognition and innate immunity. Immunobiology. 2004, 209: 39-49. 10.1016/j.imbio.2004.02.004.
51. Seternes T, Tonheim TC, Løvoll M, Bøgwald J, Dalmo RA: Specific endocytosis and degradation of naked DNA in the endocardial cells of cod (*Gadus morhua* L.). J Exp Biol. 2007, 210: 2091-2103. 10.1242/jeb.003186.
52. Zhu FG, Reich CF, Pisetsky DS: The role of the macrophage scavenger receptor in immune stimulation by bacterial DNA and synthetic oligonucleotides. Immunology. 2001, 103: 226-234. 10.1046/j.1365-2567.2001.01222.x.
53. Zhang XX, McIntosh TJ, Grinstaff MW: Functional lipids and lipoplexes for improved gene delivery. Biochimie. 2012, 94: 42-58. 10.1016/j.biochi.2011.05.005.
54. Weintraub H, Cheng PF, Conrad K: Expression of transfected DNA depends on DNA topology. Cell. 1986, 46: 115-122. 10.1016/0092-8674(86)90865-2.
55. Williams JA, Carnes AE, Hodgson CP: Plasmid DNA vaccine vector design: impact on efficacy, safety and upstream production. Biotechnol Adv. 2009, 27: 353-370. 10.1016/j.biotechadv.2009.02.003.
56. Davis HL, Millan CLB, Watkins SC: Immune-mediated destruction of transfected muscle fibers after direct gene transfer with antigen-expressing plasmid DNA. Gene Ther. 1997, 4: 181-188. 10.1038/sj.gt.3300380.
57. Tan Y, Li S, Pitt BR, Huang L: The inhibitory role of CpG immunostimulatory motifs in cationic lipid vector-mediated transgene expression in vivo. Hum Gene Ther. 1999, 10: 2153-2161. 10.1089/10430349950017149.
58. Qin LH, Ding YZ, Pahud DR, Chang E, Imperiale MJ,

Bromberg JS: Promoter attenuation in gene therapy: interferon-gamma and tumor necrosis factor-alpha inhibit transgene expression. Hum Gene Ther. 1997, 8: 2019-2029. 10.1089/hum.1997.8.17-2019.

59. McLauchlan PE, Collet B, Ingerslev E, Secombes CJ, Lorenzen N, Ellis AE: DNA vaccination against viral haemorrhagic septicaemia (VHS) in rainbow trout: size, dose, route of injection and duration of protection-early protection correlates with Mx expression. Fish Shellfish Immunol. 2003, 15: 39-50. 10.1016/S1050-4648(02)00137-7.

60. Hansen E, Fernandes K, Goldspink G, Butterworth P, Umeda PK, Chang KC: Strong expression of foreign genes following direct injection into fish muscle. FEBS Lett. 1991, 290: 73-76. 10.1016/0014-5793(91)81229-2.

61. Doh SG, Vahlsing HL, Hartikka J, Liang X, Manthorpe M:Spatial-temporal patterns of gene expression in mouse skeletal muscle after injection of lacZ plasmid DNA. Gene Ther. 1997, 4: 648-663. 10.1038/sj.gt.3300460.

62. Corbeil S, LaPatra SE, Anderson ED, Kurath G: Nanogram quantities of a DNA vaccine protect rainbow trout fry against heterologous strains of infectious hematopoietic necrosis virus. Vaccine. 2000, 18: 2817-2824. 10.1016/S0264-410X(00)00078-5.

63. Lorenzen E, Einer-Jensen K, Rasmussen JS, Kjaer TE, Collet B, Secombes CJ, Lorenzen N: The protective mechanisms induced by a fish rhabdovirus DNA vaccine depend on temperature. Vaccine. 2009, 27: 3870-3880. 10.1016/j.vaccine.2009.04.012.

64. Myhr A, Dalmo RA: Introduction of genetic engineering in aquaculture: Ecological and ethical implications for science and governance. Aquaculture. 2005, 250: 542-554. 10.1016/j.aquaculture.2004.12.032.

65. Saade F, Petrovsky N: Technologies for enhanced efficacy of DNA vaccines. Expert Rev Vaccines. 2012, 11: 189-209. 10.1586/erv.11.188.

66. Caipang CMA, Hirono I, Aoki T: Modulation of the early immune response against viruses by a teleostean interferon regulatory factor-1 (IRF-1). Comp Biochem Physiol A Mol Integr Physiol. 2009, 152: 440-446. 10.1016/j.cbpa.2008.11.022.

67. Tonheim T, Dalmo RA, Bøgwald J, Seternes T: Specific uptake of plasmid DNA without reporter gene expression in Atlantic salmon (*Salmo salar* L.) kidney after intramuscular administration. Fish Shellfish Immunol. 2008, 24: 90-101. 10.1016/j.fsi.2007.09.006.

68. Tinsley-Bown AM, Fretwell R, Dowsett AB, Davis SL, Farrar GH: Formulation of poly(D,L-lactic-co-glycolic acid) microparticles for rapid plasmid DNA delivery. J Control Release. 2000, 66: 229-241. 10.1016/S0168-3659(99)00275-8.

69. Tian J, Sun X, Chen X, Yuc J, Qu L, Wang L: The formulation and immunisation of oral poly (DL-lactide-co-glycolide) microcapsules containing a plasmid vaccine against lymphocystis disease virus in Japanese flounder (*Paralichthys olivaceus*). Int Immunopharmacol. 2008, 8: 900-908. 10.1016/j.intimp.2008.02.006.

70. Tian J, Yu J: Poly(lactic-co-glycolic acid) nanoparticles as candidate DNA vaccine carrier for oral immunization of Japanese flounder (*Paralichthys olivaceus*) against lymphocystis disease virus. Fish Shellfish Immunol. 2011, 30: 109-117. 10.1016/j.fsi.2010.09.016.

71. Hølvold LB, Fredriksen BN, Bøgwald J, Dalmo RA: Transgene and immune gene expression following intramuscular injection of Atlantic salmon (*Salmo salar* L.) with DNA-releasing PLGA nano- and microparticles. Fish Shellfish Immunol. 2013, 35: 890-899. 10.1016/j.fsi.2013.06.030.

72. Malyala P, O'Hagan DT, Singh M: Enhancing the therapeutic efficacy of CpG oligonucleotides using biodegradable microparticles. Adv Drug Deliv Rev. 2009, 61: 218-225. 10.1016/j.addr.2008.12.009.

73. Nguyen DN, Green JJ, Chan JM, Longer R, Anderson DG: Polymeric materials for gene delivery and DNA vaccination. Adv Mater. 2009, 21: 847-867. 10.1002/adma.200801478.

74. Cohen H, Levy RJ, Gao J, Fishbein I, Kousaev V, Sosnowski S, Slomkowski S, Golomb G: Sustained delivery and expression of DNA encapsulated in polymeric nanoparticles. Gene Ther. 2000, 7: 1896-1905. 10.1038/sj.gt.3301318.

75. Prabha S, Zhou WZ, Panyam J, Labhasetwar V: Size-dependency of nanoparticle-mediated gene transfection: studies with fractionated nanoparticles. Int J Pharm. 2002, 244: 105-115. 10.1016/S0378-5173(02)00315-0.

76. Sharp FA, Ruane D, Claass B, Creagh E, Harris J, Malyala P, Singh M, O'Hagan DT, Petrilli V, Tschopp J, O'Neill LA, Lavelle EC: Uptake of particulate vaccine adjuvants by dendritic cells activates the NALP3 inflammasome. Proc Natl Acad Sci U S A. 2009, 106: 870-875. 10.1073/pnas.0804897106.

77. Hedley ML, Curley J, Urban R: Microspheres containing plasmid-encoded antigens elicit cytotoxic T-cell responses. Nat Med. 1998, 4: 365-368. 10.1038/nm0398-365.

78. Singh M, Briones M, Ott G, O'Hagan D: Cationic microparticles: a potent delivery system for DNA vaccines. Proc Natl Acad Sci U S A. 2000, 97: 811-816. 10.1073/pnas.97.2.811.

79. Jang JH, Shea LD: Intramuscular delivery of DNA releasing microspheres: microsphere properties and transgene expression. J Control Release. 2006, 112: 120-128. 10.1016/j.jconrel.2006.01.013.

80. Labhasetwar V, Bonadio J, Goldstein SA, Levy JR: Gene transfection using biodegradable nanospheres: results in tissue culture and a rat osteotomy model. Colloid Surface B. 1999, 16: 281-290. 10.1016/S0927-7765(99)00079-X.

81. Tian JY, Sun XQ, Chen XG: Formation and oral administration of alginate microspheres loaded with pDNAcoding for lymphocystis disease virus (LCDV) to Japanese flounder. Fish Shellfish Immunol. 2008, 24: 592-599. 10.1016/j.fsi.2008.01.009.

82. Leal CAG, Carvalho-Castro GA, Sacchetin PSC, Lopes CO, Moraes AM, Figueiredo HCP: Oral and parenteral vaccines against *Flavobacterium columnare*: evaluation of humoral immune response by ELISA and in vivo efficiency in Nile tilapia (*Oreochromis niloticus*).

Aquacult Int. 2010, 18: 657-666. 10.1007/s10499-009-9287-x.

83. Tian J, Yu J, Sun X: Chitosan microspheres as candidate plasmid vaccine carrier for oral immunisation of Japanese flounder (*Paralichthys olivaceus*). Vet Immunol Immunopathol. 2008, 126: 220-229. 10.1016/j.vetimm.2008.07.002.

84. Kumar SR, Ahmed VPI, Parameswaran V, Sudhakaran R, Babu VS, Hameed ASS: Potential use of chitosan nanoparticles for oral delivery of DNA vaccine in Asian sea bass (*Lates calcarifer*) to protect from *Vibrio* (*Listonella*) *anguillarum*. Fish Shellfish Immunol. 2008, 25: 47-56. 10.1016/j.fsi.2007.12.004.

85. Leon-Rodriguez L, Luzardo-Alvarez A, Blanco-Mendez J, Lamas J, Leiro J: Biodegradable microparticles covalently linked to surface antigens of the scuticociliate parasite *P. dicentrarchi* promote innate immune responses *in vitro*. Fish Shellfish Immunol. 2013, 34: 236-243. 10.1016/j.fsi.2012.10.029.

86. Harikrishnan R, Kim JS, Balasundaram C, Heo MS: Vaccination effect of liposomes entrapped whole cell bacterial vaccine on immune response and disease protection in *Epinephelus bruneus* against *Vibrio harveyi*. Aquaculture. 2012, 342: 69-74.

87. Behera T, Swain P: Antigen adsorbed surface modified poly-epsilon-caprolactone microspheres stimulates both adaptive and innate immune response in fish. Vaccine. 2012, 30: 5278-5284. 10.1016/j.vaccine.2012.05.028.

88. Behera T, Swain P: Antigen adsorbed calcium phosphate nanoparticles stimulate both innate and adaptive immune response in fish, *Labeo rohita* H. Cell Immunol. 2011, 271: 350-359. 10.1016/j.cellimm.2011.07.015.

89. Myhr AI, Dalmo RA: Nanotechnology and risk: what are the issues?. Nanoethics: The Ethical and Social Implications of Nanotechnology. Edited by: Allhoff FLP, Moor J, Weckert J. 2007, New Jersey: John Wiley & Sons Inc, 149-161.

90. Gvili K, Benny O, Danino D, Machluf M: Poly(D,L-lactide-co-glycolide acid) nanoparticles for DNA delivery: waiving preparation complexity and increasing efficiency. Biopolymers. 2007, 85: 379-391. 10.1002/bip.20697.

91. Nielsen KN, Fredriksen BN, Myhr AI: Mapping uncertainties in the upstream: the case of PLGA nanoparticles in salmon vaccines. Nanoethics. 2011, 5: 57-71. 10.1007/s11569-011-0111-5.

92. Lorenzen N, LaPatra SE: DNA vaccines for aquacultured fish. Rev Sci Tech. 2005, 24: 201-213.

93. Gillund F, Tonheim T, Seternes T, Dalmo RA, Myhr AI: DNA vaccination in aquaculture –Expert judgements of impact on environment and fish health. Aquaculture. 2008, 284: 25-34. 10.1016/j.aquaculture.2008.07.044.

94. Salonius K, Simard N, Harland R, Ulmer JB: The road to licensure of a DNA vaccine. Curr Opin Investig Drugs. 2007, 8: 635-641.

95. EMEA, Committee for Veterinary Medicinal Products: Note for guidance: DNA vaccines non-amplifiable in Eukaryotic cells for veterinary use. 2001, [http://www.ema.europa.eu/docs/en_GB/document_library/Scientific_guideline/2009/10/WC500004614.pdf]

96. EMEA, Committee for the Medicinal Products for Human Use: Concept paper on guidance for DNA vaccines. 2012, [http://www.ema.europa.eu/docs/en_GB/document_library/Scientific_guideline/2012/03/WC500124898.pdf]

97. Food and Drug Administration (FDA), Center for Biologics Evaluation and Research: Guidance for Industry. Considerations for Plasmid DNA Vaccines for Infectious Disease indications. 2007, [http://www.fda.gov/downloads/biologicsbloodvaccines/guidancecomplianceregulatoryinformation/guidances/vaccines/ucm091968.pdf]

98. Foss GS, Rogne S: Gene medication or genetic modification? The devil is in the details. Nat Biotechnol. 2003, 21: 1280-1281. 10.1038/nbt1103-1280.

CHAPTER 3

ORAL VACCINATION OF FISH – ANTIGEN PREPARATIONS, UPTAKE, AND IMMUNE INDUCTION

Stephen Mutoloki, Hetron Mweemba Munang'andu, and Øystein Evensen

Keywords: Oral vaccination, fishes, oral tolerance, antigen production, local and systemic immune responses

BACKGROUND

Mucosal surfaces constitute the largest body area of living organisms in constant contact with the external environment and are responsible for the maintenance of immunological homeostasis. The gut is the most attractive route for antigen delivery in fish for several reasons. It offers an easy way of administering antigens; delivery is less stressful, both small and large sized fish can be vaccinated and this approach is associated with no side effects. Despite these advantages, there are currently very few oral vaccines registered for use in the aquaculture industry. For example, out of 17 commercially available vaccines against viruses reported in 2014, only 2 were oral preparations (1). This low number was attributed to poor performance of oral vaccines in general compared to their injection counterparts (2), possibly as a result of antigen degradation during passage through the hostile stomach environment prior to reaching the second segment of the intestine where absorption takes place. Despite this, the search for oral vaccines that meet the desired induction of the immune response and protection has continued. The struggle in developing good oral vaccines is complicated by the lack of proper understanding of what constitutes good immunological induction leading to protection on one side, and the induction of tolerance on the other.

For the most part, efforts to develop vaccines whether for parenteral or oral administration have been largely directed at bacterial and viral infections. Recently, however, other challenges facing the aquaculture industry, namely, ectoparasites have prompted efforts to address in particular sea lice and amoeba infections in salmonids.

ANTIGEN UPTAKE AND DISTRIBUTION

The gut of teleosts can be subdivided into two groups in terms of the uptake of macromolecules: (1) those devoid of stomachs, i.e., pre-larval stages of fish as well as certain species that do not develop stomachs throughout their life time, e.g., cyprinids; (2) fish with stomachs and therefore possess a segment of low pH and are capable of pre-digestion (3). The gut can further be divided into three segments based on the morphology of enterocytes namely segment 1: 60–75% of the total gut length depending on species with cells considered absorptive; segment 2: 10–15% of the gut length and with enterocytes characterized by large supranuclear vacuoles and high pinocytotic activity; and segment 3: less well-characterized but with enterocytes that have osmoregulatory function and short microvilli [summarized in Ref. (4,5)].

The most uptake of macromolecules in the gut of fish takes place in the second segment although in a few species like the cod, this segment has not been identified (6). In general, particulate and soluble antigens are taken up by different mechanisms: particles such as horse radish peroxidase (HRP) in carp are thought to be taken up by receptor-mediated endocytosis (7). HRP was observed in the endolysosomal compartment followed by intercellular spaces where after it was observed systemically (8). Ferritin (soluble), on the other hand, was taken up into supranuclear vacuoles (endosomes), a route believed to be used also by other soluble factors such as bacterial LPS (9).

Antigens administered via the gut are invariably taken up by cells lining this organ such as enterocytes and M cells in the case of higher vertebrates. In fish, enterocytes have been shown to take up antigens (10, 11). M-like cells, on the other hand, have been shown to have functional endolysosomal organelles

Citation: Mutoloki, S., Munang'andu, H. M., and Evensen, Ø. (2015). Oral vaccination of fish – antigen preparations, uptake, and immune induction. Frontiers in Immunology, 6:519. DOI: 10.3389/fimmu.2015.00519.

and not able to phagocytoze inactivated bacteria in fish (5). It is nevertheless thought that if aided with certain signals like poly D, l-lactide-co-glycolic acid (PLGA) microparticles, M-like cells may still be able to take up antigens (12). It is generally considered that the best immune induction will be that which mimics a natural infection. Different pathogens are internalized by different routes/receptors under natural infections and it is not unlikely that some of the pathogen for which oral vaccines are sought will not reach or target the enterocytes. To what extent this will be an issue in the efficacy of oral vaccines remains to be shown. This is especially relevant given the compartmentalization of immune responses as discussed below. As fish reside in the aquatic environment, it is likely that most pathogens will invade via mucosal routes, including the gills, skin (fin bases) (13), and also via the gut (14–16).

IMMUNE INDUCTION FOLLOWING ORAL VACCINATION

It is well-known that lymphoid structures associated with the gut of fish are different from those of mammals. Fish do not have lymph nodes or Peyer's patches but instead possess a less organized, diffuse gut-associated lymphoid tissue (GALT), which is functionally different from that of mammals (17). Nevertheless, the fish GALT is capable of local immune responses. Indeed, oral administration of antigens result in the up-regulation of genes related to recruitment of immune cells (18) and local antibody production (19, 20). Protection is, however, variable with some reporting adequate (3, 21–23) and others inadequate (24). As primary vaccines, orally delivered vaccines especially containing inactivated whole antigens have traditionally not featured well, often resulting in suboptimal protection against several pathogens. When used for boosting, however, oral vaccines have been shown to be capable of enhancing or extending protection (25) although the antibody response is transient, typically lasting about 3 months (3, 9, 12, 26). Interestingly, varying the dosage regime, for example, by administering the vaccine 3 days/week for 2 months instead of 5 days/month gave different results with the former being more effective (24). The basis for this difference has not been examined but may reflect a delicate balance that has to be maintained in augmenting the immune response by oral boosting and should be subject of further studies.

Other factors that influence efficacy of oral vaccines include the nature of antigens, formulation, and dosage. Bacterial whole antigens generally perform better than viral antigens. Formulation is important for some vaccines in general and for oral vaccines, it is even more so given that they act as a means to protect antigens through the stomach as further discussed below.

Dosage is key to any vaccine regime and for orally administered vaccines, it is difficult to determine the dose at individual level. Examination of feed residues and weight gain at population level following feeding gives an idea of the average in-take in the population. Then comes the distribution of antigens per weight unit of feed or per pellet, which is also difficult to determine. This is a topic addressed to a very little extent in published studies and potentially has a great impact on the outcome of mucosal immunization modalities.

Compartmentalization of the immune system is yet another factor to consider, a well-known phenomenon in mammals. The basis for this is the trafficking of immune competent cells aided by homing receptors (27, 28). Further to this, whether an asymmetric immune response is induced in fish as seen in mice (29) is not known. If so, the concept would be that immune responses induced in the gut would induce local and systemic immune responses, while parenteral delivery will not yield protective immune responses on mucosal surfaces (Figure 1). In a recent study, it was shown that rainbow trout infected with a gut parasite (*Ceratomyxa shasta*) had local (gut) IgT responses but not IgM responses, while systemically, IgM levels were high but IgT was not detected in circulation (30). Reciprocal modalities were not tested but in a recent study it was shown that parenteral vaccination efficiently stimulates systemic responses but is a poor inducer of mucosal immunity (12, 31). Furthermore, oral administration of antigens results in stimulation of both systemic and mucosal responses (32, 33). Together, these findings align with a hypothesis also for fish that the immune response shows an asymmetrical pattern but further studies are needed to better understand the importance and implications.

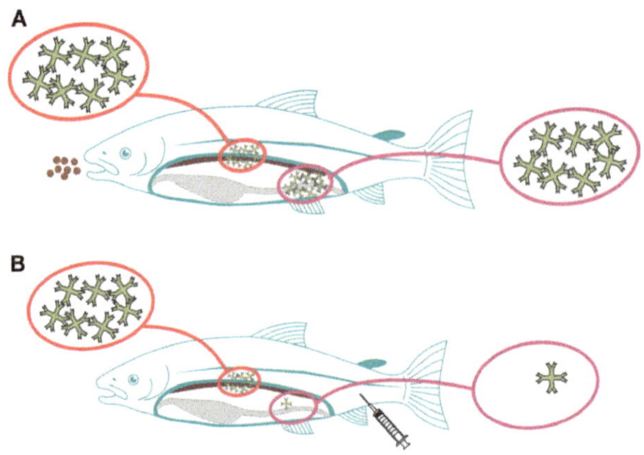

Figure 1. Proposed asymmetry for immune responses induced via mucosal (gut) versus parenteral routes in fish. When antigens are delivered via the gut, local and systemic immune responses will be elicited, symbolized by high amounts of circulating IgM (A). When the antigens are delivered parenterally, systemic responses will be strong, while local (gut) responses will be almost absent (B).

Mucosal immunoglobulins (such as IgT or IgZ) are relative new discoveries in fish immunology. For this reason, their functional role is not well-understood/studied and in general, very few studies (perhaps none) show that IgT plays a role in protecting mucosal surfaces. Although Zhang and co-workers (30) provides some guidance, it does not show that IgT is protective. It could be that IgT merely plays a role in regulating the commensal flora thus keeping it in check, preventing the proliferation of any single bacterial species beyond a level that could potentially result in pathology and disease, in line with current thinking in mammalian immunology (34). There are a few studies on the potential importance of IgT in the gills (intraepithelial location) against *I. multifilis* infection but the functional importance of IgT+ B cells is actually not known (35).

ORAL TOLERANCE

Oral tolerance is defined as the hypo-responsiveness to a fed antigen (36) and is a result of the suppression of the cellular and/or humoral immune response (37). It is a phenomenon that has been well-known for ages and in fish, it has been recognized as the suppression of antibodies (12, 24, 26) and is easily induced. In higher vertebrates, the causes of tolerance are multiple including low doses that favor the induction of Tregs; and higher doses associated with anergy (36). Repeated administration of small amounts of antigens, vaccination of too young (immunocompetent) fish, low temperatures (lower end of the permissive limit), type of antigens and administration regime as well as genetics are factors that have been implicated in the induction of tolerance in fish (11, 18, 24, 38).

Mechanisms of tolerance involve the induction of Tregs associated with up-regulation of FoxP3 and production of TGF-β (Figure 2). Other cells are also involved and these include dendritic cells, macrophages, and epithelial cells. In fish, the mechanisms have not been elucidated in any detail; it is for the most part illustrated by decreased antibody response following repeated antigen exposure (24, 26, 32). It is only recently that it has been shown that the suppression of antibody production was accompanied by the induction of FoxP3, TGF-β, and IL-10 (32), strengthening the belief that the mechanism is comparable to that of higher vertebrates.

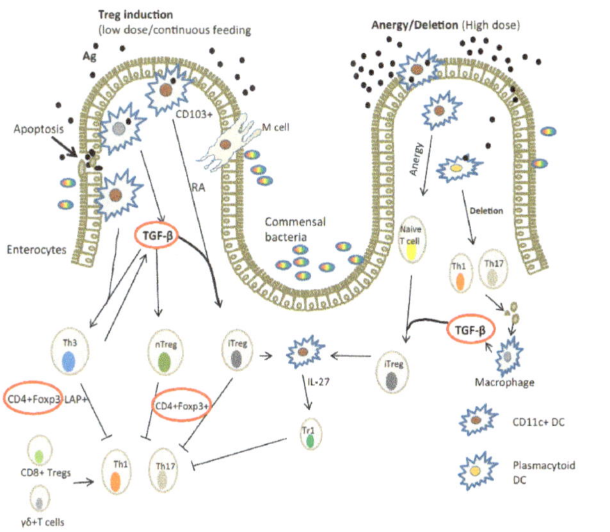

Figure 2. Mechanism of induction of oral tolerance in the gut in mammals [adapted from Ref. (36)]. There are several ways in which oral antigens may be taken up through the gut epithelium: by the enterocytes; sampled by DCs (macrophage-like cells in fish) that penetrate the lumen or via M cells. The gut environment favors tolerance probably to allow for gut microflora. DCs are known to drive Treg differentiation from FoxP3, TGF-β, and IL-10 from gut epithelial cells. Lower antigen doses tend to induce TGF-β while high doses lead to anergy. In Atlantic salmon, the expression of IL-10, TGF-β, Foxp3 (circled red) associated with suppressed antibody responses have been demonstrated (32). Key: TGF, transforming growth factor; RA, retinoic acid; DC, dendritic cells; LAP, latency associated peptide; Foxp3, forkhead box protein; IL, interleukin.

RECENT DEVELOPMENTS TO IMPROVE THE PERFORMANCE OF ORAL VACCINES IN FISH

The disadvantages associated with oral vaccines include their demand for high antigens requirements, the need to protect antigens as they pass through the stomach as well as the formulation of vaccines to improve the stimulation of protective immunity. The following are the strides that have been taken by the scientific community to bring the frontier of oral vaccinology in fish forward.

ANTIGEN PRODUCTION

A prerequisite to the production of any vaccine is the ability to scale-up antigens easily and at a low cost. While bacteria and bacteria-based products, such as subunit antigens, are quite easily propagated by fermentation, scaling up for virus antigens can be challenging and thus this section focusses mainly on virus antigens.

Virus Antigen Preparation

Virus antigens used in the production of vaccines in general and oral vaccines in particular range from native whole pathogens (whole virus preparations) to subunit or synthetic products.

Native Whole Pathogen Antigens

At present, most commercial vaccines against viral diseases for parenteral delivery in fish are produced by inactivating whole pathogens. For fish vaccines, viral antigens are typically produced by propagation in cell culture and this is limited by the yield obtained. As an example, infectious pancreatic necrosis virus (IPNV) produces yields ranging between 10^6 and $10^{10}TCID_{50}ml^{-1}$ if grown in RTG 2, CHSE, or other permissible cell lines (39–41). This is complicated by the fact that yields may be inconsistent and differ greatly between workers and laboratories. To resolve the problems stated above, several solutions have been attempted in recent years, including (1) discovery of cell lines with short regeneration time and high antigen yields; and (2) development of cell culture systems that can house higher cell numbers per unit volume. Using IPNV as an example, Asian Grouper strain K cells (AGK cells) have been produced from the skin of Orange spotted grouper (*Epinephelus coioides*) that yield high amounts of virus (approximately $10^{10}TCID_{50}ml^{-1}$) (42) while having a short cell-generation turn-around time, i.e., frequent splits and high split ratios (43).

For cell culture systems, flasks that can accommodate more cells per unit volume than ordinary flasks have recently been developed by commercial companies. An example is the multilayered hyper flasks from Corning® with scaled-up volume (up to 10 times) on which cells can be grown (www.sigmaaldrich.com). Another example is the BelloCell® from CESCO Bioengineering Inc. (44). The principle of the BelloCell is that of alternating nutrient and gaseous exchange. This results in maximum cell growth and increased cell population. Indeed, improved cell growth has been reported using this method and it has potential for use in fish vaccination, although currently there are no reports that suggest that it has yet been used so far. Other developments include the three dimensional cell culture systems, such as Hydromatrix™, MaxGel™, and Mouse ECM from Sigma® that mimic *in vivo* growth conditions of cells. As with BelloCells, the systems result in better growth of cells and will likely lead to increased antigen yields. Finally, the contribution of other improvements in the cell culture cultivation process, for example, refinement in the culture media production and process control cannot be underestimated.

Subunit Antigens Produced in Bacteria and Yeast Cell Systems

Subunit antigens are produced from heterologous protein expression systems and offer the safest and most attractive means of antigen production. Unlike live pathogens or DNA vaccines, subunit vaccines do not pose the risk of invading the host or integrating with the host DNA. In fish vaccinology, the most widely used protein expression systems for the production of subunit antigens are *E. coli* and yeast. Commercial oral vaccines have been used previously or are in production on the basis of these techniques (45–47). Nevertheless, other systems for example the Baculovirus expression system in eukaryotic cells has also been used especially at experimental level (33).

Subunit Antigens in Plant-Based Systems

During the last decade, the use of whole plants as antigen production systems has received a lot of attention owing to the advantages that they offer, such as ease of scaling up, reduced costs, and good safety margins (48). These systems have also been referred to as *molecular* farming and are the utilization of whole plants or plant cells/tissues cultured *in vitro* for the production of recombinant proteins. Plants have advantages compared to traditional platforms of recombinant protein production in the sense that they are less expensive to establish and maintain, they lack undesirable components such as hyper glycosylated proteins as found in yeast, and extraneous agents are less of a problem. Scalability is relatively easy. Finally, plants are higher eukaryotes and can fold and assemble complex/multimeric proteins and also perform post-translational modifications (48). Despite these many advantages, there are currently no recombinant proteins produced commercially. Many proof-of-principle studies have been done and several companies are investigating and exploring the commercial feasibility of such production systems. The focus of such efforts has been directed at a small number of well-characterized plants of which Tobacco cultivars Bright Yellow (BY-2) and *Nicotiana tabacum* 1 (NT-1) are most popular (49). It is noteworthy, however, that plant systems lack intrinsic benefits of cultured cells, for example, the difficulty associated with the control of growth conditions and batch to batch inconsistency (49). Regardless, this is likely to be the trend for the future also for the aquaculture research community (48).

DNA-Based Antigens

DNA-based vaccines in the aquaculture industry have met some success with one injectable vaccine against infectious hematopoietic necrosis in Atlantic salmon being licensed for commercial use in Canada (50). Although there is no commercial vaccine yet licensed for use as an oral vaccine, induction of protection against IPNV has been reported using this method at experimental level (21). The biggest problem with DNA vaccines in the aquaculture industry at the moment is the licensing requirements related to integration into genome studies, in principle related to safety to the end-consumer more than the vaccinated animal (51, 52).

Bacterial Antigens

In general, there has been more success with bacterial vaccines of the aquatic industry compared to their viral counterparts, for example, if we look at the relative number of commercial vaccines for salmonids in Norway (53). There are several reason for this, while all viruses are intracellular pathogens, most of the bacteria important to the fish industry are extracellular, with only a few exceptions for example *Piscirickettsia salmonis* whose vaccine development has equally been a challenge (54). The effective immunological response against extracellular bacterial pathogens is predominantly humoral, thus vaccines that act by inducing antibody responses will normally suffice to protect the animals against challenge. By contrast, protection against intracellular pathogens requires a combination of immunological responses, including humoral, cell-mediated, and cytotoxic responses. To produce a vaccine which induces all these responses is what makes it a challenge. By contrast, factors that contribute to the relative ease in development of bacterial vaccines compared to viral vaccines are their relatively large size and the type of antigens/immunogens they possess.

Oral bacterial vaccines have similar challenges as viral vaccines, for example, degradation of antigens as they pass through the stomach and requirements for large volumes. A study by Villumsen and others showed that protection of fish was conferred if the vaccine was administered anally compared to the oral route, with the authors suggesting that orally fed antigens were digested in the stomach (55). Indeed, commercial bacterial vaccines for oral administration, e.g., enteric red mouth vaccine (Aquavac ERM Oral vet from MSD animal health) are only intended for use as booster vaccines (56). Consequently, several studies aimed at finding the best solution to protect bacterial antigens in the stomach of fish have been conducted (57) and these are discussed in the sections below.

Some bacteria, specifically *Lactobacillus* expressing various proteins of IPNV have been applied in studies of oral vaccination of fish (58, 59). Although the results are contrasting, this is nevertheless an interesting concept that warrants further follow-up, as does the delivery of a combination of bacteria and viruses as multivalent oral vaccines. This may help synergize responses to viral antigens as has been observed for injectable vaccines.

ENCAPSULATION TECHNIQUES

Encapsulation refers to incorporation of materials, including food ingredients, cells, or others, into small capsules and is accomplished by several different techniques (60). Encapsulated materials, e.g., vaccine antigens can then be mixed with food for oral administration. In vaccine development for fish, there are essentially three methods by which this is done: (1) finished feed is top-dressed with vaccine powder by using adhesive agents, e.g., edible oil or gelatin; (2) finished feed is sprayed with the vaccine if the latter is in liquid form; and (3) mixing the antigen with the feed in the production process (61). The first two methods (top dressing) are quite simple to apply but have the disadvantage of uneven distribution in the feed and also the threat that the antigens are directly exposed to hostile stomach environment upon feeding, leading to degradation. By contrast, mixing the antigen with the feed gives the advantage of uniform distribution of the antigens in the feed. Since most fish feed is produced through an extrusion process at high temperature and pressure antigens would have to be added to the pellet at later stages, either in a vacuum infusion coating process. As a means of protecting the antigens against the hostile stomach environment, several encapsulations techniques have been developed and tried as discussed below.

Microalgae

Microalgae are potentially future candidates of recombinant vaccine production (62) both for higher vertebrates and the fish industry. This system is said to have among other advantages the ease of scalability, rapid transformation, and consistent transgene expression levels (62). In oral vaccination of fish, however, there are no reports on the use of this technology.

Alginate Particles

Alginates are another carrier that promises to revolutionize oral vaccine development. They occur naturally in brown algae and have been targets for trapping macromolecules and cells (63). In the aquaculture industry, they have been tested quite extensively (Table 1).

Table 1. Summary of previous studies of oral vaccination of fish using alginate microparticles.

Target	Fish species	Result	Reference
Commercial *Vibrio anguillarum* vaccine in two types of alginate microspheres	Carp and rainbow trout	Antibodies produced by carp only; protection not assessed	(38)
Plasmid DNA expressing major capsid protein of lymphocystis disease virus	Japanese flounder	Antigens detected in various tissues between 10 and 90 days post immunization	(64)
Aeromonas salmonicida recombinant A layer proteins in alginate beads	Goldfish	Antibodies produced but no difference in disease susceptibility between the control and treatment groups after challenge	(24)
Lactococcus garvieae bacterin	Rainbow trout	RPS of 50 achieved	(65)
L. garvieae bacterin	Rainbow trout	Achieved RPS of 53 after 30 days following challenge; after a boost at 61 days, RPS increased to 61 at 120 days	Altun et al. (66)
Flavobacterium columnare bacterium in alginate microparticles; compared oral and parenteral deliveries	Nile tilapia	No protection or significant antibody production in oral group	(67)
Infectious pancreatic necrosis virus VP2 plasmid DNA in alginate microspheres; administered by intubation	Brown trout and rainbow trout	RPS of 84 (brown trout) and between 67 and 83 (rainbow trout) after 30 days	(68)
IPNV VP2 plasmid DNA in feed pellets	Rainbow trout	Induction of transcriptional responses; RPS of 78–85	(25, 69, 70)
Inactivated IPNV virus encapsulated in alginate beads; feed pellets	Atlantic salmon	Induced antibody response; no protection measured	(32)

As can be seen from the table above, alginates seem to work well with DNA plasmids, giving RPS values of 67% or more in all species tested. Whether this technology can be transferred to other viral diseases of fish would be interesting.

Nanoparticles

In recent years, increased knowledge has accumulated over the use of different forms of nanoparticles in oral vaccination of fish. Rajesh et al. (71) examined the use of chitosan nanoparticles for orally delivering a DNA vaccine against *Vibrio anguillarum* in sea bass. While they demonstrated that fish took up the antigens, fish were, however, not protected and a relative RPS rate of 46% was recorded. A better protection against *V. parahaemolyticus* was recorded in black seabream (*Acanthropagus schelegelii* Bleeker) also vaccinated with a DNA vaccine loaded in nanoparticles, resulting in 72.3% RPS 3 weeks post vaccination.

For viral diseases, encapsulating DNA vaccines against infectious hematopoietic necrosis virus with Poly (d,l-lactic-co-glycolic acid) (PLGA) nanoparticles and adding to feed pellets showed that rainbow trout took-up the vaccine in the lower intestine within 96 h of feeding and also induced low levels of gene expression and specific antibodies but this was not sufficient to protect the fish against lethal challenge (72). When a DNA vaccine against lymphocystis loaded in PLGA was used to feed Japanese flounder, several innate immune parameters were induced suggesting that the system could be used as a carrier for plasmid DNA vaccines (73). Recently, Rivas-Aravena and co-workers (74) reported enhanced protection of Atlantic salmon fed with chitosan nanoparticles-based oral vaccines loaded with a DNA coding an alphavirus replicase (as an adjuvant) while the target antigen was ISAV. The authors reported 77% protection.

Biofilms

Biofilms are defined as communities of microbes adherent on a surface and usually held together by a polymetric extracellular matrix (75, 76). In so doing, they form a protective coat which has been taken advantage of in vaccine studies of fish to prevent the degradation of antigens through the stomach. Work in fish vaccination using biofilms is not as extensive as literature dates back to 2000–2004 and was centered on *A. hydrophila* (2, 77). In general, all the studies reported significant protection of fish fed with biofilms compared to free cells. No additional work has been reported thereafter.

Artemia

First described in the mid 1700s by Schlosser, Artemia have gained value as food for fish as well as for other uses (78). Artemia naturally consume the bacteria that they are immersed with thereby readily encapsulating them and it is this property that is exploited in vaccine development. This approach has been examined in oral vaccination of fish targeting different bacteria not only *V. anguillarum* (38) but also recombinant *E. coli.* (46). Where the immune response was evaluated, there were indications of immunosuppression following very early vaccination in carp while an increased antibody response was reported for older carp immunized at 58 days. Similarly, older sea bream developed highly increased antibodies 21 days following a boost (38).

Oral Vaccines Against Sea Lice and Amoeba

As fish farming expands, so does the number of important ectoparasites (79). In the salmonid industry at present, the most important ectoparasites include sea lice (*Lepeophtheirus salmonis* and *Caligus rogercrossey*) and amoeba (*Neoparamoeba perurans*) the causative agents of amoebic gill disease. Conventional treatments against the former have mainly involved the use of chemotherapy with products, such as organophosphates (80) and avermectins (81) but as with many other drugs, resistance is an increasing problem (82, 83). Amoebic gill disease is more problematic to treat but repeated freshwater baths (84) and also hydrogen peroxide (85) are used.

When it comes to vaccination, to the knowledge of the authors, there are no reports that address oral vaccination against sea lice (*L. salmonis*) in Atlantic salmon, or (*C. rogercrossey*) in Coho salmon. Since both parasites attach to and complete their life cycle on the skin of salmon, it is assumed that local responses would be important in preventing infections but so far no attempts have been made to induce immune responses via the mucosa.

Similarly, there are apparently no records on testing oral vaccines against amoeba in fish. Interestingly, in rodents,*Yersinia enterocolitica* expressing an amoeboid outer protein as a fusion protein has been shown to induce some degree of protection (86, 87). This could be an avenue to explore also for fish as an antigen delivery model against amoeba.

Summary of Status and Future Directions

There is a consensus among scientists working in this field that antigen uptake in teleosts takes place in the second gut segment. In order for some antigens to reach this site, protection against degradation in the stomach is necessary. The second gut segment is possibly the site where immune induction is initiated locally for antigens taken up from the intestines. However, the cell types, molecular elements, and even immune organs involved are yet to be definitively elucidated. Central to this is the role of IgT or IgZ, for example, do they neutralize antigens or merely regulate the relative number of bacteria in the gut microbiome? Could it be that bacteria coated with IgT represent proliferating, potentially disease-causing bacteria? Furthermore, what roles do IgT or IgZ have vis-à-vis IgM in protective immunity? What functional significance does compartmentalization of the different isotypes have? Thus, the mechanism of immune induction (local and systemic) and protection following oral vaccination is yet to be elucidated. Reagents that allow functional studies of IgT are in very few hands at the moment, and this has partly caused the slow pace in understanding its role in the protection of fish against pathogens. Further, it will be important to understand if antigens delivered at mucosal surfaces (oral, skin, gills, and/or nasal) will elicit both local and systemic responses. Studies here are not conclusive. While IHN virus delivered nasally elicit systemic immune responses (88) it has also been shown that live and inactivated IPN virus are taken up when delivered orally and anally (10). IPNV delivered orally give a boost response to circulating IgM (10). Despite these findings, there is a need to better understand if inactivated and live (replicating) antigens differ in their ability to induce systemic, protective responses when antigens are delivered locally. The potential of differences between pathogens also need to be explored.

When it comes to vaccine formulation, several candidates that can serve as vehicles for antigens, for example, alginates have been identified and shown to be capable of protecting antigens against degradation in the stomach. The contribution of these vehicles toward augmentation of the immune response, however, remains poorly understood. Similarly, the effects of adjuvants in this field have not been well explored, except for a few studies, e.g., recombinant TNFa (89). This is an area that is likely to take focus, as shall the continued exploration of more effective encapsulation techniques.

In terms of antigen preparations, a number of products are commercially available on the market that makes it easier and cheaper to produce larger volumes of antigens, especially those of virus nature. Here, the use of plants gives an interesting impetus and need following up.

Finally, when it comes to DNA vaccines, the future is not easy to predict. There is no doubt that progress will continue for injectable vaccines especially where other approaches have little or no efficacy. For oral vaccines as well, DNA preparations will come. However, acceptability in some regions such as Europe will depend on changes in legislation (90).

CONCLUSION

The last decade has seen an increase in the number of studies addressing oral vaccination of fish. The discovery of new methods of efficiently producing antigens particularly of viral antigens and the concept of using plant systems for the production of recombinant antigen presents new exciting future possibilities. Unraveling the immune response to antigens in the oral compartment and systemically vis-à-vis oral tolerance remains a challenge and requires more efforts. Finally, there is need to understand better a potential asymmetry in the immune responses of fish elicited by antigens delivered via the gut (or mucosal surfaces in general) versus parenteral delivery of antigens, and the importance of this for protecting the primary barriers of infection.

CONFLICT OF INTEREST STATEMENT

The authors declare that the research was conducted in the absence of any commercial or financial relationships that could be construed as a potential conflict of interest.

ACKNOWLEDGMENTS

Work leading to preparation of this paper was funded by the TARGETFISH, Targeted Disease Prophylaxis in European Fish Farming, EU Grant 311993. "Nanoparticle encapsulation of plant-based vaccine against piscine reovirus infection in salmon" RCN grant no. 239140, and "Sea Lice Research Centre", RCN grant no. 203513. We acknowledge the assistance from Dr. Ida Skaar, Norwegian Veterinary Institute, Oslo, Norway in preparing Figure 1.

REFERENCES

1. Dhar AK, Manna SK, Thomas Allnutt FC. Viral vaccines for farmed finfish. *Virusdisease* (2014) 25(1):1–17.
2. Quentel C, Vigneulle M. Antigen uptake and immune responses after oral vaccination. *Dev Biol Stand* (1997) 90:69–78.

3. Stroband HW, van der Meer JH, immermans LPM. Regional functional differentiation in the gut of the grasscarp, *Ctenopharyngodon idella* (Val.).*Histochemistry* (1979) 64:235–49. doi:10.1007/BF00495024
4. Rombout JHWM, Bot HE, Taverne-Thiele JJ. Immunological importance of the second gut segment of carp. *J Fish biol* (1989) 35(2):167–78. doi:10.1111/j.1095-8649.1989.tb02966.x
5. Rombout JHWM, Yang G, Kiron V. Adaptive immune responses at mucosal surfaces of teleost fish. *Fish Shellfish Immunol* (2014) 40(2):634–43. doi:10.1016/j.fsi.2014.08.020
6. Inami M, Taverne-Thiele AJ, Schrøder MB, Kiron V, Rombout JHWM. Immunological differences in intestine and rectum of *Atlantic cod* (*Gadus morhua L.*). *Fish Shellfish Immunol* (2009) 26(5):751–9. doi:10.1016/j.fsi.2009.03.007
7. Rombout JH, Lamers CH, Helfrich MH, Dekker A, Taverne-Thiele JJ. Uptake and transport of intact macromolecules in the intestinal epithelium of carp (*Cyprinus carpio* L.) and the possible immunological implications. *Cell Tissue Res* (1985) 239(3):519–30. doi:10.1007/BF00219230
8. McLean E, Ash R. The time-course of appearance and net accumulation of horseradish peroxidase (HRP) presented orally to juvenile carp *Cyprinus carpio* (L.). *Comp Biochem Physiol A Comp Physiol* (1986) 84(4):687–90. doi:10.1016/0300-9629(86)90388-9
9. Tatner MF, Horne MT. Susceptibility and immunity to *Vibrio anguillarum* in post-hatching rainbow trout fry, *Salmo gairdneri* Richardson 1836.*Dev Comp Immunol* (1983) 7(3):465–72. doi:10.1016/0145-305X(83)90031-9
10. Chen L, Evensen Ø, Mutoloki S. IPNV antigen uptake and distribution in Atlantic salmon following oral administration. *Viruses* (2015) 7(5):2507–17. doi:10.3390/v7052507
11. Joosten PH, Engelsma MY, van der Zee MD, Rombout JH. Induction of oral tolerance in carp (*Cyprinus carpio* L.) after feeding protein antigens.*Vet Immunol Immunopathol* (1997) 60(1–2):187–96. doi:10.1016/S0165-2427(97)00124-4
12. Firdaus-Nawi M, Zamri-Saad M, Nik-Haiha NY, Zuki MA, Effendy AW. Histological assessments of intestinal immuno-morphology of tiger grouper juvenile, *Epinephelus fuscoguttatus*. *Springerplus* (2013) 2:611. doi:10.1186/2193-1801-2-611
13. Harmache A, LeBerre M, Droineau S, Giovannini M, Bremont M. Bioluminescence imaging of live infected salmonids reveals that the fin bases are the major portal of entry for *Novirhabdovirus*. *J Virol* (2006) 80(7):3655–9. doi:10.1128/JVI.80.7.3655-3659.2006
14. Birkbeck TH, Ringø E. Pathogenesis and the GI tract of growing fish. In: Holzapfel W, Naughton P, editors. *Microbial Ecology of the Growing Animal*. Amsterdam: Elsevier (2004). p. 208–34.
15. Ling SH, Wang XH, Lim TM, Leung KY. Green fluorescent protein-tagged *Edwardsiella tarda* reveals portal of entry in fish. *FEMS Microbiol Lett*(2001) 194(2):239–43. doi:10.1111/j.1574-6968.2001.tb09476.x
16. Sakai DK. Invasive routes of *Aeromonas salmonicida* subsp. salmonicida. *Sci Rep Hokkaido Salmon Hatchery* (1979) 34(61):89.
17. Rombout JH, Abelli L, Picchietti S, Scapigliati G, Kiron V. Teleost intestinal immunology. *Fish Shellfish Immunol* (2011) 31(5):616–26. doi:10.1016/j.fsi.2010.09.001
18. Rombout JHWM, Kiron V. Mucosal vaccination of fish. In: Gudding R, Lillehaug A, Evensen O, editors. *Fish Vaccination*. Oxford: Willey Blackwell (2014). p. 56–67.
19. Anuradha K, Foo HL, Mariana NS, Loh TC, Yusoff K, Hassan MD, et al. Live recombinant *Lactococcus* lactis vaccine expressing aerolysin genes D1 and D4 for protection against *Aeromonas hydrophila* in tilapia (*Oreochromis niloticus*). *J Appl Microbiol* (2010) 109(5):1632–42. doi:10.1111/j.1365-2672.2010.04789.x
20. Siriyappagouder P, Shankar KM, Naveen Kumar BT, Patil R, Byadgi OV. Evaluation of biofilm of *Aeromonas hydrophila* for oral vaccination of*Channa striatus*. *Fish Shellfish Immunol* (2014) 41(2):581–5. doi:10.1016/j.fsi.2014.09.021
21. Li L, Lin SL, Deng L, Liu ZG. Potential use of chitosan nanoparticles for oral delivery of DNA vaccine in black seabream *Acanthopagrus schlegelii*Bleeker to protect from *Vibrio parahaemolyticus*. *J Fish Dis* (2013) 36(12):987–95. doi:10.1111/jfd.12032
22. Pabst R, Russell MW, Brandtzaeg P. Tissue distribution of lymphocytes and plasma cells and the role of the gut. *Trends Immunol* (2008)29(5):206–8. doi:10.1016/j.it.2008.02.006
23. Tobar JA, Jerez S, Caruffo M, Bravo C, Contreras F, Bucarey SA, et al. Oral vaccination of Atlantic salmon (*Salmo salar*) against salmonid rickettsial septicaemia. *Vaccine* (2011) 29(12):2336–40. doi:10.1016/j.vaccine.2010.12.107
24. Maurice S, Nussinovitch AF, Jaffe NF, Shoseyov OF, Gertler A. Oral immunization of *Carassius auratus* with modified recombinant a-layer proteins entrapped in alginate beads. *Vaccine* (2004) 23:450–9. doi:10.1016/j.vaccine.2004.06.022
25. Ballesteros NA, Rodriguez Saint-Jean S, Perez-Prieto SI. Food pellets as an effective delivery method for a DNA vaccine against infectious pancreatic necrosis virus in rainbow trout (*Oncorhynchus mykiss*, Walbaum). *Fish Shellfish Immunol* (2014) 37(2):220–8. doi:10.1016/j.fsi.2014.02.003
26. Rombout JHWM, van den Berg AA, van den Berg CTGA, Witte P, Egberts E. Immunological importance of the second gut segment of carp. III. Systemic and/or mucosal immune responses after immunization with soluble or particulate antigen. *J Fish Biol* (1989) 35(2):179–86. doi:10.1111/j.1095-8649.1989.tb02967.x
27. Brandtzaeg P, Kiyono H, Pabst R, Russell MW. Terminology: nomenclature of mucosa-associated lymphoid tissue. *Mucosal Immunol* (2008)1(1):31–7. doi:10.1038/mi.2007.9

28. Brandtzaeg P, Pabst R. Let's go mucosal: communication on slippery ground. *Trends Immunol* (2004) 25(11):570–7. doi:10.1016/j.it.2004.09.005

29. Belyakov IM, Derby MA, Ahlers JD, Kelsall BL, Earl P, Moss B, et al. Mucosal immunization with HIV-1 peptide vaccine induces mucosal and systemic cytotoxic T lymphocytes and protective immunity in mice against intrarectal recombinant HIV-vaccinia challenge. *Proc Natl Acad Sci U S A* (1998) 95(4):1709–14. doi:10.1073/pnas.95.4.1709

30. Zhang YA, Salinas I, Li J, Parra D, Bjork S, Xu Z, et al. IgT, a primitive immunoglobulin class specialized in mucosal immunity. *Nat Immunol* (2010)11(9):827–35. doi:10.1038/ni.1913

31. Valdenegro-Vega VA, Crosbie P, Vincent B, Cain KD, Nowak BF. Effect of immunization route on mucosal and systemic immune response in Atlantic salmon (*Salmo salar*). *Vet Immunol Immunopathol* (2013) 151(1–2):113–23. doi:10.1016/j.vetimm.2012.10.010

32. Chen L, Klaric G, Wadsworth S, Jayasinghe S, Kuo TY, Evensen O, et al. Augmentation of the antibody response of Atlantic salmon by oral administration of alginate-encapsulated IPNV antigens. *PLoS One* (2014) 9(10):e109337. doi:10.1371/journal.pone.0109337

33. Xue R, Liu L, Cao G, Xu S, Li J, Zou Y, et al. Oral vaccination of BacFish-vp6 against grass carp reovirus evoking antibody response in grass carp.*Fish Shellfish Immunol* (2013) 34(1):348–55. doi:10.1016/j.fsi.2012.11.024

34. Macpherson AJ, Koller Y, McCoy KD. The bilateral responsiveness between intestinal microbes and IgA. *Trends Immunol* (2015) 36:460–70. doi:10.1016/j.it.2015.06.006

35. von Gersdorff JL, Heinecke RD, Skjodt K, Rasmussen KJ, Buchmann K. Experimental evidence for direct in situ binding of IgM and IgT to early trophonts of *Ichthyophthirius multifiliis* (Fouquet) in the gills of rainbow trout, *Oncorhynchus mykiss* (Walbaum). *J Fish Dis* (2011) 34(10):749–55. doi:10.1111/j.1365-2761.2011.01291.x

36. Weiner HL, da Cunha AP, Quintana F, Wu H. Oral tolerance. *Immunol Rev* (2011) 241:241–59. doi:10.1111/j.1600-065X.2011.01017.x

37. Weiner HL. Oral tolerance: immune mechanisms and the generation of Th3-type TGF-beta-secreting regulatory cells. *Microbes Infect* (2001)3(11):947–54. doi:10.1016/S1286-4579(01)01456-3

38. Joosten PHM, Avilés-Trigueros M, Sorgeloos P, Rombout JHWM. Oral vaccination of juvenile carp (*Cyprinus carpio*) and gilthead seabream (*Sparus aurata*) with bioencapsulated *Vibrio anguillarum* bacterin. *Fish Shellfish Immunol* (1995) 5(4):289–99. doi:10.1006/fsim.1995.0028

39. Lannan CN, Winton JR, Fryer JL. Fish cell lines: establishment and characterization of nine cell lines from salmonids. *In vitro* (1984) 20(9):671–6. doi:10.1007/BF02618871

40. Munang'andu HM, Fredriksen BN, Mutoloki S, Brudeseth B, Kuo TY, Marjara IS, et al. Comparison of vaccine efficacy for different antigen delivery systems for infectious pancreatic necrosis virus vaccines in Atlantic salmon (*Salmo salar* L.) in a cohabitation challenge model. *Vaccine* (2012)30(27):4007–16. doi:10.1016/j.vaccine.2012.04.039

41. Skjesol A, Aamo T, Hegseth MN, Robertsen B, Jorgensen JB. The interplay between infectious pancreatic necrosis virus (IPNV) and the IFN system: IFN signaling is inhibited by IPNV infection. *Virus Res* (2009) 143(1):53–60. doi:10.1016/j.virusres.2009.03.004

42. Munang'andu HM, Fredriksen BN, Mutoloki S, Dalmo RA, Evensen O. Antigen dose and humoral immune response correspond with protection for inactivated infectious pancreatic necrosis virus vaccines in Atlantic salmon (*Salmo salar* L). *Vet Res* (2013) 44(1):7. doi:10.1186/1297-9716-44-7

43. Chen L, Evensen O, Mutoloki S. Delayed protein shut down and cytopathic changes lead to high yields of infectious pancreatic necrosis virus cultured in Asian grouper cells. *J Virol Methods* (2014) 195:228–35. doi:10.1016/j.jviromet.2013.10.020

44. Lu JT, Chung YC, Chan ZR, Hu YC. A novel oscillating bioreactor bellocell: implications for insect cell culture and recombinant protein production.*Biotechnol Lett* (2005) 27(15):1059–65. doi:10.1007/s10529-005-8450-3

45. Christie KE. Immunization with viral antigens: infectious pancreatic necrosis. *Dev Biol Stand* (1997) 90:191–9.

46. Lin JH, Yu CC, Lin CC, Yang HL. An oral delivery system for recombinant subunit vaccine to fish. *Dev Biol (Basel)* (2005) 121:175–80.

47. Tamaru Y, Ohtsuka M, Kato K, Manabe S, Kuroda K, Sanada M, et al. Application of the arming system for the expression of the 380R antigen from red sea bream iridovirus (RSIV) on the surface of yeast cells: a first step for the development of an oral vaccine. *Biotechnol Prog* (2006) 22(4):949–53. doi:10.1021/bp060130x

48. Clarke JL, Waheed MT, Lossl AG, Martinussen I, Daniell H. How can plant genetic engineering contribute to cost-effective fish vaccine development for promoting sustainable aquaculture? *Plant Mol Biol* (2013) 83(1–2):33–40. doi:10.1007/s11103-013-0081-9

49. Hellwig S, Drossard J, Twyman RM, Fischer R. Plant cell cultures for the production of recombinant proteins. *Nat Biotechnol* (2004) 22(11):1415–22. doi:10.1038/nbt1027

50. Salonius K, Simard N, Harland R, Ulmer JB. The road to licensure of a DNA vaccine. *Curr Opin Investig Drugs* (2007) 8(8):635–41.

51. Evensen Ø. Future of fish vaccinology. In: Gudding R, Lillehaug A, Evensen O editors. *Fish Vaccination.* Oxford: Wiley Blackwell (2014). p. 162–71.

52. Evensen O, Leong JA. DNA vaccines against viral diseases of farmed fish. *Fish Shellfish Immunol* (2013) 35(6):1751–8. doi:10.1016/j.fsi.2013.10.021

53. Pharmaq AS. *Pharmaq – Products.* (2015). Available from http://www.pharmaq.no/products/

54. Sommerset I, Krossoy B, Biering E, Frost P. Vaccines for

fish in aquaculture. *Expert Rev Vaccines* (2005) 4(1):89–101. doi:10.1586/14760584.4.1.89

55. Villumsen KR, Neumann L, Ohtani M, Strom HK, Raida MK. Oral and anal vaccination confers full protection against enteric Redmouth disease (ERM) in rainbow trout. *PLoS One* (2014) 9(4):e93845. doi:10.1371/journal.pone.0093845
56. Lillehaug A. Vaccination strategies and procedures. In: Gudding R, Lillehaug A, Evensen O, editors. *Fish Vaccination*. Oxford: Willey Blackwell (2014). p. 140–52.
57. Plant KP, Lapatra SE. Advances in fish vaccine delivery. *Dev Comp Immunol* (2011) 35(12):1256–62. doi:10.1016/j.dci.2011.03.007
58. Min L, Li-Li Z, Jun-Wei G, Xin-Yuan Q, Yi-Jing L, Di-Qiu L. Immunogenicity of *Lactobacillus*-expressing VP2 and VP3 of the infectious pancreatic necrosis virus (IPNV) in rainbow trout. *Fish Shellfish Immunol* (2012) 32(1):196–203. doi:10.1016/j.fsi.2011.11.015
59. Zhao LL, Liu M, Ge JW, Qiao XY, Li YJ, Liu DQ. Expression of infectious pancreatic necrosis virus (IPNV) VP2-VP3 fusion protein in *Lactobacillus casei* and immunogenicity in rainbow trouts. *Vaccine* (2012) 30(10):1823–9. doi:10.1016/j.vaccine.2011.12.132
60. Gibbs BF, Kermasha S, Alli I, Mulligan CN. Encapsulation in the food industry: a review. *Int J Food Sci Nutr* (1999) 50(3):213–24. doi:10.1080/096374899101256
61. Komar C, Enright WJ, Grisez L, Tan Z. Understanding fish vaccination. *Aqua Culture Asia Pacific Magazine* (2004). p. 27–9.
62. Specht EA, Mayfield SP. Algae-based oral recombinant vaccines. *Front Microbiol* (2014) 5:60. doi:10.3389/fmicb.2014.00060
63. Chan L, Lee H, Heng P. Production of alginate microspheres by internal gelation using an emulsification method. *Int J Pharm* (2002) 242(1–2):259–62. doi:10.1016/S0378-5173(02)00170-9
64. Tian JY, Sun XQ, Chen XG. Formation and oral administration of alginate microspheres loaded with pDNA coding for lymphocystis disease virus (LCDV) to Japanese flounder. *Fish Shellfish Immunol* (2008) 24(5):592–9. doi:10.1016/j.fsi.2008.01.009
65. Romalde JL, Luzardo-Alvarez A, Ravelo C, Toranzo AE, Blanco-Wendez J. Oral immunization using alginate microparticles as a useful strategy for booster vaccination against fish lactoccocosis. *Aquaculture* (2004) 236(1–4):119–29. doi:10.1016/j.aquaculture.2004.02.028
66. Altun S, Kubilay A, Ekici S, Didinen BI, Diler O. Oral vaccination against Lactococcosis in rainbow trout (Oncorhynchus mykiss) using sodium alginate and poly (lactide-co-glycolide) carrier. *Kafkas Univ. Vet. Fak Derg*. (2010) 16(B):211–7.
67. Leal CAG, Carvalho-Castro GA, Sacchetin PSC, Lopes CO, Moraes AM, Figueiredo HCP. Oral and parenteral vaccines against *Flavobacterium columnare*: evaluation of humoral immune response by ELISA and in vivo efficiency in *Nile tilapia* (*Oreochromis niloticus*). *Aquacult Int* (2010)18(4):657–66. doi:10.1007/s10499-009-9287-x
68. de las Heras AI, Rodriguez Saint-Jean S, Perez-Prieto SI. Immunogenic and protective effects of an oral DNA vaccine against infectious pancreatic necrosis virus in fish. *Fish Shellfish Immunol* (2010) 28(4):562–70. doi:10.1016/j.fsi.2009.12.006
69. Ballesteros NA, Saint-Jean SS, Encinas PA, Perez-Prieto SI, Coll JM. Oral immunization of rainbow trout to infectious pancreatic necrosis virus (Ipnv) induces different immune gene expression profiles in head kidney and pyloric ceca. *Fish Shellfish Immunol* (2012) 33(2):174–85. doi:10.1016/j.fsi.2012.03.016
70. Ballesteros NA, Saint-Jean SS, Perez-Prieto SI, Coll JM. Trout oral VP2 DNA vaccination mimics transcriptional responses occurring after infection with infectious pancreatic necrosis virus (IPNV). *Fish Shellfish Immunol* (2012) 33(6):1249–57. doi:10.1016/j.fsi.2012.09.004
71. Rajesh KS, Ishaq Ahmed VP, Parameswaran V, Sudhakaran R, Sarath BV, Sahul Hameed AS. Potential use of chitosan nanoparticles for oral delivery of DNA vaccine in Asian sea bass (*Lates calcarifer*) to protect from *Vibrio* (*Listonella*) *anguillarum*. *Fish Shellfish Immunol* (2008) 25(1–2):47–56. doi:10.1016/j.fsi.2007.12.004
72. Adomako M, St-Hilaire S, Zheng Y, Eley J, Marcum RD, Sealey W, et al. Oral DNA vaccination of rainbow trout, *Oncorhynchus mykiss* (Walbaum), against infectious haematopoietic necrosis virus using PLGA [Poly(D,L-Lactic-Co-Glycolic Acid)] nanoparticles. *J Fish Dis* (2012) 35(3):203–14. doi:10.1111/j.1365-2761.2011.01338.x
73. Tian J, Yu J. Poly(lactic-co-glycolic acid) nanoparticles as candidate DNA vaccine carrier for oral immunization of Japanese flounder (*Paralichthys olivaceus*) against lymphocystis disease virus. *Fish Shellfish Immunol* (2011) 30(1):109–17. doi:10.1016/j.fsi.2010.09.016
74. Rivas-Aravena A, Fuentes Y, Cartagena J, Brito T, Poggio V, La Torre J, et al. Development of a nanoparticle-based oral vaccine for Atlantic salmon against ISAV using an alphavirus replicon as adjuvant. *Fish Shellfish Immunol* (2015) 45(1):157–66. doi:10.1016/j.fsi.2015.03.033
75. Costerton JW, Geesey GG, Cheng KJ. How bacteria stick. *Sci Am* (1978) 238:86–95. doi:10.1038/scientificamerican0178-86
76. Donlan RM, Costerton JW. Biofilms: survival mechanisms of clinically relevant microorganisms. *Clin Microbiol Rev* (2002) 15(2):167–93. doi:10.1128/CMR.15.2.167-193.2002
77. Azad IS, Shankar KM, Mohan CV, Kalita B. Uptake and processing of biofilm and free-cell vaccines of *Aeromonas hydrophila* in Indian major carps and common carp following oral vaccination – antigen localization by a monoclonal antibody. *Dis Aquat Organ* (2000) 43(2):103–8. doi:10.3354/dao043103
78. van Stappen G. Introduction, biology and ecology of Artemia. In: Lavens P, Sorgeloos P, editors. *Manual on the Production and Use of Live Food for Aquaculture*. Rome: FAO (1996). p. 79–250.
79. Scholz T. Parasites in cultured and feral fish. *Vet*

Parasitol (1999) 84(3–4):317–35. doi:10.1016/S0304-4017(99)00039-4

80. Canty MN, Hagger JA, Moore RT, Cooper L, Galloway TS. Sublethal impact of short term exposure to the organophosphate pesticide azamethiphos in the marine mollusc *Mytilus edulis*. *Mar Pollut Bull* (2007) 54(4):396–402. doi:10.1016/j.marpolbul.2006.11.013

81. MacKinnon BM. *Sea Lice: a Review: World Aquaculture*. (Vol. 28). (1997). p. 5–10.

82. Kaur K, Helgesen KO, Bakke MJ, Horsberg TE. Mechanism behind resistance against the organophosphate azamethiphos in Salmon lice (*Lepeophtheirus salmonis*). *PLoS One* (2015) 10(4):e0124220. doi:10.1371/journal.pone.0124220

83. McNair CM. Ectoparasites of medical and veterinary importance: drug resistance and the need for alternative control methods. *J Pharm Pharmacol*(2015) 67(3):351–63. doi:10.1111/jphp.12368

84. Findlay VL, Helders BL, Munday BL, Gurney R. Demonstration of resistance to reinfection with *Paramoeba* sp. by Atlantic salmon, *Salmo salar* L.*J Fish Dis* (2006) 18(6):639–42. doi:10.1111/j.1365-2761.1995.tb00370.x

85. Adams MB, Crosbie PB, Nowak BF. Preliminary success using hydrogen peroxide to treat Atlantic salmon, *Salmo salar* L., affected with experimentally induced amoebic gill disease (AGD). *J Fish Dis* (2012) 35(11):839–48. doi:10.1111/j.1365-2761.2012.01422.x

86. Lotter H, Russmann H, Heesemann J, Tannich E. Oral vaccination with recombinant *Yersinia enterocolitica* expressing hybrid type III proteins protects gerbils from amebic liver abscess. *Infect Immun* (2004) 72(12):7318–21. doi:10.1128/IAI.72.12.7318-7321.2004

87. Lotter H, Russmann H, Heesemann J, Tannich E. Attenuated recombinant *Yersinia* as live oral vaccine carrier to protect against amoebiasis. *Int J Med Microbiol* (2008) 298(1–2):79–86. doi:10.1016/j.ijmm.2007.07.001

88. Salinas I, Lapatra SE, Erhardt EB. Nasal vaccination of young rainbow trout (*Oncorhynchus mykiss*) against infectious hematopoietic necrosis and enteric red mouth disease. *Dev Comp Immunol* (2015) 53(1):105–11. doi:10.1016/j.dci.2015.05.015

89. Galindo-Villegas J, Mulero I, Garcia-Alcazar A, Munoz I, Penalver-Mellado M, Streitenberger S, et al. Recombinant TNFalpha as oral vaccine adjuvant protects European sea bass against vibriosis: insights into the role of the CCL25/CCR9 axis. *Fish Shellfish Immunol* (2013) 35(4):1260–71. doi:10.1016/j.fsi.2013.07.046

90. Biering E, Salonius K. DNA vaccines. In: Gudding R, Lillehaug A, Evensen Ø, editors. *Fish Vaccination*. Oxford: Wiley Blackewell (2014). p. 47–55.

CHAPTER 4

VACCINE ADJUVANTS IN FISH VACCINES

Hanna L. Thim, Stéphane Villoing, Marian McLoughlin, Karen Elina Christie, Søren Grove, Petter Frost, and Jorunn B. Jørgensen

Keywords: Salmonid alphavirus, VHSV, vaccination, adjuvant, DNA vaccine, CpG, polyI:C, complement, neutralizing antibodies

INTRODUCTION

Viral diseases present a huge problem for the global aquaculture industry, where infectious diseases are estimated to be responsible for a production loss of ten to twenty percent each year [1]. Among these, pancreas disease (PD) caused by salmon pancreas disease virus, now more commonly referred to as salmonid alphavirus (SAV), is responsible for big economic losses throughout the Norwegian, Scottish and Irish aquaculture industry.

SAV strains are grouped into six different subtypes (SAV1-6), based on sequencing and phylogenetic analysis [2]. SAV1 isolates are found in Ireland and the U.K., while SAV3 isolates are exclusively found in Norway and affects both Atlantic salmon and rainbow trout [3,4]. PD histopathological signs are characterized by lesions in pancreatic acinar tissue, heart and, later, also in skeletal muscle [5,6]. Several studies have demonstrated protective immune responses against SAV in salmonids, both experimentally and in the field [7,8], and the protection has been shown to be associated with antibody (Ab)-mediated immunity and neutralizing Ab (nAbs) [7,9,10]. Following passive immunization of salmon parr and post smolts with SAV antisera, the fish are reported to be protected upon re-challenge, which indicates that protective immunity is conferred by NAbs [11]. Since then, various vaccination strategies against SAV have been tested both in Atlantic salmon and in rainbow trout, such as an inactivated SAV vaccine based on a Subtype 1 isolate [12], a recombinant live attenuated SAV vaccine (Subtype 2 isolate) [13] and subunit and DNA vaccines based on an SAV Subtype 3 isolate [14]. An inactivated whole-virus vaccine based on an Irish SAV Subtype 1 isolate has been used in Norway, Ireland and the U.K. since 2007 [15].

Most of the commercial finfish vaccines against viruses are, like the above-mentioned SAV vaccine, based on inactivated antigens (Ag) or recombinant subunit proteins formulated in oil emulsions. Oil-based adjuvants are based on creating a depot of Ag, which improves Ag delivery to Ag-presenting cells (APC) or by attracting effector cells to the site of injection. However, side effects due to oil adjuvants have been reported and are expressed both physiologically and morphologically [16,17,18]. Administration of these oil-based vaccines is performed by intraperitoneal (i.p.) injection, and relatively high doses are needed to achieve protection [19]. Hence, more potent adjuvants enabling the use of a lower Ag dose and reducing side effects would advantageously be used for fish vaccines. Many adjuvants are derived from pathogens and act via cell-associated germ line-encoded pattern recognition receptors (PRR) on APCs to provide a danger signal. Such adjuvants induce maturation of the APCs and enhance Ag presentation and associated co-stimulation [20]. Among the best studied PRRs is the Toll-like receptor (TLR) family, of which 17 types have been described in different fish species [21]. The piscine TLRs includes TLR9 recognizing bacterial and viral DNA [22,23] and TLR3 and TLR22, which both recognize double-stranded (ds) RNA [24,25]. Studies by us and other groups have demonstrated that ligands for these receptors, such as synthetic CpG oligonucleotides (ODNs; TLR9 ligand) and polyI:C (TLR3/22 ligand) can stimulate the production of pro-inflammatory cytokines/chemokines and Type I IFNs, which increase the host's ability to eliminate viral pathogens [26,27]. Further, our group has, in accordance with mammalian studies [28], shown that a combined treatment of CpG/polyI:C induces synergistic upregulation of a wide array of immune genes in Atlantic salmon [26] and significantly enhances protection on its own against SAV [29]. When

Citation: Hanna L. Thim, Stéphane Villoing, Marian McLoughlin, Karen Elina Christie, Søren Grove, Petter Frost, and Jorunn B. Jørgensen. Vaccine Adjuvants in Fish Vaccines, Vaccines, 2014; 2(2), 228-251. DOI:10.3390/vaccines2020228. ISSN 2076-393X.

formulated in an SAV whole-virus Ag formulation [30], the combo significantly increased antibody-mediated clearance of SAV from blood, thus preventing the development of SAV-specific heart lesions. This strongly indicates that a humoral Ab response is important for protection against SAV and that CpG/polyI:C boost this protection.

The Novirhabdovirus VHSV (viral haemorrhagic septicaemia virus) glycoprotein (G) DNA-vaccine is an intra-muscularly (i.m.) injected vaccine that has been shown to induce early non-specific, as well as long-lasting specific protection against VHSV in trout [31,32] and seems to act as a genetic adjuvant [33]. Several studies have suggested that the effects of DNA-vaccines are regulated by innate responses through PRR-signalling cascades [34], and although the mechanisms that induce early protective responses by the VHSV-G DNA construct are still unclear, innate signalling are most likely involved. The VHSV-G DNA vaccine has been shown to create a local immune-competent environment at the i.m. injection site [35], and an interesting question addressed here was whether i.m. co-injection of this DNA vaccine simultaneously with an i.p. administered TLR-agonist-adjuvanted vaccine based on an inactivated SAV Ag could provide improved protective responses. Combining several PRR agonists has previously shown synergistic effects when intended as adjuvants, resulting in enhanced and more durable responses to Ag, as well as dose sparing effects [36,37]. To our knowledge, this is the first report of immunity associated with inactivated virus Ag vaccines formulated with mixed TLR agonists and with VHSV-G (vhsG) DNA vaccine as a genetic adjuvant. In the current study, the protective effects between an i.p. injection of SAV Ag with CpG/polyI:C were compared to the same treatment combined with a simultaneous i.m. injection of the vhsG DNA vaccine to see whether additive or synergistic effects were induced. To evaluate vhsG adjuvant effects, several controls were included, whose effects have been investigated earlier [29,30]; SAV Ag was i.p. injected alone or with an oil adjuvant (Montanide ISA763A), formulations of which were co-injected i.m. with vhsG to thoroughly examine the possible effects these adjuvant combinations could have on protection and immune responses against an SAV Subtype 3 challenge.

EXPERIMENTAL

Reagents and Constructs

DNA plasmid pcDNA3-vhsG [38] was kindly provided by Dr. Niels Lorenzen and diluted in 1× PBS to a concentration of 0.2 mg/mL. The synthetic dsRNA (poly I:C; Merck, Nottingham, UK) and the phosphorothioate-modified class B CpG oligonucleotide (2006T: TCGTCGTTTTGTCGTTGTCGTT, Thermo Scientific, Ulm, Germany) were dissolved in TE buffer (10 mM Tris, 1 nM EDTA, pH 8) at 5 mg/mL and further diluted 10-fold in the final vaccine formulations (50 μg/dose). SAV Ag was prepared by propagating the SPDV/SAV strain F93-125 in cell culture, which then was formalin inactivated. To demonstrate the potential additive or synergistic effects of the adjuvants tested, the same suboptimal dose of inactivated SAV Ag was used for all formulations. Montanide ISA 763 (Seppic, France) was used to prepare the water-in-oil formulations, by dispersing the water phase (containing the formalin inactivated SAV Ag) into the vegetable oil phase (containing emulsifiers and stabilizers) and emulsified using a homogenizer with an emulsification rotor. All SAV Ag formulations were provided by MSD Animal Health (Bergen, Norway).

Fish

The experimental challenge study was performed at ILAB's challenge lab facility at the University of Bergen (Høyteknologisenteret, Bergen, Norway), which fulfil the confinement conditions required for working with GMOs and DNA vaccines. Atlantic salmon, pre-smolt (Fister) with a mean weight of approximately 29 g at time of vaccination were kept in tanks supplied with running fresh water at 11–12 °C and fed with commercial dry feed (Skretting, Stavanger, Norway) based on appetite. The fish were starved for a minimum of 48 h and anaesthetized with metacainum (0.1 mg/mL bath treatment) prior to all handling.

Vaccination and SAV Cohabitation Challenge

Fish were divided into 7 treatment groups (n = 65, 69 or 75, depending on the required sampling size) and a saline injected control group (n = 79). As described in detail in Table 1, three treatment groups were i.p. injected with 100 μL of the SAV1 whole-inactivated virus Ag formulation, formulated with or without oil and/or 50 μg CpG/polyI:C. Moreover, three treatment groups were, in parallel to the i.p. injections, injected i.m. with 10 μg of the PcDNA3-vhsG plasmid diluted in 50 μL PBS (1×). One group received PcDNA3-vhsG plasmid alone, and the control group was injected with 100 μL of PBS. Fish were marked by fin and/or maxilla clipping, and there were no mortalities observed after injection. A total of 570 fish were used and divided into 3 tanks; 1 tank for the SAV cohabitation challenge (422 fish excluding shedders), and 2 tanks for harvesting organs to monitor early immune gene expression (2 × 74 fish). At 6 weeks post vaccination (wpv), 86 Atlantic salmon were injected i.p. with 0.2 mL SAV Subtype 3, each receiving a viral dose of about 1×10^3 $TCID_{50}$ and added to the SAV challenge tank to serve as shedders (n = 86 corresponds to 20% of the final amount of fish in the tank). The shedders were marked with a red VIE label under the anal fin one week prior to challenge.

The treatment that is underlined, PcDNA3-vhsG, was intra-muscularly (i.m.) injected parallel to intraperitoneal (i.p.) injection of SAV Ag treatments. Five extra fish were sampled at 6 wpv and 3 wpc for analyses not included here. hpv; hours post vaccination; wpc; weeks post challenge. [1] The same fish sampled for nsP1 RT-qPCR as for the nAb assay; [2] the same fish were sampled for heart histopathology and for the nAb assay. SAV, salmonid alphavirus; Ag, antigen; nAb, neutralizing antibody.

Table 1. Treatments, dose regime, number of fish and schedule for the sampling of organs and blood.

Treatment	Total # of fish	Analysis (# of fish)					
		Immune gene expression 12 and 48 hpv	nsP1 RT-qPCR 3 wpc [1]	Histopathology 5 and 6 wpc [2]	nAb assay		
					6 wpv	3 wpc[1]	6wpc[2]
SAV Ag	74 + 1	8 + 8	10	15 + 15	13	15	15
SAV Ag oil	64 + 1	8 + 8	10	15 + 15	8	10	15
SAV Ag CpG (50 µg)/ poly I:C (50 µg)	74 + 1	8 + 8	10	15 + 15	13	15	15
SAV Ag PcDNA3-vhsG (10 µg)	68 + 1	10 + 10	10	15 + 15	8	10	15
SAV Ag oilPcDNA3-vhsG (10 µg)	68 + 1	10 + 10	10	15 + 15	8	10	15
SAV Ag CpG (50µg)/ poly I:C (50 µg) PcDNA3-vhsG (10 µg)	68 + 1	10 + 10	10	15 + 15	8	10	15
PcDNA3-vhsG (10 µg)	68 + 1	10 + 10	10	15 + 15	8	10	15
Saline 0.9%	78 + 1	10 + 10	10	15 + 15	13	15	15

RT-qPCR of Immune Gene Expression: RNA Isolation, cDNA Synthesis and RT-qPCR

Spleen and head kidney (HK) were harvested from 8 fish per group at 12 and 48 h post vaccination (hpv) and were stored according to the manufacturer's guidelines on RNAlater (Ambion, Applied Biosystems, Foster City, CA, USA). RNA isolation, cDNA synthesis and RT-qPCR were executed as described previously [26] with minor changes. Four hundred nanograms of total RNA were reverse transcribed (TaqMan Reverse Transcription Reagents kit; Applied Biosystems) into cDNA using random hexamer primers in 30-µL reaction volumes following the manufacturer's guidelines. Primer and probe sequences and the efficiencies of the assays used in this study are presented in Table 2. cDNA samples (2.5 µL) were analysed in duplicates (target genes) or triplicates (endogenous control) in 20-µL reactions on a 7500 Fast Real-Time PCR system. The Cq-threshold was automatically set to 0.2 for analysis of both endogenous and target genes. Relative expression and statistics were calculated using Relative Expression Software Tool (REST) 2009 [39], which is based on Pfaffl's mathematical model [40], where individual Cq-values were compared between saline-injected fish (control) and vaccine injected fish (test) and correlated to the endogenous control gene, EF1αβ, and PCR-efficiency.

SAV Neutralizing Ab Responses

Serum samples from 8 to 15 individuals per group and time point (see Table 1 for details) were collected at 6 wpv and 3 and 6 wpc (weeks post challenge) and examined for SAV-neutralizing activity. To do so, virus initially incubated with diluted sera was left to adhere to Chinook salmon embryo-214 (CHSE) cells, and after 8 days, the presence of cell-associated virus was detected by an ELISA-method, as reported earlier [30]. Individual sera from each group were pooled, and half of the pooled sera were heat inactivated (HI; 56 °C for 30 min). Assays for both HI and not heat inactivated (NHI) sera were repeated 3 times for all samples. Two-fold dilutions of either HI or NHI serum were added in duplicate to a 96-well microtiter plate with maintenance media (MM; Minimum essential medium eagle (MEM) supplemented with 2% Foetal bovine serum (FBS)), giving a final dilution range from 1:20 to 1:640 (1:160 to 1:5,120 for the CpG/polyI:C-treated groups) when 100 µL of virus supernatant SPDV (SAV Subtype 1 isolate, 6,000 $TCID_{50}$/mL) were added to the wells containing salmon serum dilutions. Neutralizing effects in serum were expressed as the highest reciprocal titres showing a >50% reduction of the positive control OD value using the following formula:

$$50\%\ reduction\ OD = \frac{(\sum\ virus\ ctrl\ OD - \sum\ background\ ctrl\ OD)}{2}$$

SAV nsP1 RT-qPCR Detection

To measure SAV levels during the viraemic phase, a quantitative real-time RT-PCR (RT-qPCR) was performed on viral RNA extracted from sera 3 wpc from 10 individuals per group, as described previously [26]. Non-structural protein 1 (nsP1) primers and the probe used for this assay are described in Table 1. Individual Cq-values were transformed to relative numbers by the following formula, where y represents the lowest Cq-value detected (*i.e.*, the highest number of nsP1 transcripts) and where x is any of the other Cq-values detected:

$$\mathrm{RelCq}(x) = 2^{(y-x)}$$

A sample was considered infected when it had a relative value between 1.0×10^0 and the cut off value of 3.0×10^{-7} ($x = 37.5$).

Histopathology

Heart samples for detecting SAV-induced lesions were collected from 15 fish/group at 5 and 6 weeks post-challenge and immediately fixed in 3.5% formaldehyde in buffered saline at pH 7.0 (4.0 g $NaH_2PO_4{\cdot}_2H_2O$, 6.5 g $Na_2HPO_4{\cdot}_2H_2O$, 100 mL 35% formaldehyde and 900 mL dH_2O). To evaluate the severity of SAV-induced heart lesions, a previously defined scoring system was used (no lesion: 0; minimal: 1; mild: 2;

moderate: 3; severe: 4), where scores of 2 and more are defined to be specifically induced by an SAV infection [41]. The lesion scoring was done by Marian McLoughlin (Aquatic Veterinary Services, Belfast, Ireland) using "blinded" heart samples.

Data Analyses

All analyses were done in GraphPad Prism 5.0 if not mentioned otherwise. Differences in protection (SAV nsP1 RT-qPCR and histology) were statistically evaluated by the Kruskal-Wallis rank sum test with $p < 0.05$ as the significance limit, followed by Dunn's post hoc test at a 5% level of significance. The histology test parameter used for statistical analysis was the severity of heart lesions, scored on the ordinal scale (0–4). Statistical analysis of the SAV nsP1 RT-qPCR used the individual Cq-values of each group as the test parameters. A modified expression of the relative percent protection score (RPPsc.) [29] was used to evaluate the level of protection against SAV induced by the tested treatments, based on the results obtained with the experimental methods (SAV-induced heart lesions histology or SAV-specific RT-qPCR assay). The advantage with this modified RPPsc. method is that the actual differences in degrees of severity of disease between the affected animals for the treated and control groups are taken into consideration. RT-qPCR data were statistically analysed by the Relative Expression Software Tool (REST 2009 v.2.0.13) [39].

Table 2. Primers and probe sequences for quantitative reverse-transcriptase PCR and PCR efficiency.

GENES	ASSAY	PRIMERS/ PROBE	SEQUENCE (5'–3')	PCR EF-FICIENCY	ACCES-SION NO.
CD4-1	Fw/Rev 200 nM Probe 200 nM	Forward Reverse Probe	GAATCTGCCGCTGCAAAGAC AGGGATTCCGGTCTGTATGATATCT [6FAM]CCCAAACCAAAAGGATTC[BHQ1]	1.78	EU409792
CD4-2a	SYBR® Green 3.75 µM	Forward Reverse	TGCAAAGAAGGCGCAGAT GAAAACCTTTAATTTAACAGG	1.76	EU409793
CD8a	Fw/Rev 450 nM Probe 250 nM	Forward Reverse Probe	CGTCTACAGCTGTGCATCAATCAA GGCTGTGGTCATTGGTGTAGTC [6FAM]CTGGGCCAGCCCCTAC[BHQ1]	1.74	AY693391
EF1aB	Fw/Rev 900 nM Probe 250 nM SYBR Green* 3.75 µM	Forward * Reverse * Probe	TGCCCCTCCAGGATGTCTAC CACGGCCCACAGGTACTG [6FAM]AAATCGGCGGTATTGG[BHQ1]	2.09/2.16 *	BG933897
IFNa1	SYBR Green 3.75 µM	Forward Reverse	CCTTTCCCTGCTGGACCA TGTCTGTAAAGGGATGTTGGGAAAA	2.0	AY2169594 AY2169595
IFNγ	Fw/Rev 900 nM Probe 250 nM	Forward Reverse Probe	AAGGGCTGTGATG TGTTTCTGTGTACTGAGCGGCATTACTCC [6FAM] TTGATGGGCTGGATGACTTTAGGA[BHQ1]	2.0	AY795563
mIgM	Fw/Rev 200 nM Probe 200 nM	Forward Reverse Probe	CCTACAAGAGGGAGACCGA GATGAAGGTGAAGGCTGTTTT [6FAM]TGACTGACTGTCCATGCAGCAACA CC[BHQ1]	2.0	
Mx1/2	Fw/Rev 900 nM Probe 250 nM	Forward Reverse Probe	GATGCTGCACCTCAAGTCCTATTA CGGATCACCATGGGAATCTGA [6FAM]CAGGATATCCAGTCAACGTT[BHQ1]	1.96	U66475/ U66476
PAX5	Fw/Rev 200 nM Probe 200 nM	Forward Reverse Probe	CCACTGCCAGGTCGA GAGTCAGCGAGGAGGTGGAGTA [6FAM]CCCCGGCTATCCACCACACG[BHQ1]	1.85	
sIgM	Fw/Rev 900 nM Probe 250 nM	Forward Reverse Probe	CTACAAGAGGGAGACCGGAG AGGGTCACCGTATTATCACTAGTTT TCCACAGCGTCCATCTGTCTTTC	1.94	BT060420

Vig-1	Fw/Rev 900 nM Probe 250 nM	Forward Reverse Probe	AGCAATGGCAGCATGATCAG TGGTTGGTGTCCTCGTCAAAG [6FAM]AGTGGTTCCAAACGTATGGCGAAT ACCTG[BHQ1]	1.94	BT047610
Q_nsP1	Fw/Rev 900 nM Probe 130 nM	Forward Reverse Probe	CCGGCCCTGAACCAGTT GTAGCCAAGTGGGAGAAAGCT [6FAM]CTGGCCACCACTTCGA[BHQ1]	-	AY604235

Fw; forward. Rev; reverse. * The same primers were used for SYBR Green as for TaqMan.

RESULTS

Immune Genes

Transcription levels of selected antiviral innate genes, as well as selected B- and T-cell markers were analysed at 12 and 48 hpv in HK (Figure 1) and spleen (Figure 2) for the vaccine formulations in relation to the control group. It is worth noting the often high standard deviations underlining the highly individual immunological response, which is common in fish. Tables with average Cq-values are included as supplementary material.

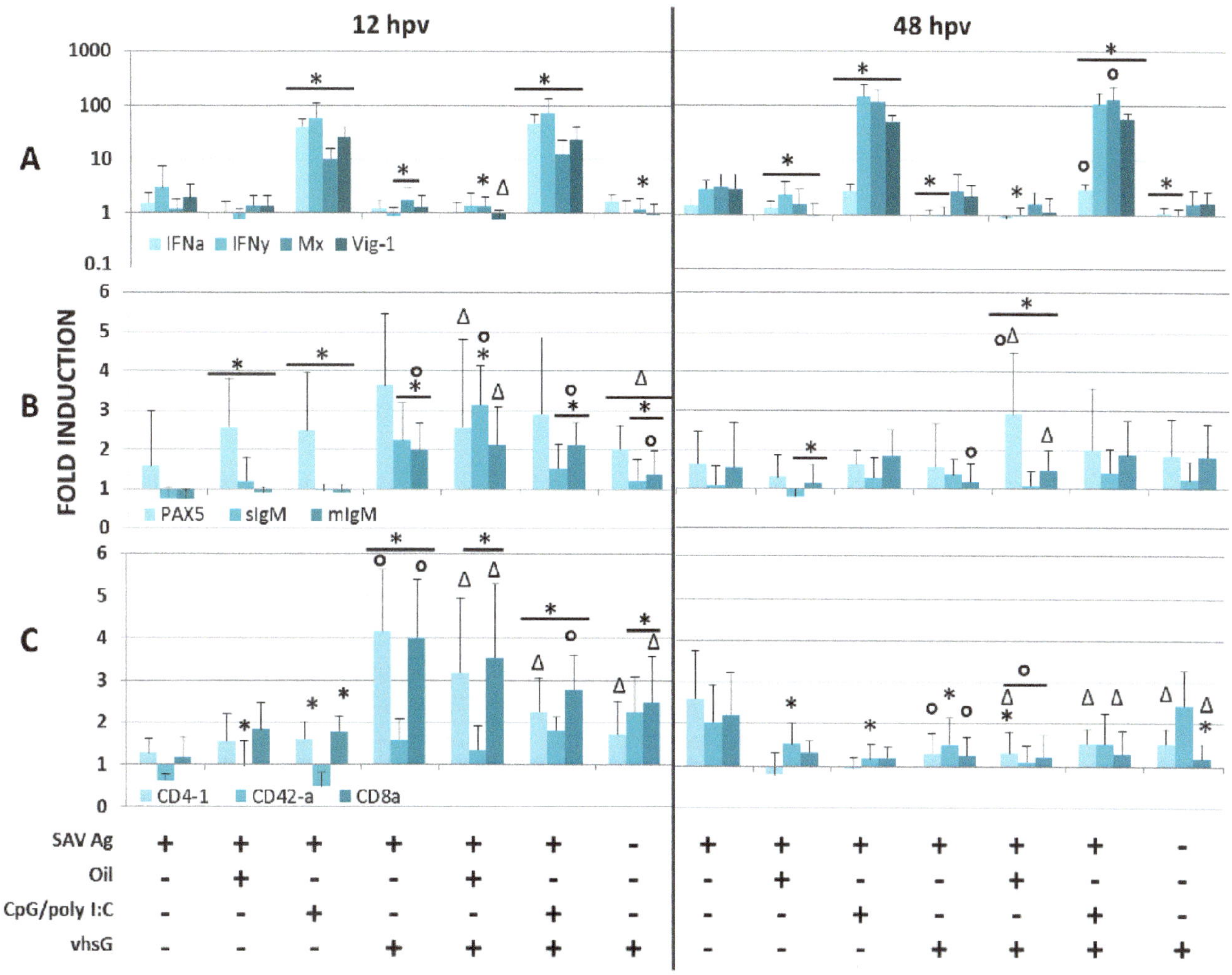

Figure 1. Relative expression of (A) antiviral, (B) B- and (C) T-cell markers in head kidney at 12 and 48 hpv for all treatments compared to saline-treated fish (expression is normalized to reference gene EF1aB). Relative expression is presented as histograms (the colour codes for each gene analysed are indicated below the histograms) calculated from fold induction by Pfaffl's method (see Materials and Methods), and significant up- or down-regulation is based on data from REST 2009. Significant differences for all treatments compared to SAV Ag are highlighted with an *, against SAV Ag vhsG as Δ and against SAV Ag CpG/polyI:C as °, + or –, respectively, indicates the presence or absence of either SAV Ag, oil, CpG/polyI:C or vhsG.

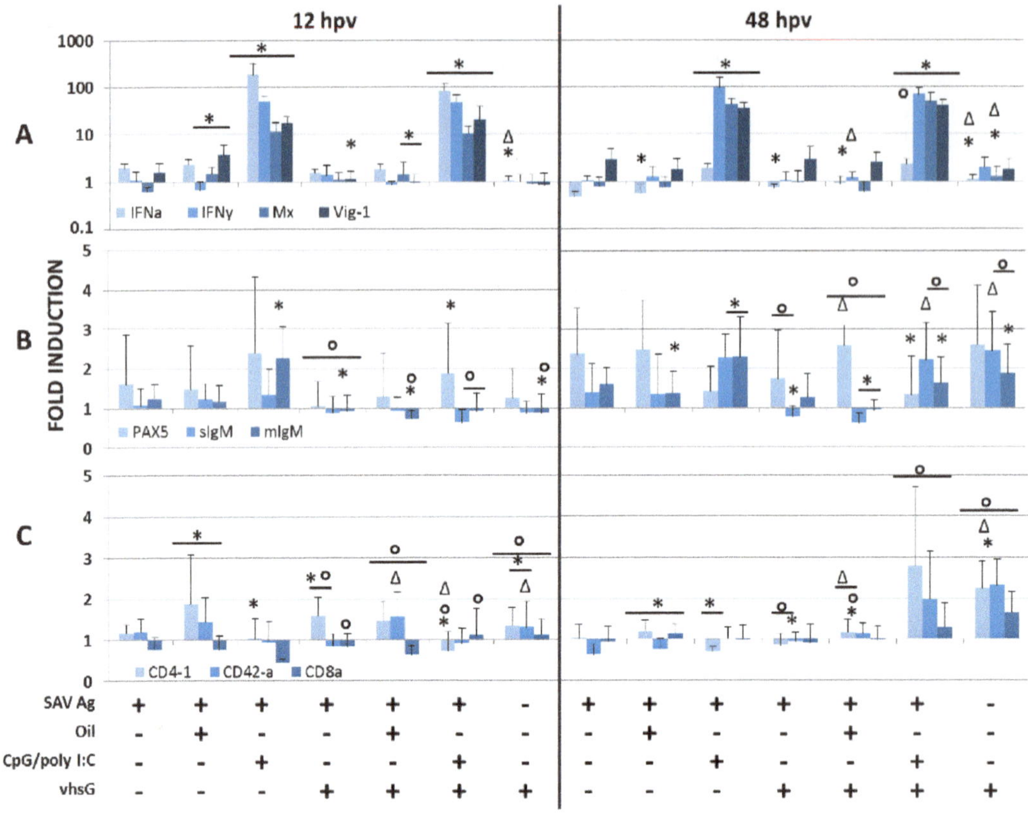

Figure 2. Relative expression of (A) antiviral, (B) B- and (C) T-cell markers in spleen at 12 and 48 hpv for all treatments compared to saline-treated fish (expression is normalized to reference gene EF1aB). Relative expression is presented as histograms (the colour codes for each gene analysed are indicated below the histograms) calculated from fold induction by Pfaffl's method (see Materials and Methods), and significant up- or down-regulation is based on data from REST 2009. Significant differences for all treatments compared to SAV Ag are highlighted with an *, against SAV Ag vhsG as Δ and against SAV Ag CpG/polyI:C as °, + or –, respectively, indicates the presence or absence of either SAV Ag, oil, CpG/polyI:C or vhsG.

Results showed that IFNa1, IFNγ and two antiviral proteins induced by IFNs (Mx and Vig-1) were strongly induced by the CpG/polyI:C-adjuvanted treatments, which confirms earlier findings [30]. In detail, for the CpG/polyI:C-adjuvanted treatments, IFNγ were highly upregulated at both time points in HK (Figure 1A; 60–150-fold) and spleen (Figure 2A; 50–100-fold), which was not seen for any of the other treatments. IFNγ is a cytokine secreted by T- and NK-cells and important for both innate and adaptive immune responses against viral infections. Upregulation of IFNa1 in HK and spleen for CpG/polyI:C-treated groups was significantly higher at 12 hpv than at 48 hpv, when a 10-fold reduction was seen. This is consistent with the fact that Type I IFN is an early induced innate antiviral actor, and in accordance, the antiviral genes, Vig-1 and Mx (induced by IFN Type I), accompanied the IFNa1-induction and were still highly expressed at 48 hpv in the CpG/polyI:C-treated groups with approximately a 40-fold induction in spleen and 100-fold for HK. Generally, the antiviral immune gene expression pattern seen for SAV Ag alone and for the formulations without CpG/polyI:C was moderate in both organs (zero- to two-fold in general). In HK, SAV Ag alone induced a low, but significant, upregulation of IFNγ, Vig-1 and Mx at 48 hpv compared to control fish. Antiviral gene responses in HK for vhsG immunized fish were in general moderate and not significantly higher than those in the other groups, except for the CpG/polyI:C-treated groups. Compared to SAV Ag CpG/polyI:C, the immune gene expression patterns for SAV Ag CpG/polyI:C vhsG were similar in both organs, except a slight, but significant, upregulation of IFNa1 and Vig-1 in HK at 48 hpv and downregulation of IFNγ in spleen at 48 hpv. In spleen at 48 hpv, IFNa1 was significantly upregulated in all vhsG-treated groups compared to SAV Ag alone (0.8- to 2.3-fold). Besides that, the expression pattern levels in spleen were similar to those in HK.

PAX5, soluble (s) and membrane-bound (m) IgM were chosen as B-cell markers. PAX5 belongs to the family of paired box transcription factors and is present during early B-cell development, but must be repressed for plasma cell differentiation to take place [42]. Expression of mIgM often occurs in parallel to PAX5, and the ratio of mIgM *vs.* sIgM is important in relation to B-cell development. Early and developing B-cells have higher levels of mIgM and no or low sIgM levels, whereas a shift in their ratio indicates the presence of antibody producing B-cell populations [43]. In general, induction of all three B-cell markers was low at both time points, and the expression pattern varied for the two immune organs. At 12 hpv, a slightly higher expression of PAX5 and mIgM was measured in HK (Figure 1B) for vhsG-treated groups with fold inductions from two to four compared to the other treatments (approximately 1.5 to 2.5). PAX5 was the only B-cell marker induced at 12 hpv in HK of fish treated with SAV Ag, SAV Ag oil or SAV Ag CpG/polyI:C. At 48 hpv, a slight increase of mIgM from 0.8- to 1.5-fold was present for these latter groups, which could indicate a later onset of early

B-cell development in groups not receiving vhsG. In spleen (Figure 2B), no clear expression pattern could be seen, and the B-cell markers were hardly detectable until 48 hpv, except for the SAV Ag CpG/polyI:C-treated group, where both PAX5 (2.4-fold) and mIgM (2.3-fold) were induced at 12 hpv. At 48 hpv, the expression of PAX5 was reduced, and both sIgM and mIgM were induced for fish treated with SAV Ag CpG/polyI:C and SAV Ag CpG/polyI:C vhsG.

Co-receptors for T-helper cells (CD4) and cytotoxic T-cells (CD8) were used as T-cell markers to further study early adaptive responses. Atlantic salmon contain two shorter CD4 domain molecules (CD4-2a and -2b) in addition to the four classical (CD4-1) domains [44]. To determine which T-cell subsets were activated, both CD4-1 and CD4-2a, in addition to CD8α transcript levels were measured in HK (Figure 1C) and spleen (Figure 2C) at 12 and 48 hpv. As for B-cell markers, the general induction of T-cell markers in both organs was low and, with a few exceptions, their mRNA levels differed little between treatments and sampling time points. As the supplementary data show, the basal levels of CD4-2a (average Cq-values of 29.7 and 27.5 at both time points for HK and spleen, respectively) were higher in both tissues and at both time points compared to CD4-1 (average Cq-values of 33.6 and 32.4 at both time points for HK and spleen, respectively), although CD4-2a levels were not affected in the two organs at either time point. CD4-1 and CD8α were the most highly upregulated in HK at 12 hpv in groups co-injected with vhsG, ~2.3- to four-fold for both markers, which declined to an average of 1.3-fold for both CD4-1 and CD8α over the next 36 h. At 48 hpv, CD4-1 and CD8α levels in HK were significantly higher for the SAV Ag alone treatment (2.6 and 2.2, respectively) compared to the other treatments (0.8–1.4 for CD4-1 and approximately 1.25 for CD8α). In spleen, the overall expression levels of T-cell markers were very low at both 12 and 48 hpv, except at 48 hpv for SAV Ag CpG/polyI:C vhsG and vhsG alone, where both CD4 markers were upregulated.

SAV Neutralizing Ab Responses

The presence of anti-SAV neutralizing responses in sera was measured at 6 wpv and at three and 6 wpc by a viral neutralization assay. It is well known that heat sensitive factors in serum may augment the neutralizing activity of Ab [45], and therefore, neutralizing activity was measured both with HI and NHI sera. For the NHI sera (Figure 3A), detectable neutralizing antibody titres (nAbTs) were present from 6 wpv for all groups, except groups treated with SAV Ag oil vhsG, vhsG alone or saline. The highest nAbTs were found in the SAV Ag CpG/polyI:C-treated group with titres of 640 before challenge that rose to 1280 and further to 2560 at three and 6 wpc, respectively. Fish treated with SAV Ag CpG/polyI:C that also received the vhsG i.m. injection showed the second highest nAbTs of 640 and 320 at 6 wpv and 3 wpc, respectively and 640 at 6 wpc. While all groups receiving a SAV Ag formulation, except the SAV Ag oil vhsG group, mounted a detectable neutralizing response before challenge, groups receiving either vhsG alone or saline had detectable nAbTs first after challenge. For vhsG alone and saline, the nAbTs are most likely induced upon exposure to the challenge virus, with titres of 80 (3 wpc) and 160 (6 wpc) for vhsG and 160 for saline at both three and 6 wpc. For HI sera, nAbTs were only present in three groups pre-challenge (Figure 3B). CpG/polyI:C-adjuvanted treatments provided the highest responses, with nAbTs ranging from 640 at 6 wpv to 1,280 at 6 wpc for SAV Ag CpG/polyI:C, and for SAV Ag CpG/polyI:C vhsG, the generation of nAbs was consistent with a titre of 160 at all sampling points. No positive sera were found among the fish injected with SAV Ag, SAV Ag oil or SAV Ag oil vhsG, while SAV Ag vhsG had detectable nAb responses both pre- and post-challenge. Fish treated with vhsG and saline, where >80% of the fish in both groups had positive SAV-specific heart lesions at 6 wpc (Figure 4B), showed detectable nAbTs at 6 wpc (80 for both treatments).

Protection

Six weeks after vaccination, all fish were challenged by cohabitation with an SAV Subtype 3 isolate, and vaccine-induced protection was measured at three, five and 6 wpc (Figure 4). No mortality was observed after challenge. At 3 wpc, during the viraemic phase [46], viral RNA from sera were isolated, and SAV nsP1 transcript levels were detected by RT-qPCR. At 3 wpc, 70% (seven out of 10 fish) of the saline-treated fish had SAV positive sera, thus indicating a successful challenge (Figure 4A). Four of the six water-formulated SAV Ag treatments, namely SAV Ag, SAV Ag CpG/polyI:C, SAV Ag vhsG and SAV Ag CpG/polyI:C vhsG, provided full protection (RPPsc. = 100%) against SAV at 3 wpc, with non-detectable Cq-values. For the two oil-formulated groups, SAV Ag oil and SAV Ag oil vhsG, 20% and 10% of the fish had nsP1 positive sera, leading to RPPsc. of 71.4% and 85.7%, respectively. Furthermore, based on the prevalence of viremia determined by nsP1 RT-qPCR, there was a significant difference between all water- and oil-formulated SAV Ag treatments compared to saline-treated fish, except for SAV Ag oil (RPP.sc. = 71%). The prevalence of nsP1 positive fish in the group treated with vhsG alone (nine out of 10 serum positive fish) was significantly higher than the prevalence in the other treatment groups, except the saline group. Protection for the vhsG alone treatment was less than for the saline injected fish with an RPPsc. of −0.28%.

Heart tissue was sampled at five and 6 wpc to evaluate the severity of SAV-induced heart lesions by histological scoring. Protection based on the reduction of the severity of SAV-induced heart lesions was comparable to the protection shown through the reduction of the prevalence of viremia at 3 wpc for all treatments. Here, only histology data for 6 wpc are presented (Figure 4B), given that the prevalence of fish with SAV-specific heart lesions in the control group was highest at 6 wpc (80% *vs.* 60% at 5 wpc), where 10 out of 12 individuals had severe heart lesions. Further, at 6 wpc, lesions in the group vaccinated with SAV Ag oil vhsG were reduced, and only one out of 15 fish showed moderate lesions (RPPsc. 83.3%). The group treated with SAV Ag oil had an RPPsc. of 58.3%, and five fish showed mild to severe lesions, compared to an RPPsc. of 100% at 5 wpc (not shown), with only two fish showing minimal heart lesions. However, there was no significant difference in protection for that treatment between both time points.

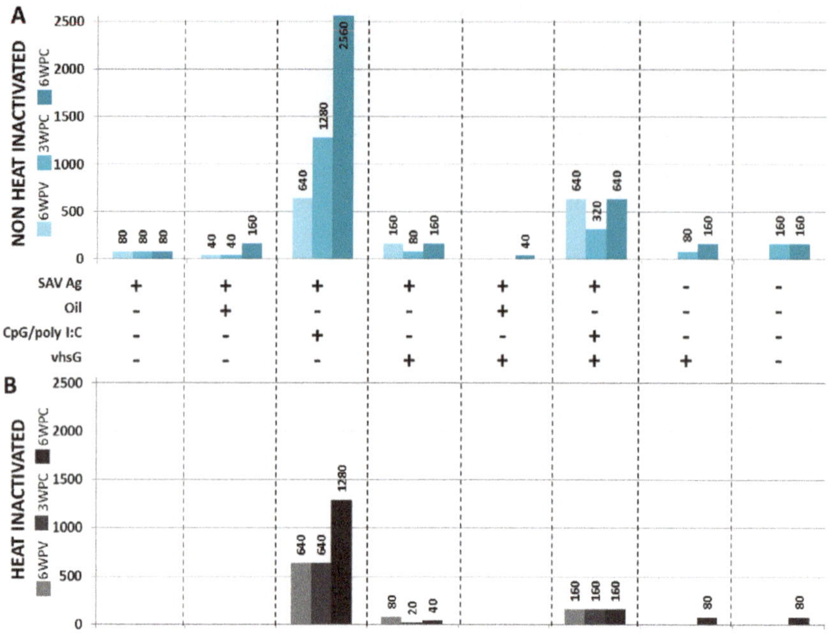

Figure 3. Vaccine-induced anti-SAV neutralizing titres from not heat inactivated (A) and heat inactivated (B) sera, collected at 6 wpv, 3 wpc and 6 wpc (the colour codes for each time point are next to the *y*-axis). Titres representing a 50% reduction, calculated as described in the Materials and Methods, are shown above the histogram corresponding to each treatment. + or −, respectively, indicates the presence or absence of either SAV Ag, oil, CpG/polyI:C or vhsG.

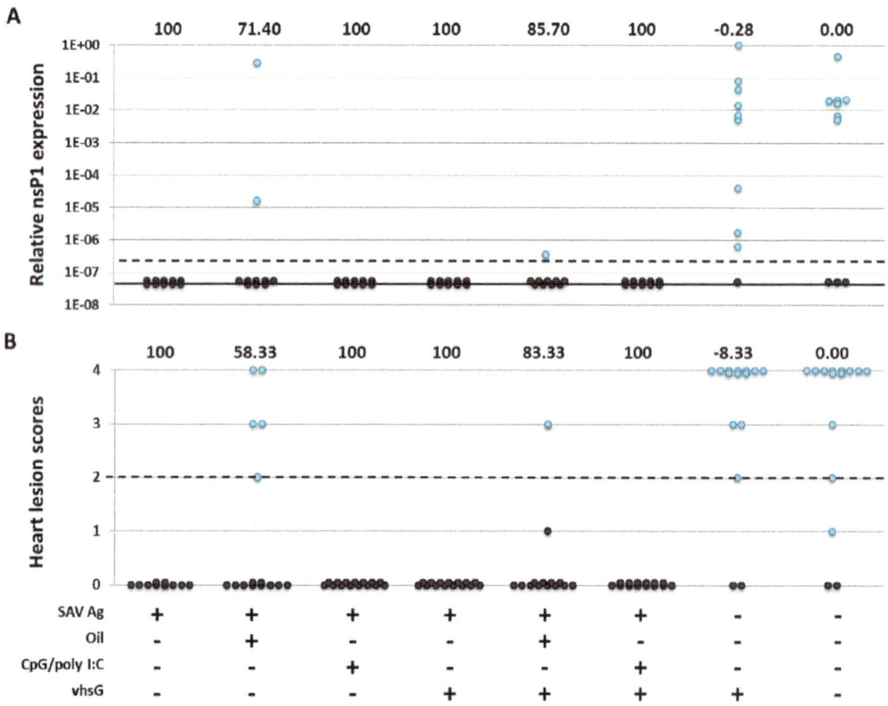

Figure 4. Protection against pancreas disease (PD) in vaccinated and control groups based on (A) the reduction of the prevalence of viremia and (B) the reduction in severity of SAV-specific heart lesions. (A) Relative SAV nsP1 expression at 3 wpc measured by SAV nsP1 RT-qPCR for each treatment group. Individual Cq-values were transformed to RelCq numbers, as described in the Materials and Methods. One (1.0×10^{0}) indicates the highest presence of nsP1 transcripts. Sera below the dotted line (cut off; 3.0×10^{-7}) were considered negative, and sera below the solid line had undetected Cq-values. Relative SAV nsP1 values are presented as black ($<3.0 \times 10^{-7}$) and blue ($\geq 3.0 \times 10^{-7}$) dots. (B) The distribution of individual heart lesion scores assessed by histology at 6 wpc for each treatment group. A score of ≥ 2 was set as the cut off (indicated by the dotted line). Individual heart lesion scores are presented as black (<2) and blue (≥ 2) dots. Relative percent protection score (RPPsc.) values corresponding to each group are shown above Graphs A and B. + or −, respectively, indicate the presence or absence of either the SAV Ag, oil, CpG/poly I:C or vhsG DNA construct.

DISCUSSION

How innate and adaptive immunity interact upon vaccination affects the outcome of protection. Especially, TLRs, included in the PRR-family, have emerged as key components of the innate immune system, activating signals critically involved in the initiation and maintenance of adaptive responses. Thus, TLR stimuli can be exploited as powerful adjuvants to elicit both primary and anamnestic immune responses. In light of this, our group used a combination of a selected CpG ODN [47] and a synthetic dsRNA (polyI:C), known to activate PRR-family members specialized in viral and nucleic acid detection (TLR-3, -9, -22 and RIG I), as a model to study the mechanisms of adjuvant action in bony fish. We showed that this TLR-ligand adjuvant combination induced a strong modulation of core response genes and increased levels of nAbs [30].

Here, to follow up and to extend this model, the CpG/poly I:C combo was i.p. injected alone or combined with an i.m. injection of the VHSV G DNA genetic adjuvant. Montanide ISA763A was included as a control adjuvant, i.p. injected in combination with the genetic adjuvant. This was to determine if any enhanced protective responses were induced, since oil-adjuvanted formulations previously had provided a lower protection, most likely due to depot effects [29,30]. An inactivated SAV whole-virus formulation was used as the test Ag and injected i.p. with the same suboptimal dose for all formulations, hypothesized to give about 70% protection based on previous unpublished results. However, in the present SAV challenge, fish injected with SAV Ag alone were fully protected (RPPsc. of 100%), thus preventing the detection of differences in protection between non-oil-based treatments. SAV Ag formulated with Montanide ISA763A oil had a lower efficacy (RPPsc. of 58.3%–85.7%) and produced lower nAb titres (25%–70% reduction) compared to the formulations receiving the equivalent water-based formulation, a difference that could be explained by the slower release of the Ag from the vaccine depot for the oil-adjuvanted formulations.

Correlating to earlier reports, the present CpG/polyI:C formulation induced a distinctively higher upregulation of early innate antiviral transcripts (IFNa1, Vig-1 and Mx) and also higher titres of nAbs compared to all other treatments [26,29,30]. nAbs are thought to be the primary correlate of protection against SAV [7,9,10], and the results suggest that an efficient clearance of virus mediated by Ab is possible. Interestingly, IFNγ expression levels were highly elevated in the CpG/poly I:C-adjuvanted groups. IFNγ, the hallmark of Th1 responses in mammals, is produced mainly by activated T-cells, NK cells and NKT cells; however, other cells, including macrophages/DCs, as well as B-cells are known to express IFNγ upon CpG-stimulation [48]. In support of this, a recent paper showed that salmon MHCII-positive mononuclear phagocytes, as well as B-cells and putative T-cells showed highly upregulated IFNγ transcript levels upon CpG-stimulation [49]. Since IFNγ has been shown to upregulate TLR9 expression in salmon leukocytes [22], the increased levels of this cytokine may represent a positive feed-back loop, where the secreted IFNγ upregulates the TLR9 receptor, and thereby, the cell's responsiveness to its own agonist are increased.

In higher vertebrates, B-cells activated via PRRs and/or by Ag cross-linking of the B-cell receptor rapidly respond, proliferate and differentiate into IgM secreting short-lived plasma cells (SLPC). T-cell help is needed to induce long-term B-cell memory, represented by memory B-cells or long-lived plasma-cells (LLPC) [50,51]. Bony fish Ab secreting cells (ASCs) are known to possess comparable B-cell subpopulations [52]; however, the understanding of the mechanisms by which these subpopulations are produced and distributed is scarce. Since germinal centres and antibody isotype shifting are not found in bony fish, the classical T-helper function in fish can be questioned. It is thus possible that CpG ODNs (through PRRs) can activate salmon B-cells directly, so that once activated, they start to proliferate and mature into ASCs. This would allow polyclonal activation of the entire B-cell pool, which has been demonstrated for mammalian B-cells [53,54]. Salmon B-cells express TLR9 and are responsive to their own agonist, CpG DNA [55]. Here, the CpG/poly I:C-adjuvanted vaccines were injected into the peritoneum, and recent studies have revealed that IgM-positive cells dominate the peritoneal cavity of unstimulated rainbow trout [56]. Upon inflammatory stimulation, these IgM positive cells were found at high levels 72 h post-injection. Since bony fish B-cells are phagocytic [57], it is possible that the vaccine can be engulfed by B-cells in the peritoneum and directly activated by CpG ODNs to differentiate into ASCs or, alternatively, migrate into secondary immunological organs, such as HK or spleen, and at those sites become ASC. This interesting possibility should be further elucidated in future studies.

Moreover, Desvignes *et al.* [9] have previously suggested that the complement in general might aid in the clearance of SAV. Other studies have since then shown that salmonid Abs are dependent on complement activity to neutralize VSHV and IHNV [58,59], which both are enveloped viruses. SAV is also an enveloped virus; hence, it is possible that complement factors are involved in SAV neutralization, as well. A previous study indeed showed induced levels of complement component C4 (classical pathway) in salmon treated either with saline or with CpG/polyI:C after SAV challenge [29]. In this study, a major fraction of the nAb titres were heat sensitive, underlining the involvement of the complement system in the clearance of SAV. The addition of naive salmon serum as the complement source was evaluated in the virus neutralization assay using HI sera samples. Interestingly, for the sera with low titres, *i.e.*, SAV Ag CpG/poly I:C vhsG, the fresh complement increased nAbs, while for the high titre group (SAV Ag CpG/poly I:C), there was no increase in nAbs (results not shown). A possible explanation for the variation in the results for the different groups could be that in the groups with very high nAbT, the neutralizing activity by the Abs *per se* is very efficient, and therefore, the complement does not provide any additional effect. In the groups with lower nAbT, the results indicate that combining the complement and Abs increases neutralization, supporting a role of complement-mediated neutralization, for example, by the classical complement pathway. Further studies aimed at elucidating the significance of the complement for the neutralizing responses need to be addressed.

B- and T-cell markers were, in general, not, or very modestly, induced for all formulations tested, which could be due to the early sampling time points. The results show that fish treated with SAV Ag CpG/polyI:C had a slight, but notable, increase of PAX5 transcripts levels at 12 hpv in both HK and spleen. In spleen, PAX5 induction occurred in parallel to that of mIgM at 12 hpv, while at 48 hpv, PAX5 expression was reduced and mIgM remained stable. This indicates that an early B-cell development takes place [43]. In rainbow trout, activated B-cells differentiate into plasmablasts and plasma cells (PC), both in HK and spleen, and are distributed through the blood system to peripheral tissues [43,52], where hydroxyurea-resistant PC can migrate back to the anterior kidney and may persist there as LLPCs. In mammals, LLPCs are generated by migration to a supportive niche in the bone marrow [50]. The ability of LLPCs to produce Ab for months to years without the stimulating Ag relies on specialized cues. One suggested cue is Type I IFN, which, when injected as an adjuvant in mice, has been shown to induce both short- (10 dpv) and long-lived (26 wpv) Ab production [60]. It has been suggested that the signals induced by Type I IFN affect either migration to survival niches or differentiation of PC [61]. It may well be possible that CpG/polyI:C through its strong induction of Type I IFN could enhance the generation of a, if present, similar long-lived Ab production in salmon, and our intent is to further investigate this aspect.

The immunological mechanisms behind the full protection provided by SAV Ag alone have been reasoned to depend on a T-cell independent (TI) nAb response, as that formulation did not induced an antiviral immune gene response at 5 dpv [30]. Here, a moderate induction of IFNa1, Mx, Vig-1 and IFNγ was evident in HK of fish treated with SAV Ag alone at 12 hpv. Inactivated viral Ags based on other enveloped viruses have been shown to trigger IFN responses [62,63]. Here, the actual mechanism responsible for the IFN responses observed *in vivo* with SAV Ag alone is unknown; nonetheless, by analogy with other alphaviruses, it can be hypothesized that the inactivated SAV particles present in the Ag preparation might induce the Type I IFN response seen, possibly through binding to the mannose receptor of *N*-glycans present on SAV E2, as observed for other alphaviruses [64] and other enveloped viruses [63]. The effect that the moderate induction of innate responses has on cellular immunity later in the course of the challenge is complex to interpret, owing to the limited knowledge on cellular immunity against viral infections in fish. Interestingly, NHI sera from fish treated with SAV Ag alone displayed nAb activity, while there was no detectable nAbT in HI sera for the group receiving SAV Ag alone. This variation between NHI and HI sera emphasizes the importance of the complement in the clearance of SAV and also the suggested ability of CpG to activate B-cells polyclonal through TLR 9, based on the high generation of heat-stable nAbs evident here for SAV Ag CpG/polyI:C compared to SAV Ag alone.

To our knowledge, very few studies have been performed on teleost to investigate the immunostimulatory and protective effects of an i.p. vaccine injected simultaneously with a DNA-based adjuvant injected i.m. The efficacy of the DNA vaccines based on the glycoprotein G from VHSV and IHNV are well-documented; these vaccines have been shown to induce a long-lasting protection against these viruses [65,66], and a commercial IHNV-G DNA vaccine is approved for use in Canadian aquaculture [67]. An early induced cross-protection after VHSV G DNA vaccination has been seen following infection with nodavirus in turbot (*Scophthalmus maximus*) [68] and with the heterologous Novirhabdovirus IHNV in rainbow trout. This study also showed that the DNA expressed G protein does not confer protection against bacterial diseases [69], hence emphasizing that the early non-specific protection provided is purely antiviral. In one study on rainbow trout [70], two or four CpG motifs were incorporated into the plasmid backbone along with VHSV G open reading frame, and this modified DNA vaccine gave significantly higher immune responses (Mx and IFNγ) and a significantly higher production of anti VHSV nAb compared to when plasmid without CpG motifs were administered. Based on these findings, it was hypothesized that the G protein might be able to contribute with a non-specific antiviral protection also against an SAV infection. As evident here, VHSV G glycoprotein DNA vaccination showed no additive effects on early protective responses against SAV. Furthermore, co-injecting vhsG with CpG/poly I:C or oil-adjuvanted SAV Ag neither caused additive nor synergistic effects on the immune gene expression. Muscle samples (n = 8) harvested at 48 hpv from vhsG co-injected groups were analysed for vhsG-transcripts, and the levels varied between individual fish, where some individuals had undetectable vhsG mRNA levels (results not shown). The low transcription of vhsG in muscle and the weak effect on innate immune gene expression could be explained by the relative early sampling time points. Indeed, studies on rainbow trout have shown that it can take as long as 14 dpv (earlier samplings at two and 7 dpv did not show any expression) before ISGs are significantly upregulated in spleen [70], and similar results in the liver of rainbow trout [71] and kidney of Japanese flounder (*Paralichthys olivaceus*) [72] have also been described. Protective immune responses observed after challenge in G DNA-vaccinated rainbow trout have been related to increased Mx expression and other ISGs [31,35,73], preceded by the upregulation of IFN Type I and II [35,74]. This early non-specific protection has been followed by a more specific long-lasting anti-viral response, based on both humoral (nAb) and cellular protective mechanisms (MHCII, T-cells) [73,75,76]. Interestingly, both B- and T-cell markers analysed for all vhsG formulations in this study had the highest expression compared to all other groups and were most distinct at 12 hpv in HK. This could indicate that the glycoprotein G has been expressed, and the high standard deviations indicate a high variation between individuals related to either immune gene expression profile and/or of vhsG expression.

In regards to humoral immunity, diverse responses were seen in the different groups i.m. injected with vhsG. While the SAV Ag oil vhsG treatment showed reduced nAbs titres compared to SAV Ag oil, the water-based SAV Ag formulation co-injected with vhsG had higher titres than SAV Ag alone at both 6 wpv and 6 wpc. Finally, a reduction in neutralizing responses was present when vhsG was co-injected with the SAV Ag CpG/polyI:C formulation, and the NHI nAbTs were reduced by half or more at all three time points compared to SAV Ag CpG/polyI:C. These data show that vhsG induced a slightly higher production of nAbs for the SAV Ag-treated fish co-injected with vhsG, and it also suggests the presence of factor(s) that cause a reduction in humoral responses when CpG/polyI:C-adjuvanted SAV Ag is co-injected with vhsG. For both mammals and teleosts, it has been shown that polyvalent vaccination often negatively affects the generation of specific Ab [77,78,79]. Skinner *et al.* [78] have shown that concurrent i.p. vaccination of a polyvalent oil vaccine with i.m. injection of a rhabdovirus-specific DNA vaccine delayed seroconversion of IHNV-specific nAb compared to DNA vaccination alone. The negative effects on nAbT when SAV

Ag CpG/poly I:C was co-injected with the vhsG vaccine can thus have several explanations, such as antigenic competition or Ag immunodominance. However, without the possibility of analysing the presence of vhsG-specific Ab (due to the limited amount of sera remaining), neither explanation can be claimed for certain.

CONCLUSIONS

To summarize, the vaccine based on a suboptimal dose of inactivated SAV whole virus Ag formulated with CpG/polyI:C induced the highest nAb responses, followed by the combined treatment of SAV Ag vhsG and, finally, the oil-adjuvanted SAV Ag formulation. The expression of several innate antiviral immune genes showed consistently elevated levels in the groups injected with CpG/polyI:C compared to the other adjuvants tested. B- and T-cell markers were, in general, not, or very modestly, induced for all formulations tested. For groups receiving the vhsG DNA vaccine, no antiviral immune gene expression was detected at these early time points, but as indicated by Martinez-Alonso *et al.* [70], a later induction could be possible. Pre-challenge humoral responses for SAV Ag co-injected with vhsG had slightly higher levels of both heat-sensitive and heat-stable neutralizing factors compared to SAV Ag alone, suggesting a very moderate adjuvant effect of vhsG. No enhancement of nAbs responses was evident when co-injecting vhsG with the TLR-ligand-adjuvanted SAV Ag formulation. Instead, a negative influence was observed, which may result from Ag competition.

Overall, these data show that CpG/polyI:C is a potent TLR-ligand combo, which could be used in future salmonid vaccination strategies against SAV. Further work should be aimed at investigating the duration of the efficacy of these TLR-ligands as adjuvants, while at the same time facilitating their administration. One potential strategy may be to protect the TLR-ligand/Ag formulation from degradation by encapsulating them in protective vehicles, such as nanoparticles, as recently described in [80]. Considering the potency of the TLR-ligand combo tested here, one can anticipate that CpG/polyI:C used as an adjuvant in an SAV Ag formulation could provide an Ag dose-sparing effect.

ACKNOWLEDGMENTS

The authors would like to thank Niels Lorenzen at the National Veterinary Laboratory (Danish Technical University, Copenhagen, Denmark) for providing the VHSV-G DNA plasmid construct and Barbara Spasowska and Glenn Sundnes at MSD Animal Health Innovation AS for their excellent work with the organization of all fish samples.

AUTHOR CONTRIBUTIONS

HLT carried out all of the experiments for this study except the nsP1qPCR, participated in the data-analyses and drafted the manuscript. SV participated in the design of this study, nsP1qPCR, data-analyses and revision of the manuscript. MMcL carried out the histopathology analyses, KEC contributed to the virus neutralization assay. SG provided some of the qPCR assays. PF participated in the design and revision of the manuscript. JBJ participated in the design, interpretation of the data and revision of the manuscript.

REFERENCES

1. Thorarinsson, R.; Powell, D.B. Effects of disease risk, vaccine efficacy, and market price on the economics of fish vaccination. *Aquaculture* 2006, *256*, 42–49.
2. Fringuelli, E.; Rowley, H.M.; Wilson, J.C.; Hunter, R.; Rodger, H.; Graham, D.A. Phylogenetic analyses and molecular epidemiology of european salmonid alphaviruses (sav) based on partial e2 and nsp3 gene nucleotide sequences. *J. Fish Dis.* 2008, *31*, 811–823.
3. Hodneland, K.; Bratland, A.; Christie, K.E.; Endresen, C.; Nylund, A. New subtype of salmonid alphavirus (sav), togaviridae, from atlantic salmon salmo salar and rainbow trout oncorhynchus mykiss in norway. *Dis. Aquat. Org.*2005, *66*, 113–120.
4. Taksdal, T.; Olsen, A.B.; Bjerkas, I.; Hjortaas, M.J.; Dannevig, B.H.; Graham, D.A.; McLoughlin, M.F. Pancreas disease in farmed atlantic salmon, salmo salar l, and rainbow trout, oncorhynchus mykiss (walbaum), in norway. *J. Fish Dis.*2007, *30*, 545–558.
5. McLoughlin, M.F.; Graham, D.A. Alphavirus infections in salmonids—A review. *J. Fish Dis.* 2007, *30*, 511–531.
6. Graham, D.A.; Frost, P.; McLaughlin, K.; Rowley, H.M.; Gabestad, I.; Gordon, A.; McLoughlin, M.F. A comparative study of marine salmonid alphavirus subtypes 1–6 using an experimental cohabitation challenge model. *J. Fish Dis.*2011, *34*, 273–286.
7. Biering, E.; Villoing, S.; Sommerset, I.; Christie, K.E. Update on viral vaccines for fish. *Dev. Biol.* 2005, *121*, 97–113.
8. Houghton, G. Acquired protection in atlantic salmon salmo salar parr and post-smolts against pancreas disease. *Dis. Aquat. Org.* 1994, *18*, 109–118.
9. Desvignes, L.; Quentel, C.; Lamour, F.; Le, V.A. Pathogenesis and immune response in atlantic salmon (salmo salar l.) parr experimentally infected with salmon pancreas disease virus (spdv). *Fish Shellfish Immunol.* 2002, *12*, 77–95.
10. McLoughlin, M.F.; Nelson, R.T.; Rowley, H.M.; Cox, D.I.; Grant, A.N. Experimental pancreas disease in atlantic salmon salmo salar postsmolts induced by salmon pancreas disease virus (spdv). *Dis. Aquat. Org.* 1996, *26*, 117–124.
11. Houghton, G.; Ellis, A.E. Pancreas disease in atlantic salmon: Serum neutralisation and passive immunisation. *Fish Shellfish Immunol.* 1996, *6*, 465–472.
12. Lopez-Doriga, M.V.; Smail, D.A.; Smith, R.J.; Domenech, A.; Castric, J.; Smith, P.D.; Ellis, A.E. Isolation of salmon pancreas disease virus (spdv) in cell culture and its ability to protect against infection by the "wild-type" agent. *Fish Shellfish Immunol.* 2001, *11*, 505–522.
13. Moriette, C.; Leberre, M.; Lamoureux, A.; Lai, T.L.; Bremont, M. Recovery of a recombinant salmonid alphavirus fully attenuated and protective for rainbow trout. *J. Virol.* 2006, *80*, 4088–4098.
14. Xu, C.; Mutoloki, S.; Evensen, O. Superior protection conferred by inactivated whole virus vaccine over subunit and DNA vaccines against salmonid alphavirus

infection in atlantic salmon (salmo salar l.). *Vaccine* 2012, *30*, 3918–3928.

15. Tavornpanich, S.; Viljugrein, H.; Stene, A.; Brun, E. Estimation of the reproduction number of salmon pancreas disease virus subtype 3 in homogeneously mixed populations of norwegian farmed atlantic salmon. *Prev. Vet. Med.* 2013, *111*, 329–332.
16. Aunsmo, A.; Ovretveit, S.; Breck, O.; Valle, P.S.; Larssen, R.B.; Sandberg, M. Modelling sources of variation and risk factors for spinal deformity in farmed atlantic salmon using hierarchical- and cross-classified multilevel models. *Prev. Vet. Med.* 2009, *90*, 137–145.
17. Haugarvoll, E.; Bjerkas, I.; Szabo, N.J.; Satoh, M.; Koppang, E.O. Manifestations of systemic autoimmunity in vaccinated salmon. *Vaccine* 2010, *28*, 4961–4969.
18. Skinner, L.A.; Schulte, P.M.; LaPatra, S.E.; Balfry, S.K.; McKinley, R.S. Growth and performance of atlantic salmon, salmo salar l., following administration of a rhabdovirus DNA vaccine alone or concurrently with an oil-adjuvanted, polyvalent vaccine. *J. Fish Dis.* 2008, *31*, 687–697.
19. Sommerset, I.; Krossøy, B.; Biering, E.; Frost, P. Vaccines for fish in aquaculture. *Expert Rev. Vaccines* 2005, *4*, 89–101.
20. McKee, A.S.; Munks, M.W.; Marrack, P. How do adjuvants work? Important considerations for new generation adjuvants. *Immunity* 2007, *27*, 687–690.
21. Palti, Y. Toll-like receptors in bony fish: From genomics to function. *Dev. Comp. Immunol.* 2011, *35*, 1263–1272.
22. Skjæveland, I.; Iliev, D.B.; Zou, J.; Jørgensen, T.; Jørgensen, J.B. A tlr9 homolog that is up-regulated by ifn-[gamma] in atlantic salmon (salmo salar). *Dev. Comp. Immunol.* 2008, *32*, 603–607.
23. Iliev, D.B.; Skjaeveland, I.; Jorgensen, J.B. Cpg oligonucleotides bind tlr9 and rrm-containing proteins in atlantic salmon (salmo salar). *BMC Immunol.* 2013, *14*, 12.
24. Cheng, Y.S.; Xu, F. Anticancer function of polyinosinic-polycytidylic acid. *Cancer Biol. Ther.* 2011, *10*, 1219–1223.
25. Matsuo, A.; Oshiumi, H.; Tsujita, T.; Mitani, H.; Kasai, H.; Yoshimizu, M.; Matsumoto, M.; Seya, T. Teleost tlr22 recognizes rna duplex to induce ifn and protect cells from birnaviruses. *J. Immunol.* 2008, *181*, 3474–3485.
26. Strandskog, G.; Skjaeveland, I.; Ellingsen, T.; Jørgensen, J.B. Double-stranded rna- and cpg DNA-induced immune responses in atlantic salmon: Comparison and synergies. *Vaccine* 2008, *26*, 4704–4715.
27. He, H.; Genovese, K.J.; Swaggerty, C.L.; Nisbet, D.J.; Kogut, M.H. *In vivo* priming heterophil innate immune functions and increasing resistance to salmonella enteritidis infection in neonatal chickens by immune stimulatory cpg oligodeoxynucleotides. *Vet. Immunol. Immunopathol.* 2007, *117*, 275–283.
28. He, H.; Genovese, K.J.; Swaggerty, C.L.; MacKinnon, K.M.; Kogut, M.H. Co-stimulation with tlr3 and tlr21 ligands synergistically up-regulates th1-cytokine ifn-gamma and regulatory cytokine il-10 expression in chicken monocytes.*Dev. Comp. Immunol.* 2012, *36*, 756–760.
29. Strandskog, G.; Villoing, S.; Iliev, D.B.; Thim, H.L.; Christie, K.E.; Jørgensen, J.B. Formulations combining cpg containing oliogonucleotides and poly i:C enhance the magnitude of immune responses and protection against pancreas disease in atlantic salmon. *Dev. Comp. Immunol.* 2011, *35*, 1116–1127.
30. Thim, H.L.; Iliev, D.B.; Christie, K.E.; Villoing, S.; McLoughlin, M.F.; Strandskog, G.; Jorgensen, J.B. Immunoprotective activity of a salmonid alphavirus vaccine: Comparison of the immune responses induced by inactivated whole virus antigen formulations based on cpg class b oligonucleotides and poly I:C alone or combined with an oil adjuvant.*Vaccine* 2012, *30*, 4828–4834.
31. Kim, C.H.; Johnson, M.C.; Drennan, J.D.; Simon, B.E.; Thomann, E.; Leong, J.A. DNA vaccines encoding viral glycoproteins induce nonspecific immunity and mx protein synthesis in fish. *J. Virol.* 2000, *74*, 7048–7054.
32. Lorenzen, E.; Einer-Jensen, K.; Martinussen, T.; LaPatra, S.E.; Lorenzen, N. DNA vaccination of rainbow trout against viral hemorrhagic septicemia virus: A dose-response and time-course study. *J. Aquat. Anim. Health* 2000, *12*, 167–180.
33. Von Gersdorff Jørgensen, L.; Sigh, J.; Kania, P.W.; Holten-Andersen, L.; Buchmann, K.; Clark, T.; Rasmussen, J.S.; Einer-Jensen, K.; Lorenzen, N. Approaches towards DNA vaccination against a skin ciliate parasite in fish. *PLoS One*2012, *7*, e48129.
34. Kobiyama, K.; Jounai, N.; Aoshi, T.; Tozuka, M.; Takeshita, F.; Coban, C.; Ishii, K. Innate immune signaling by, and genetic adjuvants for DNA vaccination. *Vaccines* 2013, *1*, 278–292.
35. Purcell, M.K.; Nichols, K.M.; Winton, J.R.; Kurath, G.; Thorgaard, G.H.; Wheeler, P.; Hansen, J.D.; Herwig, R.P.; Park, L.K. Comprehensive gene expression profiling following DNA vaccination of rainbow trout against infectious hematopoietic necrosis virus. *Mol. Immunol.* 2006, *43*, 2089–2106.
36. Querec, T.; Bennouna, S.; Alkan, S.; Laouar, Y.; Gorden, K.; Flavell, R.; Akira, S.; Ahmed, R.; Pulendran, B. Yellow fever vaccine yf-17d activates multiple dendritic cell subsets via tlr2, 7, 8, and 9 to stimulate polyvalent immunity. *J. Exp. Med.* 2006, *203*, 413–424.
37. Querec, T.D.; Akondy, R.S.; Lee, E.K.; Cao, W.P.; Nakaya, H.I.; Teuwen, D.; Pirani, A.; Gernert, K.; Deng, J.S.; Marzolf, B.; *et al.* Systems biology approach predicts immunogenicity of the yellow fever vaccine in humans. *Nat. Immunol.*2009, *10*, 116–125.
38. Lorenzen, N.; Lorenzen, E.; Einer-Jensen, K.; LaPatra, S.E. DNA vaccines as a tool for analysing the protective immune response against rhabdoviruses in rainbow trout. *Fish Shellfish Immunol.* 2002, *12*, 439–453.
39. Pfaffl, M.W.; Horgan, G.W.; Dempfle, L. Relative expression software tool (rest) for group-wise comparison and statistical analysis of relative expression results in real-time pcr. *Nucleic Acids Res.* 2002, *30*, e36.
40. Pfaffl, M.W. A new mathematical model for relative

quantification in real-time rt-pcr. *Nucleic Acids Res.* 2001, *29*, e45.

41. Christie, K.E.; Graham, D.A.; McLoughlin, M.F.; Villoing, S.; Todd, D.; Knappskog, D. Experimental infection of atlantic salmon salmo salar pre-smolts by i.p. Injection with new irish and norwegian salmonid alphavirus (sav) isolates: A comparative study. *Dis. Aquat. Org.* 2007, *75*, 13–22.
42. Hagman, J.; Lukin, K. "Hands-on" regulation of b cell development by the transcription factor pax5. *Immunity* 2007,*27*, 8–10.
43. Zwollo, P.; Haines, A.; Rosato, P.; Gumulak-Smith, J. Molecular and cellular analysis of b-cell populations in the rainbow trout using pax5 and immunoglobulin markers. *Dev. Comp. Immunol.* 2008, *32*, 1482–1496.
44. Moore, L.J.; Dijkstra, J.M.; Koppang, E.O.; Hordvik, I. Cd4 homologues in atlantic salmon. *Fish Shellfish Immunol.* 2009,*26*, 10–18.
45. Meyer, K.; Basu, A.; Przysiecki, C.T.; Lagging, L.M.; di Bisceglie, A.M.; Conley, A.J.; Ray, R. Complement-mediated enhancement of antibody function for neutralization of pseudotype virus containing hepatitis c virus e2 chimeric glycoprotein. *J. Virol.* 2002, *76*, 2150–2158.
46. Grove, S.; Austbo, L.; Hodneland, K.; Frost, P.; Lovoll, M.; McLoughlin, M.; Thim, H.L.; Braaen, S.; Konig, M.; Syed, M.; *et al.* Immune parameters correlating with reduced susceptibility to pancreas disease in experimentally challenged atlantic salmon (salmo salar). *Fish Shellfish Immunol.* 2013, *34*, 789–798.
47. Strandskog, G.; Ellingsen, T.; Jørgensen, J.B. Characterization of three distinct cpg oligonucleotide classes which differ in ability to induce ifn alpha/beta activity and cell proliferation in atlantic salmon (salmo salar l.) leukocytes. *Dev. Comp. Immunol.* 2007, *31*, 39–51.
48. Malmgaard, L. Induction and regulation of ifns during viral infections. *J. Interferon Cytokine Res.* 2004, *24*, 439–454.
49. Iliev, D.B.; Thim, H.; Lagos, L.; Olsen, R.; Jørgensen, J. Homing of antigen-presenting cells (apcs) in head kidney and spleen—Salmon head kidney hosts diverse apc types. *Front. Immunol.* 2013, *4*, 137.
50. Chu, V.T.; Berek, C. The establishment of the plasma cell survival niche in the bone marrow. *Immunol. Rev.* 2013, *251*, 177–188.
51. Tangye, S.G. Staying alive: Regulation of plasma cell survival. *Trends Immunol.* 2011, *32*, 595–602.
52. Ye, J.; Kaattari, I.; Kaattari, S. Plasmablasts and plasma cells: Reconsidering teleost immune system organization. *Dev. Comp. Immunol.* 2011, *35*, 1273–1281.
53. Bernasconi, N.L.; Onai, N.; Lanzavecchia, A. A role for toll-like receptors in acquired immunity: Up-regulation of tlr9 by bcr triggering in naive b cells and constitutive expression in memory B cells. *Blood* 2003, *101*, 4500–4504.
54. Capolunghi, F.; Cascioli, S.; Giorda, E.; Rosado, M.M.; Plebani, A.; Auriti, C.; Seganti, G.; Zuntini, R.; Ferrari, S.; Cagliuso, M.; *et al.* Cpg drives human transitional b cells to terminal differentiation and production of natural antibodies. *J. Immunol.* 2008, *180*, 800–808.
55. Thim, H.L.; Lagos, L.; Iliev, D.B.; Jensen, I.; Jørgensen, J.B. CpG Activation of Salmonid B-Cells Can Induce Long Lived Protection. Presented at HAVBRUK, Tromsø, Norway, March 2014.
56. Korytář, T.; Jaros, J.; Verleih, M.; Rebl, A.; Kotterba, G.; Kühn, C.; Goldammer, T.; Köllner, B. Novel insights into the peritoneal inflammation of rainbow trout (oncorhynchus mykiss). *Fish Shellfish Immunol.* 2013, *35*, 1192–1199.
57. Li, J.; Barreda, D.R.; Zhang, Y.A.; Boshra, H.; Gelman, A.E.; Lapatra, S.; Tort, L.; Sunyer, J.O. B lymphocytes from early vertebrates have potent phagocytic and microbicidal abilities. *Nat. Immunol.* 2006, *7*, 1116–1124.
58. Encinas, P.; Gomez-Casado, E.; Fregeneda, G.; Olesen, N.J.; Lorenzen, N.; Estepa, A.; Coll, J.M. Rainbow trout surviving infections of viral haemorrhagic septicemia virus (vhsv) show lasting antibodies to recombinant g protein fragments. *Fish Shellfish Immunol.* 2011, *30*, 929–935.
59. Fregeneda-Grandes, J.; Skall, H.; Olesen, N. Antibody response of rainbow trout with single or double infections involving viral haemorrhagic septicaemia virus and infectious haematopoietic necrosis virus. *Dis. Aquat. Org.* 2009,*83*, 23–29.
60. Le Bon, A.; Schiavoni, G.; D'Agostino, G.; Gresser, I.; Belardelli, F.; Tough, D.F. Type I interferons potently enhance humoral immunity and can promote isotype switching by stimulating dendritic cells *in vivo*. *Immunity* 2001, *14*, 461–470.
61. Manz, R.A.; Arce, S.; Cassese, G.; Hauser, A.E.; Hiepe, F.; Radbruch, A. Humoral immunity and long-lived plasma cells. *Curr. Opin. Immunol.* 2002, *14*, 517–521.
62. De Wit, M.C.; Horzinek, M.C.; Haagmans, B.L.; Schijns, V.E.J.C. Host-dependent type 1 cytokine responses driven by inactivated viruses may fail to default in the absence of il-12 or ifn-α/β. *J. Gen. Virol.* 2004, *85*, 795–803.
63. Milone, M.C.; Fitzgerald-Bocarsly, P. The mannose receptor mediates induction of ifn-α in peripheral blood dendritic cells by enveloped rna and DNA viruses. *J. Immunol* 1998, *161*, 2391–2399.
64. Shabman, R.S.; Morrison, T.E.; Moore, C.; White, L.; Suthar, M.S.; Hueston, L.; Rulli, N.; Lidbury, B.; Ting, J.P.-Y.; Mahalingam, S.; *et al.* Differential induction of type I interferon responses in myeloid dendritic cells by mosquito and mammalian-cell-derived alphaviruses. *J. Virol.* 2007, *81*, 237–247.
65. Lorenzen, N.; LaPatra, S.E. DNA vaccines for aquacultured fish. *Rev. Sci. Technol.* 2005, *24*, 201–213.
66. Einer-Jensen, K.; Delgado, L.; Lorenzen, E.; Bovo, G.; Evensen, O.; Lapatra, S.; Lorenzen, N. Dual DNA vaccination of rainbow trout (oncorhynchus mykiss) against two different rhabdoviruses, vhsv and ihnv, induces specific divalent protection. *Vaccine* 2009, *27*, 1248–1253.

67. Alonso, M.; Leong, J.A. Licensed DNA vaccines against infectious hematopoietic necrosis virus (ihnv). *Recent Pat. DNA Gene Seq.* 2013, *7*, 62–65.

68. Sommerset, I.; Lorenzen, E.; Lorenzen, N.; Bleie, H.; Nerland, A.H. A DNA vaccine directed against a rainbow trout rhabdovirus induces early protection against a nodavirus challenge in turbot. *Vaccine* 2003, *21*, 4661–4667.

69. Lorenzen, N.; Lorenzen, E.; Einer-Jensen, K.; LaPatra, S.E. Immunity induced shortly after DNA vaccination of rainbow trout against rhabdoviruses protects against heterologous virus but not against bacterial pathogens. *Dev. Comp. Immunol.* 2002, *26*, 173–179.

70. Martinez-Alonso, S.; Martinez-Lopez, A.; Estepa, A.; Cuesta, A.; Tafalla, C. The introduction of multi-copy cpg motifs into an antiviral DNA vaccine strongly up-regulates its immunogenicity in fish. *Vaccine* 2011, *29*, 1289–1296.

71. McLauchlan, P.E.; Collet, B.; Ingerslev, E.; Secombes, C.J.; Lorenzen, N.; Ellis, A.E. DNA vaccination against viral haemorrhagic septicaemia (vhs) in rainbow trout: Size, dose, route of injection and duration of protection-early protection correlates with mx expression. *Fish Shellfish Immunol.* 2003, *15*, 39–50.

72. Byon, J.Y.; Ohira, T.; Hirono, I.; Aoki, T. Use of a cdna microarray to study immunity against viral hemorrhagic septicemia (vhs) in japanese flounder (paralichthys olivaceus) following DNA vaccination. *Fish Shellfish Immunol.*2005, *18*, 135–147.

73. Boudinot, P.; Blanco, M.; de Kinkelin, P.; Benmansour, A. Combined DNA immunization with the glycoprotein gene of viral hemorrhagic septicemia virus and infectious hematopoietic necrosis virus induces double-specific protective immunity and nonspecific response in rainbow trout. *Virology* 1998, *249*, 297–306.

74. Zou, J.; Tafalla, C.; Truckle, J.; Secombes, C.J. Identification of a second group of type i ifns in fish sheds light on ifn evolution in vertebrates. *J.Immunol.* 2007, *179*, 3859–3871.

75. Boudinot, P.; Bernard, D.; Boubekeur, S.; Thoulouze, M.I.; Bremont, M.; Benmansour, A. The glycoprotein of a fish rhabdovirus profiles the virus-specific t-cell repertoire in rainbow trout. *J. Gen. Virol.* 2004, *85*, 3099–3108.

76. Lorenzen, N.; Lapatra, S.E. Immunity to rhabdoviruses in rainbow trout: The antibody response. *Fish Shellfish Immunol.* 1999, *9*, 345–360.

77. Raadsma, H.W.; O'Meara, T.J.; Egerton, J.R.; Lehrbach, P.R.; Schwartzkoff, C.L. Protective antibody titres and antigenic competition in multivalent dichelobacter nodosus fimbrial vaccines using characterised rdna antigens. *Vet. Immunol. Immunopathol.* 1994, *40*, 253–274.

78. Skinner, L.A.; LaPatra, S.E.; Adams, A.; Thompson, K.D.; Balfry, S.K.; McKinley, R.S.; Schulte, P.M. Concurrent injection of a rhabdovirus-specific DNA vaccine with a polyvalent, oil-adjuvanted vaccine delays the specific anti-viral immune response in atlantic salmon, salmo salar l. *Fish Shellfish Immunol.* 2010, *28*, 579–586.

79. Taborda, C.P.; Rivera, J.; Zaragoza, O.; Casadevall, A. More is not necessarily better: Prozone-like effects in passive immunization with igg. *J. Immunol.* 2003, *170*, 3621–3630.

80. De Titta, A.; Ballester, M.; Julier, Z.; Nembrini, C.; Jeanbart, L.; van der Vlies, A.J.; Swartz, M.A.; Hubbell, J.A. Nanoparticle conjugation of cpg enhances adjuvancy for cellular immunity and memory recall at low dose. *Proc. Natl. Acad. Sci. USA* 2013.

CHAPTER
5 VACCINES FOR FINFISH AQUACULTURE

Brian Dixon

Keywords: Antibody, antigen presentation, cloning, cytokine, immunity, major histocompatibility complex receptors

THE NEED FOR FINFISH VACCINES

In 2007 the global market for aquaculture was worth US$ 60 billion, of which US$ 20 billion was the value of finfish aquaculture. Losses in finfish aquaculture can cause massive economic losses. A Canadian example was the recent outbreak of IPN virus in the Mainstream Dixon Bay farm in British Columbia which resulted in the destruction of 500.000 fish, a loss of well over a million dollars in revenue for that one company in one year. There are 375 marine finfish and 822 freshwater finfish licenses currently in Canada, for a total of 1.197 finfish licenses (http://www.dfo-mpo.gc.ca/aquaculture/faq-eng.htm). A typical farm has between 6 to 24 cages that each contain 35.000 to 50.000 fish, so the 500.000 animals at the Dixon Bay site noted above is average. That means that there are an approximate total of nearly 600 million fish grown annually in Canada. If 5% of those farms suffer severe outbreaks that cause the destruction of stock, which is approximately 30 million fish, it will cost of 60 million dollars in losses. In fact the data suggest 5% loss to parasites and diseases is conservative. The data for Norwegian farms over a 14 year period (http://www.bellona.org/aquaculture/artikler/Lost_production/) suggest an average of 8.6% losses, of which 85% or 7.31% was due to fish deaths, the majority of which is caused by pathogens and parasites. If we use the 2007 numbers from above, 7% losses to disease would have a value of 1.4 billion dollars. Thus by far the biggest target area for preventing very costly losses to the finfish aquaculture industry is to prevent fish from dying of diseases. Better vaccines could prevent a large proportion of those losses as treatment will prevent the destruction of animals.

What do we need to know to design better vaccines for fish? The most important piece of knowledge needed here is detailed information about way fish immunity develops and eliminates pathogens. This will tell us what types of pathogenic molecules should be included in vaccines and exactly what type of response they will elicit. Additionally, this detailed knowledge will inform us of the appropriate choices for adjuvants which are the critical component of vaccines that initiate innate immune responses, without which our peptide or protein will not be recognized. While teleost immune systems possess most of the same cells and molecules that mammalian immune systems do, we cannot rely on knowledge of mammalian immunity to guide the production of vaccines for teleost fish. The main reason for this is, as René Stet and I argued in 2003, that while teleost fish possess all of the same components as our immune systems, they use them very differently. Indeed the data we have accumulated since then has only confirmed our argument that teleost fish and mammalian immune systems have evolved along divergent pathways from the same starting point (Figure 1).

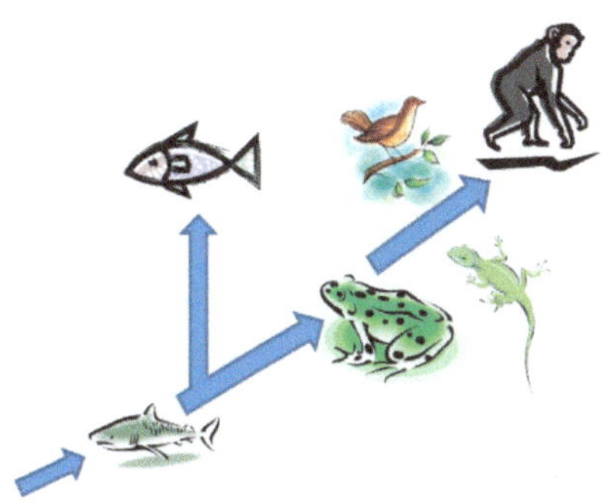

Figure 1. A schematic representation of the evolution of immune systems. The immunoglobulin superfamily protein based immune sys-

Citation: Brian Dixon. Vaccines for finfish aquaculture Electronic Journal of Biotechnology, 2012; Vol. 15, No. 5. DOI: 10.2225/vol15-issue5-fulltext-18. ISSN: 0717-3458.

tem found in all vertebrates arose in an ancestor of the sharks after its divergence from the lineage of lampreys and hagfish, which possess an adaptive immune system based on leucine rich repeat proteins. Teleost fish underwent a series of genome duplications during their evolution, which allowed them to use the same basic set of genes and cells in a completely different manner from the lobe-finned fishes which were the ancestors of amphibians and hence the whole tetrapod lineage.

Current Knowledge of Fish Immune System Genes

Fish immune systems have been studied at the molecular level for over 20 years. Molecular biology allowed the cloning of immunoglobulin genes starting in 1989, with the cloning of the catfish genes (Ghaffari and Lobb, 1989). Cloning of the carp Major Histocompatibility (MH) genes, which encode receptors essential for the initiation of adaptive immune responses, was first reported in 1990 (Hashimoto et al. 1990). In both these cases, numerous papers detailing the cloning of these both of these genes from other species followed over the following five years (using MH genes as an example, seeDixon et al. (1995) for a review of the numerous species studied between 1990 and 1995).

Cytokines, generally small soluble proteins that are secreted by one cell in order to send a signal to another, play many roles in the physiological functioning of an animal. For example, Interleukin-1β (IL-β), which is a key cytokine expressed early in immune responses by macrophages and endothelial cells in order to activate the immune response by initiating the fever response and activating both T cells and macrophage. However, it also plays a role in communicating between the endocrine system and the immune system during the suppression of immune response due to stress while also exhibiting a host of other properties, including, stimulating myeloid cell production in the bone marrow, angiogenesis (Bar et al. 2003) and blood vessel dilation (Jiménez-Altayó et al. 2006). Tumour necrosis factor α (TNFα) shares many of the same immunological properties as IL1β. Other key cytokines are Interferon γ (IFNγ) which is released by T cells and NK cells in order to activate macrophages, and Il-6, which is produced by T cells, macrophages and endothelial cells in order to activate the acute phase/fever response, while also inducing the growth of B and T cells. Despite their importance, the first fish cytokine genes took longer to isolate, with the first being TGF-beta in 1998 (Hardie et al. 1998) followed by several others, including IL-1b in 1999 (Zou et al. 1999), and the first chemokine, chemoattractants that traffic lymphocytes from one location to another for example to sites of injury and infection, in 1998 (Dixon et al. 1998). One interesting feature of cytokines in teleost fish is that most of those cloned to date exist as duplicated genes in the genome of most species. For example, both type I interferons (Chang et al. 2009) and type II interferons are present as duplicate genes. It is not entirely clear at this point what the functional significance of this gene duplication is as both are usually upregulated upon immune stimulation, for example IFN-γ, however there is differential regulation of these genes, suggesting that each isoform may have different functions (Purcell et al. 2009).

Molecular techniques such as subtractive hybridization (SSH) added numerous cloned genes for all classes of immune molecules to the list of those isolated throughout the 1990s (for example just from my own work see: three forms of invariant chain: Fujiki et al. 2003a, IL-6: Fujiki et al. 2003b, CD9: Fujiki et al. 2002, C5a receptor: Fujiki et al. 2003c). Finally in the 2000-s there were genome projects for model species such as zebrafish and pufferfish, as well as large scale EST projects for key aquaculture species such as Atlantic Salmon and Cod. Thus, large lists of the genes which make up the immune system of several fish species are now available including those that encode receptors, cytokines and effector molecules. These lists are limited however, either due to poor annotation or a lack of identification of a significant proportion of the sequences produced. In some cases over 30% of the sequences in the databases are unknown. Additionally, the one key piece of information that has emerged from data mining of these databases is that many key immune system genes, which are single copy in mammals, are duplicated in teleost fish (invariant chain (Fujiki et al. 2003a; Chang et al. 2009; Purcell et al. 2009) complicating the task of developing an understanding of teleost immune systems. But what functional data is available regarding those genes?

Knowledge of Fish Immune System Gene Expression Patterns

The early discoveries of fish immune system genes, either by targeted cloning or by selective techniques such as SSH, were often accompanied by the characterization of gene expression patterns by northern blot or reverse transcriptase PCR analysis of various fish tissues in unstimulated and immune stimulated conditions (again just drawing from my own work, some examples can be seen in the Fujiki et al. references (Fujiki et al. 2002; Fujiki et al. 2003a; Fujiki et al. 2003b; Fujiki et al. 2003c) or Kales et al. (2004). Thus basic expression patterns and conditions for some genes are known. For the large numbers of genes produced by genome and EST projects however, the information is less reliable. This may be due to the gene duplication and lack of annotation issue noted above, or for technical reasons, such as the use of oligonucleotide based microarrays for expression analysis, upon which the oligonucelotides used fail to differentiate between alternate forms of the genes (for example between isotypes of antibody or duplicates of interleukin genes). That being said some useful information can be mined from these databases regarding the expression pattern of some specific genes from microarray studies or lists of EST derived from specific tissues before and after immune stimulation.

Knowledge of the Function of Fish Immune System Gene Products

The area where there is a big gap in functional knowledge between fish and mammals is at the level of protein function. Even for some of the earliest genes to be discovered, there are still no antibodies or functional tests available, although in recent years this has become more of a focus for researchers. Antibodies to immunoglobulin genes from fish had been produced several years before the cloning of the genes (Etlinger et al. 1977; Warr et al. 1979), due to the fact that isolating the immunoglobulins from blood was not too onerous (Raison et al. 1978). Indeed, antibodies to fish immunoglobulins for various species are one commercial product that has emerged from this line of research, with companies such as Cedarlane

Labs in Burlington Ontario and Aquatic Diagnostics in Stirling, Scotland, marketing these. Antibodies to other immune system genes such as MH receptors were isolated relatively quickly after the discovery of the genes (Van Erp et al. 1996; Rodrigues et al. 1998; Kales et al. 2004; Nath et al. 2006) while antibodies to fish cytokines have been slower to emerge. These have been used for some functional studies where possible, but the lack of antibodies to most available proteins has been a limiting factor in true understanding of immune function in teleosts.

What We Need I: Understanding of How Antigen Presentation Starts Responses

One of the key specific pieces of knowledge we need to know in order to develop better finfish vaccines is how fish immune systems recognize pathogens and target them for their B and T cells. This will require knowing what particular protein fragments of pathogens are recognized by teleost T cells and antibodies in order to assess their general nature, as well as knowledge of the specific epitopes from key pathogens that are recognized. This will require functional knowledge of the accessory molecules involved in antigen presentation such as TAP, Tapasin, and the other molecules of the antigen presentation pathways.

Knowing the Peptides Targeted by the Immune System Is Essential

The immune system recognizes pathogens and starts specific responses-antibodies or T cells targeted to short specific peptide sequences of the pathogen only when T cells have been activated to produce the cytokine signals which start the response (see below). T cell activation requires both peptides of the pathogen presented in the groove of the MH receptor to be recognized by the T cell receptor and a confirmatory signal from the antigen presenting cell that confirms that the peptide is indeed pathogenic in origin.

For vaccine development, one key piece of knowledge would be the actual sequence of the peptides which can activate the T cells. This knowledge can be used to include such peptide sequences within the vaccines, guaranteeing activation of T cells specific for the pathogen. The technology for doing this has been worked out for mammalian systems (Barber et al. 1996; Gleimer et al. 2011), but challenges remain for performing this in teleost systems. Firstly one needs to know which particular MH receptor alleles are capable of binding peptides of the pathogens. This is achieved by characterizing the MH alleles of several different families of fish to assess the diversity of MH alleles present (one species of fish can contain hundreds of different alleles: for example, my lab has discovered over 100 different MH class I beta alleles in 200 arctic char individuals that we have typed to date). One then runs a disease trial with these families using the pathogen of interest and sequences the MH alleles present in the survivors (and mortalities if possible). The specific MH alleles present in high frequency in the survivors carries the peptides that protective T cells can recognize, and conversely if there are alleles present at high frequency in the mortalities, these are not effective and causes susceptibility to the pathogen. This approach was used by (Grimholt et al. 2003) to characterize Atlantic salmon MH alleles that were protective against the pathogens A. salmonicida and Infectious Salmon Anemia Virus (ISAV), and the process has been started for other countries fish stocks, such as the lab of Dr Sergio Marshall which has looked at the alleles in Chilean Atlantic salmon which protect against ISAV and *P. salmonis* (Gómez et al. 2010). This information can be used to breed protective MH alleles into the broodstock, but one must be careful to keep a diversity of alleles present as well as new pathogens emerge and it is difficult to predict which MH alleles will be protective for those diseases.

However once the main protective MH alleles have been identified and bred into the population, it should be possible to design a vaccine specific for those key MH alleles that will provide the peptides it binds and allow enhanced T cell activation. The process involves cloning the genes for the MH class II alpha and beta chain alleles of interest into mammalian expression vectors and using those vectors to express the complete MH receptor on the surface of a cell line which does not normally express that receptor (for example expressing fish MH receptors on the surface of mammalian cells). In order to be expressed on the surface the alpha and beta chain must bind to a peptide which fits into the unique groove of the receptor. One then acid washes the receptors from the surface of the cell, purifies them on a column containing MH specific antibodies under non-denaturing conditions so the alpha-beta-peptide trimer remains intact and then dissociates the peptide from the alpha and beta chains using mild detergent, purifies the peptide from the MH chains using the same chromatography column and finally sequences the peptide using mass spectrometry. Even for a single MH allele this should yield a large number of peptides as usually only 2 anchor amino acids are needed to bind the peptide to the MH receptor groove and peptides bound to MH class I grooves are usually around 9 amino acids in length, while MH class II bound peptides can be around 20 amino acids long. The number and types of peptides which can be bound by at least one Atlantic salmon allele have been modelled (Cárdenas et al. 2010). By searching the main amino acid sequences through sequence databases containing pathogen genomes or ESTs, one can identify pathogenic peptide sequences which match those bound by the protective MH alleles. Inclusion of the pathogenic proteins which include these target peptide sequences should ensure a robust immune response, thus improving vaccine efficacy in a species population-specific, indeed population-specific, manner.

What do we need to actually carry this out for finfish? For salmonids, the tools are available are now as studies linking MH alleles to specific diseases have already been carried out and my laboratory has antibodies against salmonid MH receptors. All that remains is to surpass the technical hurdles involved in expressing salmonid MH alleles on the surface of MH negative cell lines. Dr. Sergio Marshall□s laboratory and mine are currently collaborating on that process and hope to achieve this shortly. For other aquaculture species, studies linking MH alleles to disease as well as the development of MH receptor specific antibodies are still needed, but these are both easily achievable.

The placement of specific peptides in the groove of MH receptors does not depend solely on the MH receptor itself, however. The process of peptide loading for both MH class I and MH class II receptors is guided by accessory molecules (Saunders and Van Endert, 2011; Schulze and Wucherpfennigm, 2012). In the case of peptide loading into the MH class I receptor, comprising MH class I heavy chain and beta-2 microglobulin, the peptides which are eventually bound are edited by the proteasome in the cytoplasm, only selectively transported into the ER by TAP and must be actively loaded into the groove by tapasin. The chaperones calreticulin and ERP57 are bound to the MH class I: beta-2 microglobulin: tapasin: TAP complex and may also exert influence on which specific peptides are inserted into the MH receptor groove (Wearsch and Cresswell, 2008). For MH class II peptide loading, the DM accessory molecule loads peptides into the groove and may influence the nature of the peptides bound. All of these protein:protein interactions have been characterized mammalian systems, but for not teleost fish as while most of the genes have been cloned, no antibodies which can be used for immunoprecipitation or microscopic co-localization studies are commercially available. However understanding these interactions and the way they ultimately shape the peptide repertoire bound by MH class I and II receptors could reveal key information needed for understanding the types of pathogenic proteins included as targets in vaccines: choosing a protein that will be processed with high efficiency so that peptides from it are presented with high frequency by MH receptors will substantially increase vaccine efficacy.

In order to understand how MH class I receptors bind to and carry peptides to the cell surface in teleosts, if the same process in mammals is a guide, one would have to have antibodies to at least eight of these different proteins (MH class I, beta-2 microglobulin, calreticulin, calnexin, ERP57, Tapasin and both TAP subunits) just to examine the molecular interactions between proteins within the ER. For MH class II one would need to examine the interaction between the MH class II receptor, MH class II associated invariant chain within the ER and the interaction of MH class II receptors and DM in the MH class II compartment vesicles. For salmonids, the genes encoding all of the accessory proteins known to be involved in the MH class I antigen presentation pathway have all been cloned (Grimholt et al. 2002; Fuller et al. 2004; Kales et al. 2004; Landis et al. 2006), including ERP57 for which we have just submitted a manuscript (Sever Bols and Dixon, unpublished). We have developed antisera to all of these proteins and we are currently trying to characterize the protein:protein interactions involved in teleost (well, at least salmonid) MH class I presentation. For the salmonid MH class II pathway, which at first glance appears less complicated due to the fact that fewer proteins are involved, the study of the process is not quite as advanced. In part this is due to the fact that a teleost homologue of DM has not yet been identified, despite all of the recent genome projects. DM is evolutionarily related to MH class II receptors and is similar in structure and sequence. Given that teleost MH receptors are only roughly 35% identical in protein sequence to their mammalian equivalents, it may be that the teleost DM equivalent sequences are sitting in a database identified as aberrant MH class II receptor sequences. The other obstacle to the study of this process is the complicated nature of the other chaperone molecule involved in this pathway - MH class II associated invariant chain (Ii). In mammals, this molecule, while encoded by a single gene, produces four different polypeptides using alternative translation start sites and alternative splicing (Strubin et al. 1986). Genes encoding zebrafish (Yoder et al. 1999) and rainbow trout (Fujiki et al. 2003c) equivalents of Ii have been isolated, but in this case there are multiple gene copies instead of alternative splicing, each of which encodes a variant form, some of which are missing key features of the mammalian Ii protein. Each of these isoforms, however, is very similar in sequence to each other, making it difficult for the polyclonal antibodies to distinguish between them. Thus elucidating the steps of the antigen presentation pathway, particularly which Ii isoforms are binding to the MH class II receptor at particular time points or under different conditions, will prove difficult. My laboratory is currently looking at strategies needed to clone the teleost DM gene and producing monoclonal antibodies which can distinguish the various Ii isoforms.

What We Need II: Cytokines are Key Players

Once immune responses are initiated, the type of response that follows is guided by the cytokines that are produced in the early phases of the immune response. Depending on these initial signals, the immune response can upregulate mechanisms that response to viruses, extracellular bacteria or allergens and parasites such as worms. In mammals, these three pathways develop separately and in particular the extracellular bacteria (TH1) and allegen/worm (TH2) pathways are mutually exclusive (Hall et al. 2011; Bolon, 2012). Thus in developing vaccines it is important to ensure that the correct pathway is triggered and this could be achieved by assaying the cytokines expressed during the response to the vaccine. How does this translate to fish immunity? Many cytokine genes have been discovered, but the gene for the key cytokine that initiates the TH2 response has not yet been isolated. Additionally, while gene discovery has provided knowledge of fish cytokines at an academic level, it has not yet translated into very many practical products for assessing the health or immune status of fish in either a commercial or research settings. Such assays would ideally measure the levels of the protein products of specific genes, perhaps even specific isoforms of these genes as post-transcriptional modifications may be required for functionality of these proteins and are quite common mechanisms of regulating expression in mammalian immune system genes, especially the cytokine genes.

Thus assays which could detect specific amounts of cytokines in the blood or other tissues of fish or the number of cells from peripheral blood or a cell culture could be used not only for academic research into fish immune responses, which is currently becoming limited due to the lack of those assays, but also for many other aspects of fish physiology as well. Indeed with sufficient knowledge as to the exact role of these cytokines in initiating immune responses, sufficiently inexpensive and simple assays for quantifying production of these cytokines could be used commercially as a screening tool for diseases in aquaculture situations. This would be particularly useful for novel pathogens for which there is not yet a diagnostic test.

Assays for Cytokine Concentrations Are Needed

A common assay used for the assessment of cytokine levels in mammalian tissue is a quantitative or direct Enzyme Linked ImmunoSorbent Assay (ELISA), sometimes called a sandwich ELISA. This assay requires two antibodies to the target cytokine, one to coat the 96 well plates for □capturing□ the target cytokine from the sample, and a second, usually conjugated to a flourochrome or enzyme, for detection. The 'capture' antibody is coated onto the plate first, and after washing to remove any excess, the sample containing the unknown amount of the target compound is added. The target compound is bound by the capture antibody, while everything else in the sample is washed away. Finally the detection antibody is added and binds to the target cytokine. The assay usually comes with a set of standards of known concentration so that when the absorbance or fluorescence value of the unknown are plotted on a curve, the actual concentration of the target cytokine in the unknown can be determined.

For determining the number of cells in a culture dish or in the diverse population of in peripheral blood lymphocytes (PBL) that produce a target cytokine, the best assay is an ELISpot assay. This is a variation on the common ELISA, in which it is the cells thought to be producing the target cytokine that are added to a 96 well plate which has been lined with a nylon membrane (indeed cells which have been stimulated can be added to one set of wells, while control unstimulated cells can be added to another set of wells, usually in triplicate or quadruplicate). The nylon membrane has been pre-coated with an antibody specific for the target cytokine. Thus as soon as the cells produce any target cytokine it is immediately immobilized on the solid phase of the nylon membrane. The target cytokine can then be detected using an enzyme conjugated second antibody specific for that target cytokine. Enzymatic detection such a nitro blue tetrazolium is used so that dark spot develops on the nylon membrane where each target cytokine producing cell was located. The membrane can then be examined using a microscope to count the number of dark spots in each well. A comparison of the number of spots between stimulated and unstimulated cell cultures, for example, can inform the researcher whether or not the stimulus induced cytokine production. Or alternatively PBL from fish suspected to be infected could be compared to fish from a disease free environment. One advantage of the ELIspot assay is that once the image is acquired, the users can utilize any of numerous computer graphic programs that can measure the number of spots to derive numerical values.

For both the above assays, monoclonal or polyclonal antibodies can be used as either the capture or detection antibodies, although it is common to use monoclonal antibodies for capture as they have high specificity for the target cytokine. If the detection antibody is directly conjugated to the fluorescent tag or enzyme used for visualization, the two antibodies can be from the same species, but if an enzyme or fluorescently linked secondary antibody is needed, then the capture and detection antibodies must be from different species in order to prevent false positives.

It is possible that the specificity of the polyclonal antibodies for a particular cytokine might not be sufficient for accurate quantification, in which case the applicants might want to develop monoclonal antisera to that particular cytokine. This would however reduce the target species range for which the kit could be used, but it may be possible to design peptides for protein production which would react at least to both rainbow trout and Atlantic salmon.

While knowledge of the multiple isoforms of most fish cytokines is limited, it is clear that in most cases, both isoforms are upregulated during immune responses there are also differences in the degree by which these duplicate genes are upregulated (for example see interferon gamma up regulation in Purcell et al. 2009). If future research shows that there are indeed different functions for these duplicate cytokines, these assays could be modified by the inclusion of isoform specific monoclonal antibodies which would allow differentiation between the two forms.

CONCLUDING REMARKS

The discussion above makes it clear that the Road Not Taken hypothesis advanced by Rene Stet and myself clearly holds water: teleosts and tetrapods have the same components in their immune systems, but evolution has set them on different paths causing them to regulate and use those components in very different ways (Stet et al. 2003). This means that in order to make vaccines that are truly effective for teleost fish, we need to understand immune processes as they happen in teleost fish and knowledge of mammalian immune processes can only be a loose guide at best. In most cases this means moving into the post genomic era and starting the painstaking process of examining the tens of thousands of sequences in the databases in functional studies, making recombinant proteins and antibodies in order to truly understand their function in a teleost context. Current attempts at explaining teleost immune responses based solely on mRNA expression patterns using genes annotated based on sequence similarities of less than 40%, not taking into account teleost gene duplications and radically different gene regulation patterns, for example the recent microarray study of Canadian sockeye salmon (Miller et al. 2011), are doomed to be wildly inaccurate until we have the necessary teleost-specific knowledge of immune regulation and function required to understand the significance of changes in gene expression. That knowledge, however, will be hugely beneficial in designing truly effective teleost vaccines, provided it can be incorporated into vaccine production technology in a cost effective manner.

REFERENCES

1. Bar, D.; Apte, R.N.; Voronov, E.; Dinarello, C.A. and Cohen, S. (2003). A continuous delivery system of IL-1 receptor antagonist reduces angiogenesis and inhibits tumor development. *The FASEB Journal*, vol. 18, no. 1, p. 161-163.
2. Barber, L.D.; Percival, L.; Valiante, N.M.; Chen, L.; Lee, C.; Gumperz, J.E.; Phillips, J.H.; Lanier, L.L.; Bigge, J.C.; Parekh, R.B. and Parham, P. (1996). The inter-locus recombinant HLA-B*4601 has high selectivity in peptide binding and functions characteristic of HLA-C. *The Journal of Experimental Medicine*, vol. 184, no. 2, p. 735-740.
3. Bolon, B. (2012). Cellular and molecular mechanisms of autoimmune disease.*Toxicologic Pathology*, vol. 40, no. 2, p. 216-229.
4. Cárdenas, C.; Bidon-Chanal, A.; Conejeros, P.; Arenas, G.; Marshall, S. and Luque, F. (2010). Molecular modeling of Class I and II alleles of the Major Histocompatibility Complex in *Salmo salar*. *Journal of Computer-Aided Molecular Design*, vol. 24, no. 12, p. 1035-1051.

5. Chang, M.; Nie, P.; Collet, B.; Secombes, C. and Zou, J. (2009). Identification of an additional two-cysteine containing type I interferon in rainbow trout *Oncorhynchus mykiss*provides evidence of a major gene duplication event within this gene family in teleosts. *Immunogenetics*, vol. 61, no. 4, p. 315-325.

6. Dixon, B.; Van Erp, S.H.M.; Rodrigues, P.N.S.; Egberts, E. and Stet, R.M. (1995). Fish major histocompatibility complex genes: An expansion. *Developmental & Comparative Immunology*, vol. 19, no. 2, p. 109-133.

7. Dixon, B.; Shum, B.; Adams, E.J.; Magor, K.E.; Hedrick, R.P.; Muir, D. and Parham, P. (1998). CK-1, a putative chemokine of rainbow trout (*Oncorhynchus mykiss*). *Immunological Reviews*, vol. 166, no. 1, p. 341-348.

8. Etlinger, H.M.; Hodgins, H.O. and Chiller, J.M. (1977). Evolution of the lymphoid system II. Evidence for immunoglobulin determinants on all rainbow trout lymphocytes and demonstration of mixed leukocyte reaction. *European Journal of Immunology*, vol. 7, no. 12, p. 881-887.

9. Fujiki, K.; Gauley, J.; Bols, N. and Dixon, B. (2002). Cloning and characterization of cDNA clones encoding CD9 from Atlantic salmon (*Salmo salar*) and rainbow trout (*Oncorhynchus mykiss*). *Immunogenetics*, vol. 54, no. 8, p. 604-609.

10. Fujiki, K.; Liu, L.; Sundick, R.S. and Dixon, B. (2003a). Molecular cloning and characterization of rainbow trout (*Oncorhynchus mykiss*) C5a anaphylatoxin receptor. *Immunogenetics*, vol. 55, no. 9, p. 640-646.

11. Fujiki, K.; Nakao, M. and Dixon, B. (2003b). Molecular cloning and characterization of a carp (*Cyprinus carpio*) cytokine-like cDNA that shares sequence similarity with IL-6 subfamily cytokines CNTF, OSM and LIF. *Developmental & Comparative Immunology*, vol. 27, no. 2, p. 127-136.

12. Fujiki, K.; Smith, C.M.; Liu, L.; Sundick, R.S. and DIXON, B. (2003c). Alternate forms of MHC class II-associated invariant chain are not produced by alternative splicing in rainbow trout (*Oncorhynchus mykiss*) but are encoded by separate genes. *Developmental & Comparative Immunology*, vol. 27, no. 5, p. 377-391.

13. Fuller, J.R.; Pitzer, J.E.; Godwin, U.; Albertino, M.; Machon, B.D.; Kearse, K.P. and Mcconnell, T.J. (2004). Characterization of the molecular chaperone calnexin in the channel catfish, *Ictalurus punctatus*, and its association with MHC class II molecules.*Developmental & Comparative Immunology*, vol. 28, no. 6, p. 603-617.

14. Ghaffari, S.H. and Lobb, C.J. (1989). Nucleotide sequence of channel catfish heavy chain cDNA and genomic blot analyses. Implications for the phylogeny of Ig heavy chains. *The Journal of Immunology*, vol. 143, no. 8, p. 2730-2739.

15. Gleimer, M.; Wahl, A.R.; Hickman, H.D.; Abi-Rached, L.; Norman, P.J.; Guethlein, L.A.; Hammond, J.A.; Draghi, M.; Adams, E.J.; Juo, S.; Jalili, R.; Gharizadeh, B.; Ronaghi, M.; Garcia, K.C.; Hildebrand, W.H. and Parham, P. (2011). Although divergent in residues of the peptide binding site, conserved chimpanzee Patr-AL and polymorphic human HLA-A*02 have overlapping peptide-binding repertoires. *The Journal of Immunology*, vol. 186, no. 3, p. 1575-1588.

16. Gómez, D.; Conejeros, P.; Marshall, S. and Consuegra, S. (2010). MHC evolution in three salmonid species: A comparison between class II alpha and beta genes. *Immunogenetics*, vol. 62, no. 8, p. 531-542.

17. Grimholt, U.; Drabløs, F.; Jørgensen, S.; Høyheim, B. and Stet, R. (2002). The Major Histocompatibility Class I locus in Atlantic salmon (*Salmo salar* L.): polymorphism, linkage analysis and protein modelling. *Immunogenetics*, vol. 54, no. 8, p. 570-581.

18. Grimholt, U.; Larsen, S.; Nordmo, R.; Midtlyng, P.; Kjoeglum, S.; Storset, A.; Saebø, S. and Stet, R.J.M. (2003). MHC polymorphism and disease resistance in Atlantic salmon (*Salmo salar*); facing pathogens with single expressed Major Histocompatibility Class I and Class II loci. *Immunogenetics*, vol. 55, no. 4, p. 210-219.

19. Hall, B.M.; Verma, N.D; Tran, G.T. and Hodgkinson, S.J. (2011). Distinct regulatory CD4+T cell subsets; Differences between naïve and antigen specific T regulatory cells.*Current Opinion in Immunology*, vol. 23, no. 5, p. 641-647.

20. Hardie, L.J.; Laing, K.J.; Daniels, G.D.; Grabowski, P.S; Cunningham, C. and Secombes, C.J. (1998). Isolation of the first piscine transforming growth factor β gene: Analysis reveals tissue specific expression and a potential regulatory sequence in rainbow trout (*Oncorhynchus mykiss*). *Cytokine*, vol. 10, no. 8, p. 555-563.

21. Hashimoto, K.; Nakanishi, T. and Kurosawa, Y. (1990). Isolation of carp genes encoding major histocompatibility complex antigens. *Proceedings of the National Academy of Sciences of the United States of America*, vol. 87, no. 17, p. 6863-6867.

22. Jiménez-Altayó, F.; Briones, A.M.; Giraldo, J.; Planas, A.M.; Salaices, M. and Vila, E. (2006). Increased superoxide anion production by interleukin-1β impairs nitric oxide-mediated relaxation in resistance arteries. *Journal of Pharmacology and Experimental Therapeutics*, vol. 316, no. 1, p. 42-52.

23. Kales, S.; Fujiki, K. and Dixon, B. (2004). Molecular cloning and characterization of calreticulin from rainbow trout (*Oncorhynchus mykiss*). *Immunogenetics*, vol. 55, no. 10, p. 717-723.

24. Landis, E.; Palti, Y.; Dekoning, J.; Drew, R.; Phillips, R. and Hansen, J. (2006). Identification and regulatory analysis of rainbow trout tapasin and tapasin-related genes.*Immunogenetics*, vol. 58, no. 1, p. 56-69.

25. Miller, K.M.; Li, S.; Kaukinen, K.H.; Ginther, N.; Hammill, E.; Curtis, J.M.R.; Patterson, D.A.; Sierocinski, T.; Donnison, L.; Pavlidis, P.; Hinch, S.G.; Hruska, K.A.; Cooke, S.J.; English, K.K. and Farrell, A.P. (2011). Genomic signatures predict migration and spawning failure in wild Canadian salmon. *Science,* vol. 331, no. 6014, p. 214-217.

26. Nath, S.; Kales, S.; Fujiki, K. and Dixon, B. (2006). Major Histocompatibility Class II genes in rainbow trout (*Oncorhynchus mykiss*) exhibit temperature dependent downregulation. *Immunogenetics*, vol. 58, no. 5-6, p. 443-453.

27. Purcell, M.K.; Laing, K.J.; Woodson, J.C.; Thorgaard, G.H. and Hansen, J.D. (2009). Characterization of the interferon genes in homozygous rainbow trout reveals two novel genes, alternate splicing and differential regulation of duplicated genes. *Fish & Shellfish Immunology*, vol. 26, no. 2, p. 293-304.

28. Raison, R.L.; Hull, C.J. and Hildemann, W.H. (1978). Characterization of immunoglobulin from the Pacific hagfish, a primitive vertebrate. *Proceedings of the National Academy of Sciences of the United States of America*, vol. 75, no. 11, p. 5679-5682.

29. Rodrigues, P.N.S.; Dixon, B.; Roelofs, J.; Rombout, J.H.M.W.; Egberts, E.; Pohajdak, B. and Stet, R.J.M. (1998). Expression and temperature-dependent regulation of the $Beta_2$-Microglobulin (Cyca-B2m) gene in a cold-blooded vertebrate, the common carp (*Cyprinus carpio* L.). *Developmental Immunology*, vol. 5, no. 4, p. 263-275.

30. Saunders, P.M. and Van Endert, P. (2011). Running the gauntlet: From peptide generation to antigen presentation by MHC class I. *Tissue Antigens*, vol. 78, no. 3, p. 161-170.

31. Schulze, M.S.E.D. and Wucherpfennig, K.W. (2012). The mechanism of HLA-DM induced peptide exchange in the MHC class II antigen presentation pathway. *Current Opinion in Immunology*, vol. 24, no. 1, p. 105-111.

32. Stet, R.J.M.; Kruiswijk, C.P. and Dixon, B. (2003). Major Histocompatibility lineages and immune gene function in teleost fishes: The road not taken. *Critical Reviews in Immunology*, vol. 23, no. 5-6, p. 441-471

33. Strubin, M.; Berte, C. and Mach, B. (1986). Alternative splicing and alternative initiation of translation explain the four forms of the Ia antigen-associated invariant chain. *The EMBO Journal*, vol. 5, no. 13, p. 3483-3488.

34. Van Erp, S.; Dixon, B.; Figueroa, F.; Egberts, E. and Stet, R. (1996). Identification and characterization of a new Major Histocompatibility Complex Class I gene in carp (*Cyprinus carpio* L.). *Immunogenetics*, vol. 44, no. 1, p. 49-61.

35. Warr, G.W.; Deluca, D. and Griffin, B.R. (1979). Membrane immunoglobulin is present on thymic and splenic lymphocytes of the trout *Salmo gairdneri*. *The Journal of Immunology*, vol. 123, no. 2, p. 910-917.

36. Wearsch, P.A. and Cresswell, P. (2008). The quality control of MHC class I peptide loading. *Current Opinion in Cell Biology*, vol. 20, no. 6, p. 624-631.

37. Yoder, J.A.; Haire, R.N. and Litman, G.W. (1999). Cloning of two zebrafish cDNAs that share domains with the MHC class II-associated invariant chain. *Immunogenetics*, vol. 50, no. 1-2, p. 84-88.

38. Zou, J.; Cunningham, C. and Secombes, C.J. (1999). The rainbow trout *Oncorhynchus mykiss* interleukin-1β gene has a different organization to mammals and undergoes incomplete splicing. *European Journal of Biochemistry*, vol. 259, no. 3, p. 901-908.

CHAPTER 6

PHOTOBACTERIOSIS: PREVENTION AND DIAGNOSIS

Francesca Andreoni and Mauro Magnani

INTRODUCTION

Photobacteriosis or fish pasteurellosis is a septicemia caused by the gram negative, halophilic bacterium Photobacterium damselae subsp. piscicida, a member of the Vibrionaceae family that shares its species epithet with Photobacterium damselae subsp. damselae [1]. Photobacteriosis is considered one of the most dangerous bacterial diseases in aquaculture worldwide due to its wide host range, high mortality rate, and ubiquitous distribution [2]. The pathogen is able to infect a wide variety of marine fish, including the yellowtail (Seriola quinqueradiata) in Japan, gilthead sea bream (Sparus aurata), sea bass (Dicentrarchus labrax), and sole (Solea senegalensis and Solea solea) in Europe, striped bass (Morone saxatilis), white perch (Morone americana), and hybrid striped bass (Morone saxatilis (Morone chrysops)) in the USA, cobia (Rachycentron canadum) in Taiwan, and golden pompano (Trachinotus ovatus) in China [3–5].

Differences in susceptibility to the disease have been described on the basis of fish age. Larvae and juveniles are more susceptible to photobacteriosis, and acute infection induces 90–100% mortality of juvenile sea bream, whereas fish over 50 g are more resistant due to more efficient phagocytosis and killing of the bacteria by neutrophils and macrophages [6]. Bacteria that reside in different tissues and inside phagocytes cause chronic and acute forms of photobacteriosis. In its acute form, multifocal necrosis is present in the liver, spleen, and kidney and bacteria accumulate freely in phagocytes, capillaries, and interstitial spaces. Chronic lesions in the internal organs are characterized by the presence of white tubercles about 0.3–0.5 mm in diameter [7].

Citation: Francesca Andreoni and Mauro Magnani, "Photobacteriosis: Prevention and Diagnosis," Journal of Immunology Research, vol. 2014, Article ID 793817, 7 pages, 2014. DOI:10.1155/2014/793817.

Adherence and invasive capacities are essential in the first stage of infection [3]. P. damselae subsp. piscicida has been reported to be weakly or moderately adherent and invasive in various fish cell lines but has shown a high binding capacity to fish intestines [8]. The adherence seems to be mediated by a protein or glycoprotein receptor of the bacterial cell surface, and the internalization of the bacteria occurs through an actin microfilament-dependent mechanism [8], with cell metabolism playing an active role [9]. However, the precise nature of the mechanism responsible for adherence and interaction with host cell receptors and virulence factors contributing to the invasion of fish nonphagocytic cells is still unknown [9].

Several virulence mechanisms of P. damselae subsp. piscicida have been described. The polysaccharide capsular material plays an important role in the pathogenesis of the bacterium, conferring resistance to serum killing and increasing fish mortality [10]. Furthermore, the intracellular survival of the pathogen is likely to confer protection against specific and nonspecific host defenses and exogenous antimicrobial agents including antibiotics [8]. Extracellular products with phospholipase, cytotoxic, and hemolytic activities may account for the damage to the infected cells, the consequent release of the microorganisms, and the invasion of adjacent cells. In particular, a key pathogenicity factor of P. damselae subsp. piscicida is an exotoxin, the plasmid-encoded apoptosis-inducing protein of 56 kDa (AIP56), abundantly secreted by virulent strains and responsible for apoptogenic activity against sea bass macrophages and neutrophils in acute fish photobacteriosis [11]. The AIP56 toxin is a zinc-metalloprotease that acts by cleaving NF-κB p-65, with the catalytic activity located in the N-terminal domain and the C-terminal domain involved in binding and internalization into the cytosol of target cells [12]. The AIP56 induces activation of caspases 8, 9, and 3, loss of mitochondrial membrane potential, release of cytochrome c into the cytosol, and overproduction of ROS, suggesting activation of both extrinsic and intrinsic apoptotic pathways [13]. Through the activation of the cell death process involving macrophages and neutrophils, the pathogen is able to subvert the immune defenses of the host and to produce infectious disease.

Another important virulence mechanism of P. damselae subsp. piscicida is the acquisition of iron from its host by using high-affinity iron-binding siderophores, low molecular weight iron-chelating molecules that interact with bacterial membrane receptors to transport iron into the bacterium [14]. Furthermore, P. damselae subsp.piscicida is able to acquire iron from hemin and hemoglobin as unique iron sources in vitro [14], and iron limitation results in an increased binding of hemin in virulent strains [15]. The heme uptake of the bacterium includes a TonB system to transport heme into the cytoplasm and an ATP-binding cassette (ABC) system to drive it across the cytoplasmic membrane [16, 17].

Little is known about the fish immune response to the bacterium and the factors responsible for its failure to protect against P. damselae subsp. piscicida. A transcriptomic approach has recently been applied to elucidate the early immune responses of juvenile gilthead sea bream to P. damselae subsp. piscicida infection. A rapid recognition of the pathogen is shown by the upregulation of lectins, peptides with antimicrobial activity, chemokines, and chemokine receptors, as well as protein of iron and the heme metabolism as a response against bacteria that are dependent on iron. However, this defensive reaction can be either beneficial or devastating to the host [18]. Moreover, the upregulation of genes with highly specialized suppressive functions has been observed indicating an active suppression of immunity that can be induced by the host to reduce tissue damages or by the pathogen to evade the host response [18].

PREVENTION OF P. DAMSELAE SUBSP. PISCICIDA INFECTION

Antibiotics have been the first line of defense in fish aquaculture to control photobacteriosis outbreaks, but after only a few years the pathogen acquired resistance to various antibiotics. In fact, different transferable genetic elements (R plasmids) carrying genes for resistance against kanamycin, sulphonamide, tetracycline [19–22], ampicillin [22, 23], chloramphenicol [22, 24], florfenicol [25], and erythromycin [26] have been documented inP. damselae subsp. piscicida. Differences in the geographic distribution of multidrug transferable elements have been observed among several strains collected in Japan and United States [22, 27]. Furthermore, the intracellular parasitism of P. damselae subsp. piscicida within macrophages undermines the effectiveness of chemotherapy.

Taking into account all of these issues, research has been focused on the development of effective vaccines to prevent photobacteriosis and reduce the use of antibiotics in fish farming with benefits at biological and environmental point [28]. Conventional P. damselae subsp. piscicida vaccines are based on inactivated products containing cellular (heat-o formalin killed bacteria) and soluble antigens (LPS and ribosomal formulations) for immersion and injection administration (Table 1). However, they appear to be ineffective in protecting against pasteurellosis [2, 3, 28–38]. Bacterins overexpressing a 97 kDa OMP and a 52 kDa ECP protein, involved in the internalization of the bacterium, are reported to be effective in both sea bass and yellowtail when delivered by immersion eliciting a strong antibody response in the gills and mucosae that may block pathogen entry and colonization [2]. However, the only commercially available vaccine is an ECP-enriched bacterin preparation that has been employed in several European countries with mixed results ranging from good in Spain, Turkey, and Greece to poor in Italy [28, 39]. The recommended vaccination protocol consists of two bath immersions at monthly intervals starting at the larval stage when the fish is 50 mg and an oral booster immunization when fish reaches 2 g body weight [3].

Table 1: Overview of vaccines against Photobacterium damselae subsp. piscicida.

	Type of vaccine	Type of product	Vaccination procedure	Species	References
Lipoprotein	Recombinant subunit vaccine	Experimental	Injection	Sea bass	Andreoni et al. [40]
rHSP60, rENOLASE, and rGAPDH antigens, singles or in combination	Recombinant subunit vaccine	Experimental	Injection	Cobia	Ho et al. [4] and Ho et al. [41]
Formalin-killed bacterin overexpressing a 97 kDa OMP and 52 kDa ECP	Inactivated	Licensed	Immersion	Sea bass and yellowtail	Barnes et al. [2]
Formalin-killed bacterin with Escherichia coli LPS	Inactivated	Experimental	Immersion	Sea bream	Hanif et al. [38]
Live attenuated aroA mutant	Live attenuated	Experimental	Injection or immersion	Hybrid striped bass	Thune et al. [37]
Formalin-killed bacterin, ECP, and crude capsular polysaccharide (cCPS)	Inactivated	Experimental	Injection, immersion, and oral	Sea bass	Bakopoulos et al. [36]

LPS mixed with chloroform-killed bacterin	Inactivated	Experimental	Injection	Yellowtail	Kawakami et al. [35]
ECP-enriched formalin-inactivated bacterin	Inactivated	Commercialized	Immersion	Sea bass, sea bream, and sole	Magariños et al. [39]
Live attenuated bacteria	Live attenuated	Experimental	Immersion	Yellowtail	Kusuda and Hamaguchi [34]
Ribosomal antigens	Subunit vaccine	Experimental	Injection	Yellowtail	Kusuda et al. [33]
LPS formulation	Subunit vaccine	Experimental	Immersion and spray methods	Yellowtail	Fukuda and Kusuda [32]
Heat- and formalin-killed bacterin	Inactivated	Experimental	Immersion and oral	Yellowtail	Fukuda and Kusuda [29], Hamaguchi and Kusuda [30], and Kusuda and Salati [31]

Recombinant DNA technology and biotechnological approaches have thus far been used to a very limited extent for the development of bacterial vaccines for fish and effective preventive measures against fish pasteurellosis do not yet exist. Studies on the development of subunit vaccines have recently been reported in cobia from a Taiwan P. damselae subsp. piscicida isolate [4]. Immunoproteomics, using western blotting on protein analyzed with 2DE and LC-MS/MS to isolate immune-reactive proteins, has been applied to identify P. damselae subsp. piscicida antigens that were then cloned and produced as recombinant proteins. In particular, three antigens were shown to induce a protective effect in cobia and therefore were reported as potential vaccine candidates for the development of a subunit vaccine against the pathogen. However, the protection of these vaccine candidates has not been investigated in other fish species, where P. damselae subsp. piscicidacauses serious disease and high mortality, and against other P. damselae subsp. piscicida isolates [4]. Moreover, antigen combinations were studied revealing that bivalent subunit vaccines may achieve a better efficiency than monovalent or trivalent antigens [41].

In our laboratory a biotechnological approach based on the reverse vaccinology has been applied to design a vaccine against fish pasteurellosis [40]. New genomic sequences of P. damselae subsp. piscicida were the starting point for bioinformatic analysis aiming to identify new proteins localized on the bacterial surface. In fact, the primary condition in selecting a bacterial protein as a vaccine candidate is its cellular localization. Cytosolic proteins are unlikely to be immunological targets, whereas surface exposed and secreted proteins are more easily accessible to the host immune system [47]. In vitro screening of the in silico selected vaccine candidates by an inhibition adherence assay revealed that immunoglobulins from mice immunized with one of the recombinant vaccine candidates were able to affect the adherence of P. damselae subsp. piscicida to fish epithelial cells. The candidate antigen, found to be involved in the adherence and internalization of P. damselaesubsp. piscicida in CHSE-214 cells, was predicted in silico as likely lipoprotein with outer membrane localization. The N-terminal signal peptide of 20 amino acids contains the lipobox motif, 2 positively charged residues within the first 7 amino acids and a transmembrane helix of 10 residues. A database search revealed homology with hypothetical proteins and no putative conserved domain; therefore, no putative biological role could be assigned to this lipoprotein. Vaccination and challenge experiments in a laboratory trial indicated that immunization of sea bass with the recombinant antigen induced the production of specific antibodies and conferred protection against P. damselae subsp. piscicida challenge [40]. In vivo long persistence of lipoprotein antibodies was obtained with a single antigen administration in agreement with Ho et al. [4] who reported that multiple administrations do not increase protection in fish. The recombinant lipoprotein is potentially able to protect sea bass against P. damselae subsp. piscicida and could be an interesting candidate for the design of a recombinant vaccine against photobacteriosis. However, protection efficacy over time, increasing doses of the antigen, and its use in combination with different adjuvants must be further investigated in field experiments.

Due to the inconsistency of effective measures to prevent photobacteriosis, research has also focused on alternative methods to control the disease. Such methods include probiotics, to be applied in aquaculture to improve health, and a strain of Pediococcus pentosaceus, a lactic acid bacterium isolated from the intestine of adult cobia, has been investigated for its probiotic potential [48]. The acidic pH derived from metabolic acids in lactic acid bacteria culture supernatant has been shown to inhibit P. damselae subsp. piscicida growth in vitro. Dietary supplementation with P. pentosaceus in cobia enhances the growth rate and respiratory burst of peripheral blood leukocytes in fed fish. Furthermore, lactic acid bacteria feeding increased the survival rate of cobia after P. damselae subsp. piscicida immersion challenge. The mechanism affording this protection is still unclear. Although feeding with lactic acid bacteria did not increase specific antibody response after the immunization of cobia with inactivated P. damselae subsp. piscicida vaccine, it heightened the synergic protection against P. damselae subsp. piscicida challenge by 22% and could be administered by itself as a probiotic or with vaccination [48]. Furthermore, selective breeding for fish strains genetically resistant to photobacteriosis constitutes a potential strategy to reduce the probability of disease outbreak and avoid the dramatic consequences of high mortality in fish farms [49]. Quantitative trait loci mapping is applied to detect the regions of the host genome that are associated with resistance to the disease and marker-assisted selection is a useful approach used in several aquaculture species [50–52].

A study investigating quantitative trait loci for resistance to fish pasteurellosis in the gilthead sea bream identified two significant quantitative trait loci, one affecting late survival and another impacting overall survival, and a potential marker for disease resistance [49]. The identification of phase-specific quantitative trait loci in gilthead sea bream supports the hypothesis of a biphasic defense response with a primary infection by experimental exposure to the pathogen and a secondary infection with bacteria released from moribund and dead fish [49, 53]. Results of quantitative trait loci, mapped by identifying regions of the genome that explain complex traits such as survival, could also be used to gain a better understanding of the mechanisms of disease resistance and defense response. Further insights might also be gained through comparative mapping with other species susceptible to photobacteriosis.

IDENTIFICATION OF P. DAMSELAE SUBSP. PISCICIDA AND DIAGNOSIS OF INFECTION

Rapid diagnosis of fish photobacteriosis outbreaks is essential for proper management and effective control in aquaculture. Disease diagnosis is usually made using standard microbiological methods, based on pathogen culturing and isolation steps. Biochemical and serological confirmation is also necessary to characterize the bacterium and to discriminate between the two closely related subspecies, piscicida and damselae of P. damselae. The miniaturized system AIP20E is usually used for a presumptive identification of the P. damselaesubsp. piscicida. Although P. damselae generally displays a unique code of 2005004 for the piscicida [54] and 2015004 for the damselae subspecies [1], some strains exhibit aberrant reactions that can lead to misleading results [55]. Hence, differentiation of the subspecies P. damselae subsp. damselae can be achieved when three or more positive results are obtained in the lysine decarboxylase (LDC) production, motility, nitrate reduction to nitrite, gas production from glucose, thiosulfate citrate bile salts-sucrose (TCBS-1) growth, and urease tests, because all these tests yield negative results for all P. damselae subsp. piscicida strains [55, 56]. Serological methods such as agglutination or the ELISA have also been developed and commercialized [3].

To overcome the problem of time-consuming and laborious procedures, in the last few years molecular methods have been developed in order to achieve an accurate and specific identification of P. damselae subsp.piscicida and a rapid diagnosis of photobacteriosis (Table 2). The point at issue is the strong similarity at the DNA level between the two subspecies that makes it difficult to identify sequences useful for designing a subspecies-specific method [3, 42, 57]. rRNA sequences have been considered for this purpose [42], but strong similarities have been detected both in the 16S, 23S, and 5S (>99%) and the intergenic spacer regions (98–99.5%) between the two subspecies of P. damselae. Moreover, the mosaic-like structure of the latter makes them unsuitable for diagnostic purposes [42, 58]. Only a PCR-based method at species level has been developed using the 16S sequences [42].

Table 2: Methods for direct identification of Photobacterium damselae subsp. piscicida.

Assay	Target	Additional culture step	Specificity	Availability on the market	References
PCR-based detection method	16S gene	—	P. damselae		Osorio et al. [42]
Multiplex PCR assay	16S gene ureC gene	—	P. damselae subspecies		Osorio et al. [1]
Multiplex PCR assay	Pbp-1A gene ureC gene internal amplification control	—	P. damselae subspecies	Photobacterium damselae-PCR detection Kit by Diatheva	Amagliani et al. [43]
PCR technique and plating method	cps gene	TCBS-1 agar	P. damselae subspecies		Rajan et al. [44]
Enzyme immunoassay	Polyclonal antibodies against P. damselae subsp. piscicida	—	P. damselae subsp. piscicida	Aquaeia-Pp kit by BIONOR	Romalde et al. [45]
PCR-RFLP method	GenBank AY191120, AY191121 sequences	—	P. damselae, restriction analysis for subspecies identification		Zappulli et al. [46]

[a]Cross reactions with P. damselae subsp. damselae and P. histaminum.

Integrated sets of methods combine the amplification of the capsular polysaccharide gene to identify the speciesP. damselae with an additional culture step on TCBS-1 agar to differentiate P. damselae subsp. piscicida from P. damselae subsp. damselae

[44] or the amplification of two P. damselae-specific targets with restriction analysis of PCR products to obtain a unique digestion profile for P. damselae subsp. piscicida strains [46].

A multiplex PCR method based on the 16S rRNA and ureC genes has been proposed to discriminate between the two subspecies. The ureC gene is present in P. damselae subsp. damselae genome but is not found in P. damselae subsp. piscicida [1]. On the contrary, a P. damselae subsp. piscicida-specific target sequence, conserved among strains of different geographical origin but not shared by P. damselae subsp. damselae, has not yet been reported [42, 44].

An additional multiplex PCR protocol has been developed in our laboratory as a valid alternative to standard culture methods for the rapid and specific diagnosis of photobacteriosis in fish [43]. The gene coding for a penicillin binding protein 1A (GenBank accession number EU164926) was selected from a large-scale genome project as the PCR target for the identification of P. damselae subsp. piscicida because of several mismatches with the corresponding P. damselae subsp. damselae gene mainly clustered in the 3′ end of the gene. However, specificity analysis also indicated amplification of the target gene in two P. damselae subsp. damselae strains. This is due to the fact that a stronger sequence similarity to P. damselae subsp. piscicida than to other P. damselae subsp. damselae strains was found in these two P. damselae subsp. damselae strains. Hence, an additional PCR target, the ureC gene, lacking in the P. damselae subsp. piscicida genome, was introduced in the assay with the aim of differentiating each strain at the subspecies level together with an internal amplification control to obtain a clear distinction between truly negative and false negative results. The optimized multiplex PCR is able to correctly identify and discriminate both subspecies of P. damselae with a detection limit of 500 fg DNA, corresponding to 100 genomic units, twofold higher than that of immunodiagnostic systems (i.e., Bionor Aquaeia-Pp kit) [45].

CONCLUSIONS

Partial genome sequencing of several P. damselae subsp. piscicida strains has been previously reported [40, 59] and recently a draft of the complete genome sequence of P. damselae subsp. piscicida DI21 strain has been deposited in the public databases (GeneBank accession number PRJNA168653), but the complete gene annotation is not yet available. This information together with the comparative analysis of the genome sequence of different strains of P. damselae subsp. piscicida and P. damselae subsp. damselae will provide further insights laying the groundwork for the development of effective vaccines and diagnostic tools for the causative agent of fish pasteurellosis.

REFERENCES

1. C. R. Osorio, A. E. Toranzo, J. L. Romalde, and J. L. Barja, "Multiplex PCR assay for ureC and 16S rRNA genes clearly discriminates between both subspecies of Photobacterium damselae," Diseases of Aquatic Organisms, vol. 40, no. 3, pp. 177–183, 2000.
2. A. C. Barnes, N. M. S. dos Santos, and A. E. Ellis, "Update on bacterial vaccines: Photobacterium damselae subsp. piscicida," Developments in Biologicals, vol. 121, pp. 75–84, 2005.
3. J. L. Romalde, "Photobacterium damselae subsp. piscicida: an integrated view of a bacterial fish pathogen,"International Microbiology, vol. 5, no. 1, pp. 3–9, 2002.
4. L.-P. Ho, J. Han-You Lin, H.-C. Liu, H.-E. Chen, T.-Y. Chen, and H.-L. Yang, "Identification of antigens for the development of a subunit vaccine against Photobacterium damselae ssp. piscicida," Fish and Shellfish Immunology, vol. 30, no. 1, pp. 412–419, 2011.
5. R. Wang, J. Feng, Y. Su, L. Ye, and J. Wang, "Studies on the isolation of Photobacterium damselae subsp. piscicida from diseased golden pompano (Trachinotus ovatus Linnaeus) and antibacterial agents sensitivity," Veterinary Microbiology, vol. 162, no. 2–4, pp. 957–963, 2013.
6. M. Noya, B. Magariños, and J. Lamas, "Interactions between peritoneal exudate cells (PECs) of gilthead seabream (Sparus aurata) and Pasteurella piscicida. A morphological study," Aquaculture, vol. 131, no. 1-2, pp. 11–21, 1995.
7. B. Magariños, A. E. Toranzo, and J. L. Romalde, "Phenotypic and pathobiological characteristics ofPasteurella piscicida," Annual Review of Fish Diseases, vol. 6, pp. 41–64, 1996.
8. B. Magariños, J. L. Romalde, M. Noya, J. L. Barja, and A. E. Toranzo, "Adherence and invasive capacities of the fish pathogen Pasteurella piscicida," FEMS Microbiology Letters, vol. 138, no. 1, pp. 29–34, 1996.
9. F. Acosta, J. Vivas, D. Padilla et al., "Invasion and survival of Photobacterium damselae subsp. piscicida in non-phagocytic cells of gilthead sea bream, Sparus aurata L," Journal of Fish Diseases, vol. 32, no. 6, pp. 535–541, 2009.
10. B. Magariños, R. Bonet, J. L. Romalde, M. J. Martinez, F. Congregado, and A. E. Toranzo, "Influence of the capsular layer on the virulence of Pasteurella piscicida for fish," Microbial Pathogenesis, vol. 21, no. 4, pp. 289–297, 1996.
11. A. do Vale, M. T. Silva, N. M. S. dos Santos et al., "AIP56, a novel plasmid-encoded virulence factor ofPhotobacterium damselae subsp. piscicida with apoptogenic activity against sea bass macrophages and neutrophils," Molecular Microbiology, vol. 58, no. 4, pp. 1025–1038, 2005.
12. D. S. Silva, L. M. Pereira, A. R. Moreira, et al., "The apoptogenic toxin AIP56 is a metalloprotease A-B toxin that cleaves NF-κb P65," PLoS Pathogens, vol. 9, no. 2, Article ID e1003128, 2013.
13. C. Costa-Ramos, A. do Vale, P. Ludovico, N. M. S. dos Santos, and M. T. Silva, "The bacterial exotoxin AIP56 induces fish macrophage and neutrophil apoptosis using mechanisms of the extrinsic and intrinsic pathways," Fish and Shellfish Immunology, vol. 30, no. 1, pp. 173–181, 2011.
14. B. Magariños, J. L. Romalde, M. L. Lemos, J. L. Barja, and A. E. Toranzo, "Iron uptake by Pasteurella piscicida and its role in pathogenicity for fish," Applied and

Environmental Microbiology, vol. 60, no. 8, pp. 2990–2998, 1994.

15. M. L. Lemos and C. R. Osorio, "Heme, an iron supply for vibrios pathogenic for fish," BioMetals, vol. 20, no. 3-4, pp. 615–626, 2007.

16. S. J. Río, C. R. Osorio, and M. L. Lemos, "Heme uptake genes in human and fish isolates ofPhotobacterium damselae: existence of hutA pseudogenes," Archives of Microbiology, vol. 183, no. 5, pp. 347–358, 2005.

17. F. Andreoni, R. Boiani, G. Serafini et al., "Expression, purification, and characterization of the recombinant putative periplasmic hemin-binding protein (hutB) of Photobacterium damselae subsp.piscicida," Bioscience, Biotechnology and Biochemistry, vol. 73, no. 5, pp. 1180–1183, 2009.

18. C. Pellizzari, A. Krasnov, S. Afanasyev et al., "High mortality of juvenile gilthead sea bream (Sparus aurata) from photobacteriosis is associated with alternative macrophage activation and anti-inflammatory response: results of gene expression profiling of early responses in the head kidney," Fish and Shellfish Immunology, vol. 34, no. 5, pp. 1269–1278, 2013.

19. E. H. Kim and T. Aoki, "Drug resistance and broad geographical distribution of identical R plasmids ofPasteurella piscicida isolated from cultured yellowtail in Japan," Microbiology and Immunology, vol. 37, no. 2, pp. 103–109, 1993.

20. E. H. Kim and T. Aoki, "The transposon-like structure of IS26-tetracycline, and kanamycin resistance determinant derived from transferable R plasmid of fish pathogen, Pasteurella piscicida," Microbiology and Immunology, vol. 38, no. 1, pp. 31–38, 1994.

21. E. H. Kim and T. Aoki, "Sulfonamide resistance gene in a transferable R plasmid of Pasteurella piscicida,"Microbiology and Immunology, vol. 40, no. 5, pp. 397–399, 1996.

22. C. S. del Castillo, H. B. Jang, J.-I. Hikima et al., "Comparative analysis and distribution ofpP9014, a novel drug resistance IncP-1 plasmid from Photobacterium damselae subsp. piscicida," International Journal of Antimicrobial Agents, vol. 42, no. 1, pp. 10–18, 2013.

23. H. Morii, M. S. Bharadwaj, and N. Eto, "Cloning and nucleotide sequence analysis of the ampicillin resistance gene on a conjugative R plasmid from the fish pathogen Photobacterium damselae subsp.piscicida," Journal of Aquatic Animal Health, vol. 16, no. 4, pp. 197–207, 2004.

24. H. Morii, N. Hayashi, and K. Uramoto, "Cloning and nucleotide sequence analysis of the chloramphenicol resistance gene on conjugative R plasmids from the fish pathogen Photobacterium damselae subsp. piscicida," Diseases of Aquatic Organisms, vol. 53, no. 2, pp. 107–113, 2003.

25. E. Kim and T. Aoki, "Sequence analysis of the florfenicol resistance gene encoded in the transferable R-plasmid of a fish pathogen, Pasteurella piscicida," Microbiology and Immunology, vol. 40, no. 9, pp. 665–669, 1996.

26. H. Morii and Y. Ishikawa, "Cloning and nucleotide sequence analysis of the chloramphenicol and erythromycin resistance genes on a transferable R plasmids from the fish pathogen Photobacterium damselae subsp. piscicida," Bulletin of the Faculty of Fisheries, vol. 93, pp. 41–50, 2012.

27. M.-J. Kim, I. Hirono, K. Kurokawa et al., "Complete DNA sequence and analysis of the transferable multiple-drug resistance plasmids (R plasmids) from Photobacterium damselae subsp. piscicida isolates collected in Japan and the United States," Antimicrobial Agents and Chemotherapy, vol. 52, no. 2, pp. 606–611, 2008.

28. T. Håstein, R. Gudding, and Ø. Evensen, "Bacterial vaccines for fish: an update of the current situation worldwide," Developments in Biologicals, vol. 121, pp. 55–74, 2005.

29. Y. Fukuda and R. Kusuda, "Efficacy of vaccination for pseudotuberculosis in cultured yellowtail by various routes of administration," Bulletin of the Japanese Society of Scientific Fisheries, vol. 47, pp. 147–150, 1981.

30. M. Hamaguchi and R. Kusuda, "Field testing of Pasteurella piscicida formalin killed bacterin against pseudotuberculosis in cultured yellowtail, Seriola quinqueradiata," Bulletin of Marine Science and Fisferies, vol. 11, pp. 11–16, 1989.

31. R. Kusuda and F. Salati, "Major bacterial diseases affecting mariculture in Japan," Annual Review of Fish Diseases, vol. 3, no. C, pp. 69–85, 1993.

32. Y. Fukuda and R. Kusuda, "Vaccination of yellowtail against pseudotuberculosis," Fish Pathology, vol. 20, no. 2-3, pp. 421–425, 1985.

33. R. Kusuda, M. Ninomiya, M. Hamaguchi, and A. Muraoka, "Efficacy of ribosomal vaccine prepared fromPasteurella piscicida against pseudotuberculosis in cultured yellowtail," Fish Pathology, vol. 23, no. 3, pp. 191–196, 1988.

34. R. Kusuda and M. Hamaguchi, "The efficacy of attenuated live bacterin of Pasteurella piscicida against pseudotuberculosis in yellowtail," Bulletin of the European Association of Fish Pathologists, vol. 8, pp. 51–53, 1988.

35. H. Kawakami, N. Shinohara, Y. Fukuda, H. Yamashita, H. Kihara, and M. Sakai, "The efficacy of lipopolysaccharide mixed chloroform-killed cell (LPS-CKC) bacterin of Pasteurella piscicida on Yellowtail, Seriola quinqueradiata," Aquaculture, vol. 154, no. 2, pp. 95–105, 1997.

36. V. Bakopoulos, D. Volpatti, L. Gusmani, M. Galeotti, A. Adams, and G. J. Dimitriadis, "Vaccination trials of sea bass, Dicentrarchus labrax (L.), against Photobacterium damsela subsp. piscicida, using novel vaccine mixtures," Journal of Fish Diseases, vol. 26, no. 2, pp. 77–90, 2003.

37. R. L. Thune, D. H. Fernandez, J. P. Hawke, and R. Miller, "Construction of a safe, stable, efficacious vaccine against Photobacterium damselae ssp. piscicida," Diseases of Aquatic Organisms, vol. 57, no. 1-2, pp. 51–58, 2003.

38. A. Hanif, V. Bakopoulos, I. Leonardos, and G. J.

Dimitriadis, "The effect of sea bream (Sparus aurata) broodstock and larval vaccination on the susceptibility by Photobacterium damsela subsp. piscicida and on the humoral immune parameters," Fish and Shellfish Immunology, vol. 19, no. 4, pp. 345–361, 2005.

39. B. Magariños, M. Noya, J. L. Romalde, G. Pérez, and A. E. Toranzo, "Influence of fish size and vaccine formulation on the protection of gilthead seabream against Pasteurella piscicida," Bulletin of the European Association of Fish Pathologists, vol. 14, pp. 120–122, 1994.

40. F. Andreoni, R. Boiani, G. Serafini et al., "Isolation of a novel gene from Photobacterium damselae subsp. piscicida and analysis of the recombinant antigen as promising vaccine candidate," Vaccine, vol. 31, no. 5, pp. 820–826, 2013.

41. L.-P. Ho, C.-J. Chang, H.-C. Liu, H.-L. Yang, and J. H.-Y. Lin, "Evaluating the protective efficacy of antigen combinations against Photobacterium damselae ssp. piscicida infections in cobia, Rachycentron canadum L," Journal of Fish Diseases, vol. 37, no. 1, pp. 51–62, 2014.

42. C. R. Osorio, M. D. Collins, A. E. Toranzo, J. L. Barja, and J. L. Romalde, "16S rRNA gene sequence analysis of Photobacterium damselae and nested PCR method for rapid detection of the causative agent of fish pasteurellosis," Applied and Environmental Microbiology, vol. 65, no. 7, pp. 2942–2946, 1999.

43. G. Amagliani, E. Omiccioli, F. Andreoni et al., "Development of a multiplex PCR assay forPhotobacterium damselae subsp. piscicida identification in fish samples," Journal of Fish Diseases, vol. 32, no. 8, pp. 645–653, 2009.

44. P. R. Rajan, J. H.-Y. Lin, M.-S. Ho, and H.-L. Yang, "Simple and rapid detection of Photobacterium damselae ssp. piscicida by a PCR technique and plating method," Journal of Applied Microbiology, vol. 95, no. 6, pp. 1375–1380, 2003.

45. J. L. Romalde, B. Magariños, F. Lores, C. R. Osorio, and A. E. Toranzo, "Assessment of a magnetic bead-EIA based kit for rapid diagnosis of fish pasteurellosis," Journal of Microbiological Methods, vol. 38, no. 1-2, pp. 147–154, 1999.

46. V. Zappulli, T. Patarnello, P. Patarnello et al., "Direct identification of Photobacterium damselaesubspecies piscicida by PCR-RFLP analysis," Diseases of Aquatic Organisms, vol. 65, no. 1, pp. 53–61, 2005.

47. M. Scarselli, M. M. Giuliani, J. Adu-Bobie, M. Pizza, and R. Rappuoli, "The impact of genomics on vaccine design," Trends in Biotechnology, vol. 23, no. 2, pp. 84–91, 2005.

48. C.-F. Xing, H.-H. Hu, J.-B. Huang, et al., "Diet supplementation of Pediococcus pentosaceus in cobia (Rachycentron canadum) enhances growth rate, respiratory burst and resistance against photobacteriosis," Fish and Shellfish Immunology, vol. 35, no. 4, pp. 1122–1128, 2013.

49. C. Massault, R. Franch, C. Haley et al., "Quantitative trait loci for resistance to fish pasteurellosis in gilthead sea bream (Sparus aurata)," Animal Genetics, vol. 42, no. 2, pp. 191–203, 2011.

50. T. Moen, M. Baranski, A. K. Sonesson, and S. Kjøglum, "Confirmation and fine-mapping of a major QTL for resistance to infectious pancreatic necrosis in Atlantic salmon (Salmo salar): population-level associations between markers and trait," BMC Genomics, vol. 10, article 368, 2009.

51. R. D. Houston, C. S. Haley, A. Hamilton et al., "The susceptibility of Atlantic salmon fry to freshwater infectious pancreatic necrosis is largely explained by a major QTL," Heredity, vol. 105, no. 3, pp. 318–327, 2010.

52. M. R. Baerwald, J. L. Petersen, R. P. Hedrick, G. J. Schisler, and B. May, "A major effect quantitative trait locus for whirling disease resistance identified in rainbow trout (Oncorhynchus mykiss)," Heredity, vol. 106, no. 6, pp. 920–926, 2011.

53. J. Antonello, C. Massault, R. Franch et al., "Estimates of heritability and genetic correlation for body length and resistance to fish pasteurellosis in the gilthead sea bream (Sparus aurata L.)," Aquaculture, vol. 298, no. 1-2, pp. 29–35, 2009.

54. A. E. Toranzo, S. Barreiro, J. F. Casal, A. Figueras, B. Magariños, and J. L. Barja, "Pasteurellosis in cultured gilthead seabream (Sparus aurata): first report in Spain," Aquaculture, vol. 99, no. 1-2, pp. 1–15, 1991.

55. A. Thyssen, L. Grisez, R. Van Houdt, and F. Ollevier, "Phenotypic characterization of the marine pathogen Photobacterium damselae subsp. piscicida," International Journal of Systematic Bacteriology, vol. 48, no. 4, pp. 1145–1151, 1998.

56. S. Botella, M.-J. Pujalte, M.-C. Macián, M.-A. Ferrús, J. Hernández, and E. Garay, "Amplified fragment length polymorphism (AFLP) and biochemical typing of Photobacterium damselae subsp. damselae,"Journal of Applied Microbiology, vol. 93, no. 4, pp. 681–688, 2002.

57. G. Gauthier, B. Lafay, R. Ruimy et al., "Small-subunit rRNA sequences and whole DNA relatedness concur for the reassignment of Pasteurella piscicida (Snieszko et al.) Janssen and Surgalla to the genusPhotobacterium as Photobacterium damsela subsp. piscicida comb. nov," International Journal of Systematic Bacteriology, vol. 45, no. 1, pp. 139–144, 1995.

58. C. R. Osorio, M. D. Collins, J. L. Romalde, and A. E. Toranzo, "Variation in 16S-23S rRNA intergenic spacer regions in Photobacterium damselae: a mosaic-like structure," Applied and Environmental Microbiology, vol. 71, no. 2, pp. 636–645, 2005.

59. H. Naka, I. Hirono, K. Kurokawa, and T. Aoki, "Random sequencing of genomic DNA of Photobacterium damselae subsp. piscicida," Fisheries Science, vol. 71, no. 6, pp. 1209–1216, 2005.

CHAPTER 7

COMPARATIVE PATHOGENOMICS OF BACTERIA CAUSING INFECTIOUS DISEASES IN FISH

Ponnerassery S. Sudheesh, Aliya Al-Ghabshi, Nashwa Al-Mazrooei, and Saoud Al-Habsi

INTRODUCTION

Genome sequencing has provided us with powerful insights into the genetic makeup of the microbial world. The microbial genomics today has progressed from the long drawnout individual genome sequencing projects in the past to a level of technological advancement, where sequencing and comparing the genomes of several strains of a single pathogen is accomplished in a very short period of time [1, 2]. We are currently passing through a period of explosive developments in the field and an overwhelming glut in the genome sequence data of microorganisms. To date, over 1800 microbial genomes have been published and the sequencing of more than 5200 microbial genome are in different stages of completion (http://www.ncbi.nlm.nih.gov/genomes/lproks.cgi).

The genomics information has categorically disproved the earlier thinking that microbial genomes are static and has demonstrated that genomic evolutionary processes are much more flexible and dynamic than previously thought. This has led to the emergence of new ideas such as "uprooting the tree of life" and the concept of "horizontal genomics" [3–8]. This new thinking about microbial genome evolution has emerged from the observations of lineage-specific genome reduction and horizontal gene transfer (HGT), frequently occurring in bacterial genomes. Increasingly, genome sequencing projects have identified an unexpected level of diversity among bacteria, which can often be linked to recombination and gene transfer between a variety of prokaryotic organisms.

There is large variation in size and content of bacterial genomes between different genera and species, and also among strains of the same species. Known genome sizes of bacteria range from under 0.6 to 10 megabases (Mb). The smallest bacterial genomes reported are for the mycoplasmas and related bacteria, with sizes as low as 530 kilobases [9]. It has been emphasized that the adaptive capability ("versatility") of bacteria directly correlates with genome size [10].

Genome sequencing of bacterial pathogens has produced exciting information on evolutionary relationships between pathogenic and nonpathogenic species and has demonstrated how each has developed special adaptations advantageous for each of their unique infectious lifestyles. In the longer term, an understanding of their genome and biology will enable scientists to design means of disrupting their infectious lifestyles.

The genomes of bacteria are made up of circular or linear chromosomes, extrachromosomal linear or circular plasmids as well as different combinations of these molecules. The functionally related genes are clustered together in very close proximity to each other, and those genes located on the "core" part of the chromosome present a relatively uniform G+C content and a specific codon usage. Closely related bacteria generally have very similar genomes [11].

The stability and integrity of the "core" sequences of the genome, however, is often interrupted by the presence of DNA fragments with a G+C content and a codon usage markedly different from those of the "core" genome. The "flexible" gene pool or the so-called "mobilome" [12], is created by the acquisition of strain-specific "assortments" of genetic information mainly represented by mobile genetic elements (MGE), such as plasmids, bacteriophages, genomic/pathogenicity islands (GEIs/PAIs), integrons, IS elements (ISEs), and transposons (see Figure 1). The flexible genes scattered in the genome provide the microbes with an additional repertoire of arsenal, for example, resistance to antibiotics, production of toxic compounds as well as other virulence factors [13].

Citation: Ponnerassery S. Sudheesh, Aliya Al-Ghabshi, Nashwa Al-Mazrooei, and Saoud Al-Habsi, "Comparative Pathogenomics of Bacteria Causing Infectious Diseases in Fish," International Journal of Evolutionary Biology, vol. 2012, Article ID 457264, 16 pages, 2012. DOI:10.1155/2012/457264.

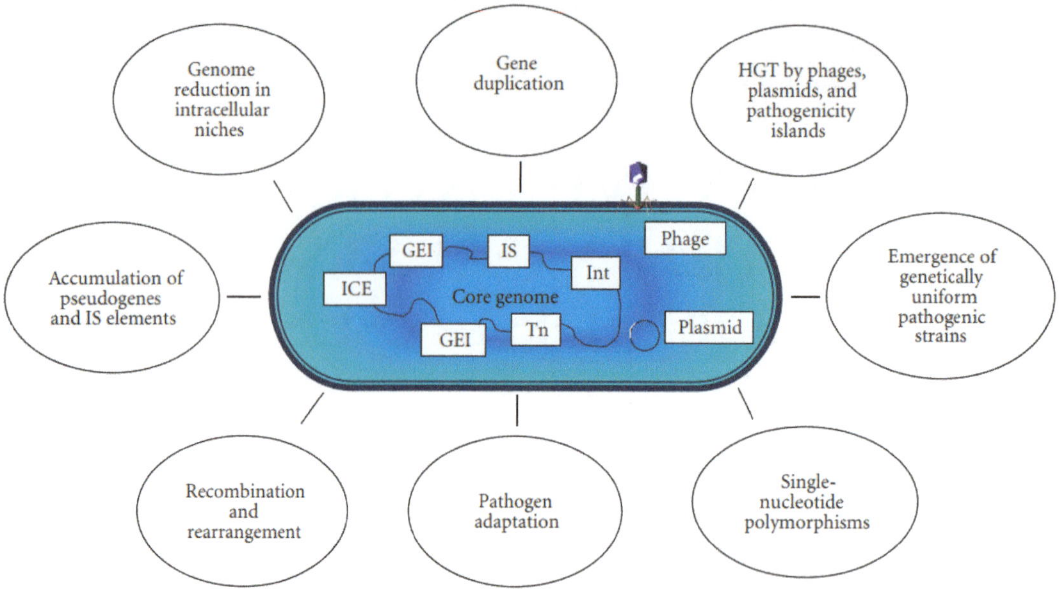

Figure 1: Major factors responsible for the pathogenomic evolution of bacteria (modified from [14, 15]; HGT: horizontal gene transfer, GEIs: genomic islands, ICEs: integrative conjugative elements, Int: integrons, Tn: conjugative transposons, IS: IS elements.

A fundamental question in biology is to define the minimum number of genes or functions to support cellular life. The size of bacterial genomes is primarily the result of two counteracting processes: the acquisition of new genes by gene duplication or by horizontal gene transfer; the deletion of nonessential genes. Genomic flux created by these gains and losses of genetic information can substantially alter gene content. This process drives divergence of bacterial species and eventually adaptation to new ecological niches [16].

Bacterial pathogens are a major cause of infectious diseases and mortality in wild fish stocks and fish reared in confined conditions. Disease problems constitute the largest single cause of economic losses in aquaculture [17]. Concurrent with the rapid growth and intensification of aquaculture, increased use of water bodies, pollution, globalization, and transboundary movement of aquatic fauna, the list of new pathogenic bacterial species isolated from fish has been steadily increasing [18]. In addition, the virulence and host range of existing pathogens has also been increasing, posing considerable challenge to fish health researchers, who are actively looking for more efficient vaccines and therapeutic drugs to combat bacterial fish diseases. The current treatment methods are ineffective and have many practical difficulties.

At the level of host-pathogen interaction, there is considerable pressure on pathogens to adapt to the harsh host environment as well as to adapt and evolve along with the ever changing external environment. The interplay between the host and the pathogen is a complex one, each driven by the need to secure the success of the species. Adaptations by one partner to exploit new environments will often stimulate the other to modify its characteristics to take advantage of the change. As a consequence of this cycle of interaction created by changing environments, new strains of pathogen will evolve. Over time, these strains may emerge as new species with characteristic disease symptoms.

The use of antibiotics to control fish diseases has met with limited success and has the potential danger of antibiotic resistance development in aquatic bacteria (World Health Organization antimicrobial resistance fact sheet 194, http://www.who.int/inf-fs/en/fact194.html) [19]. As aquaculture is one of the fastest growing food production industries in the world, demand for sustainable ways of combating fish diseases is gaining significance. There is tremendous scope for developing novel vaccines and therapeutic drugs against bacterial fish pathogens.

Genomic evolution and adaptive strategies of bacterial fish pathogens are poorly understood and lags far behind that of human and terrestrial animal pathogens. A detailed knowledge of the genome sequences of bacterial fish pathogens and how the genomes of the pathogenic species or strains evolved from nonpathogenic ancestors or counterparts will help us better understand their pathogenicity mechanisms and strategies of host adaptations. This information will help identifying novel vaccine and drug targets in the genomes of pathogens.

Recently, genome sequencing of a number of bacteria pathogenic to fish and other aquatic organisms have been completed. The genome sequence and genome characteristics of important bacterial fish pathogens completed and published to date are summarized in Table 1.

Table 1: Currently sequenced genomes of bacterial pathogens of fish.

Organisms	Size (Mb)	CDS**	Unknown/ Hypothetical genes (%)	Pseudogenes prophages, ISE/GEI	% GC	Chromosomes	Plasmids
Vibrio anguillarum 775 serotype O1	4.117	3880	26	92	44.3	2	1
Vibrio anguillarum 96F serotype O1	4.065	3766	26	38	42	2	0
Vibrio anguillarum RV22 serotype O2β	4.022	3949	26	68	43.1	2	0
Vibrio ordalii ATCC 33509	3.415	3281	—	31	43.3	2	0
Vibrio harveyi ATCC BAA-1116*	6.054	—	—	—	45.4	2	1
Vibrio vulnificus YJ016 biotype 1	5.26	5028	34	—	46.1	2	1
Vibrio splendidus strain LGP32	4.974	4498	24.8	—	43.8	2	0
Aliivibrio salmonicida strain LFI1238	4.655	4286	—	1179	38.3	2	4
Flavobacterium psy-chrophilumJIP02/86	2.862	2432	45.3	94	32.5	1	1
Flavobacterium bran-chiophilumFL-15	3.56	2867	—	54	32.9	1	—
Flavobacterium colum-nareATCC 49512*	3.2	2896	—	—	32.0	1	—
Edwardsiella tarda EIB202	3.76	3486	28	97	59.7	1	1
Edwardsiella ictaluri 93–146*	3.812	3783	—	100	57.4	1	—
Aeromonas hydrophila ATCC 7966	4.744	5195	27.7	7	61.5	1	0
Aeromonas salmonicida A449	4.702	4437	—	258	58.5	1	5
Aeromonas veronii Strain B565	4.551	4057	—	—	58.7	1	—
A. caviae Ae398	4.43	—	—	6	61.4	1	1
Renibacterium salmoni-narumATCC 33209	3.155	3507	25.3	151	56.3	1	0
Streptococcus parauberis	2.143	2641	21.3	—	35.6	1	0
Lactococcus garvieaeU-NIUD074	2.172	2101	21.8	224	38.7	1	0
Mycobacterium marinum M	6.636	5424	26	65	62.5	1	1

*Unpublished.

**Coding sequences.

The main aim of this paper is to put together and summarize the scattered genome sequencing information on important bacterial fish pathogens available in the literature to date. We sincerely believe that this paper will provide a genomic perspective on the adaptive evolutionary strategies of bacterial fish pathogens in different ecological niches and will help better understand the virulence mechanisms and pathogenesis of infections. It is hoped that this will lead to finding the most appropriate vaccine and therapeutic drug targets in the genomes and developing efficient control and treatment methods for fish diseases.

BACTERIAL PATHOGENS OF FISH

Although pathogenic species representing majority of existing bacterial taxa have been implicated in fish diseases, only a relatively small number of pathogens are responsible for important economic losses in cultured fish worldwide. Major bacterial pathogens responsible for infectious disease outbreaks in different species of fish are listed in Table 2. Major groups of bacteria causing infectious diseases in fish and the important genome characteristics of these bacteria are described in the following sections.

Table 2: Major bacterial pathogens of economically important fish.

Causative agent/species	Disease	Main host fish
Gram-negatives		
Vibrio anguillarum	Vibriosis	Salmonids, turbot, sea bass, striped bass, eel, ayu, cod, and red sea bream
Aliivibrio salmonicida (formerly Vibrio salmonicida)	Vibriosis	Atlantic salmon, cod
Vibrio vulnificus	Vibriosis	Eels, tilapia
Vibrio ordalii	Vibriosis	Salmonids
Vibrio carchariae (syn.: Vibrio harveyi)	Vibriosis, infectious gastroenteritis	Shark, abalone, red drum, sea bream, sea bass, cobia, and flounder
Moritella viscosa (formerly Vibrio viscosus)	Winter ulcer	Atlantic salmon
Photobacterium damselae subsp. piscicida (formerly Pasteurella piscicida)	Photobacteriosis (pasteurellosis)	Sea bream, sea bass, sole, striped bass, and yellowtail
Pasteurella skyensis	Pasteurellosis	Salmonids and turbot
Tenacibaculum maritimum (formerly Flexibacter maritimus)	Flexibacteriosis	Turbot, salmonids, sole, sea bass, gilthead sea bream, red sea bream, and flounder
Flavobacterium psychrophilum	Coldwater disease	Salmonids, carp, eel, tench, perch, ayu
Flavobacterium branchiophila	Bacterial gill disease	A broad range of cultured cold water and warm water salmonid and nonsalmonid fishes
Flavobacterium columnare	Columnaris disease	cyprinids, salmonids, silurids, eel, and sturgeon
Pseudomonas anguilliseptica	Pseudomonadiasis, winter disease	Sea bream, eel, turbot, and ayu
Aeromonas salmonicida	Furunculosis	salmon, trout, goldfish, koi and a variety of other fish species
Aeromonas hydrophila Aeromonas veronii Biovar Sobria Aeromonas sobria Biovar Sobria (Motile aeromonads)	Motile aeromonas septicemia (MAS), hemorrhagic septicemia, ulcer disease or red-sore disease, and epizootic ulcerative syndrome (EUS)	A wide variety of salmonid and nonsalmonid fish, sturgeon, tilapia, catfish, striped bass, and eel
Edwardsiella ictaluri	Enteric septicemia	Catfish and tilapia
Edwardsiella tarda	Edwardsiellosis	Salmon, carps, tilapia, catfish, striped bass, flounder, and yellowtail
Yersinia ruckeri	Enteric redmouth	Salmonids, eel, minnows, sturgeon, and crustaceans
Piscirickettsia salmonis	Piscirickettsiosis	Salmonids
Gram-positives		

Lactococcus garvieae (formerly Enterococcus seriolicida)	Streptococcosis or lactococcosis	Yellowtail and eel
Streptococcus iniae	Streptococcosis	Yellowtail, flounder, sea bass, and barramundi
Streptococcus parauberis	Streptococcosis	Turbot
Streptococcus phocae	Streptococcosis	Atlantic salmon
Renibacterium salmoninarum	Bacterial kidney disease	Salmonids
Mycobacterium marinum	Mycobacteriosis	Sea bass, turbot, and Atlantic salmon

VIBRIOS

Bacteria in the genus Vibrio are mainly pathogenic to marine and brackish water fish. However, they are occasionally reported in freshwater species as well [20, 21]. The distribution of vibriosis is worldwide and causes great economic loss to the aquaculture industry [22]. Vibriosis, one of the major bacterial diseases affecting fish, bivalves, and crustaceans, is mainly caused by pathogenic species such as Vibrio anguillarum, V. harveyii (Syn. V. carchariae), V. ordalii, and Aliivibrio salmonicida (formerly Vibrio salmonicida) [23, 24].

Other species such as V. vulnificus [25, 26] and Moritella viscosa (formerly Vibrio viscosus) [27] have been implicated in fish diseases such as septicemia and winter ulcer, respectively; more pathogenic species have been isolated frequently and reported in the literature [28].

Genome sequences of four major fish pathogenic vibrios, V. anguillarum, V. ordalii, Aliivibrio salmonicida, and V.vulnificus have been completed and published [29–31]. Generally, they have two chromosomes, one larger and one smaller. The majority of genes that encode cell functions and pathogenic factors are located in the large one. The small chromosome usually contains genes for environmental adaptation.

Vibrio anguillarum is the most studied aetiological agent of vibriosis [32]. V. anguillarum typically causes a hemorrhagic septicemia. The O1 and O2 serotypes are the virulent strains frequently isolated from diseased fish [33, 34]. Many O1 serotype strains harbor 65 kb pJM1-type plasmids, which carry the siderophore anguibactin biosynthesis and transport genes, a main virulence factor of V. anguillarum, while one of the O1 serotype strains and other serotypes, such as all of the O2 strains, are plasmidless [28, 35, 36]. The O1 serotype strains cause disease in salmonid fish, whereas O2 β strains are usually isolated from cod and other nonsalmonids [28, 32].

Vibrio ordalii is a very close relative of V. anguillarum [37] and was previously recognized as V. anguillarumbiotype 2. Vibriosis caused by these two species are strikingly different based on histological evidences [38]. V. anguillarum has a special affinity for blood and loose connective tissue, whereas V. ordali is mostly present as aggregates in skeletal and cardiac muscles. V. ordalii has a lesser affinity for blood and develops bacteremia only at late stages of disease.

Genomic sequences of three different strains of V. anguillarum (the strain 775 containing plasmid pJM1, serotype O1 strain 96F, and plasmidless serotype O2 β strain RV22) and V. ordali have recently been published [31]. The pJM1 plasmid in the strain 775 contains 65 genes including the anguibactin biosynthesis and transport genes that are unique for the strain.

V. anguillarum 775 contains more transposase genes (about 53) than 96F (about 23), RV22 (about 42), and V. ordalii (about 18).

The genome comparison of V. anguillarum serotypes has revealed some interesting differences in the genomic composition, indicating horizontal acquisition of virulence genes and the evolution of different potential virulence mechanisms among the closely related serotypes [31]. The V. anguillarum 96F strain has a type III secretion system 2 (T3SS2) cluster, which is absent in the 775 strain. The T3SS2 genes are highly conserved with other T3SS2 genes reported in V. parahaemolyticus, V. cholera, and V. mimicus [39–41]. In the 775 strain, three transposase genes are present at the T3SS2 chromosomal location, one of which probably originated from the pJM1, indicating that the gene cluster is inactivated by a transposition, deletion, or inversion event [31]. The 775 strain also contains 10 genomic islands including integrase, transposase, and some novel sequences conferring genomic plasticity to adapt to specific ecological niches.

The strain RV22 genome contains the toxin-antitoxin systems, and genes encoding the accessory V. choleraeenterotoxin (Ace) and the Zonula occludens toxin (Zot), which is not present in the 775 strain. The yersiniabactin-like siderophore cluster, which is highly conserved in many Vibrio species and Photobacterium damselae subspecies piscicida [42], is present in strain RV22 and V. ordalii.

A striking feature of V. ordali genome is its significant reduction in size (3.4 Mb) compared to the V. anguillarum strains 775 (4.1 Mb), 96F (4.0 Mb), and RV22 (4.0 Mb). V. ordali lacks the ABC transporter genes, the type VI secretion systems, and the gene for microbial collagenase. The Syp biofilm formation cluster, which is conserved in many Vibrio species such as V. fischeri, V. vulnificus, and V. parahaemolyticus [43, 44], is present only in V. ordalii. Thus, it is probable that the transition of V. anguillarum to V. ordalii is mediated by genome reductive evolution to become an endosymbiotic organism; V. ordali has the smallest genome of all vibrios.

Vibrio vulnificus includes three distinct biotypes. Biotype 1 strains cause human disease, while biotype 2 infects primarily eels, and biotype 3 infections has been associated with persons handling Tilapia, although the source and reservoir of biotype 3 have yet to be identified [45]. In another classification the terms clade 1 and clade 2 are used based on the multilocus sequence typing (MLST) [46]. Biotype 1 strains are present in both clades, whereas biotype 2 strains are present only in clade 1, and biotype 3 strains appear to be a hybrid between clades 1 and 2. Clade 1 strains are most often isolated from environmental samples, while clade 2 strains are mostly associated with human disease and are considered more virulent. Recent comparative genomic analysis of these biotypes or clades has clearly differentiated them based on the possession of an array of clade-specific unique genes including the presence of a virulence-associated genomic island XII in the highly virulent strains [30].

Aliivibrio salmonicida (formerly Vibrio salmonicida) causes coldwater vibriosis in marine fish such as farmed Atlantic salmon (Salmo salar), sea-farmed rainbow trout (Oncorhynchus mykiss), and captive Atlantic cod (Gadus morhua) [47]. The Gram-negative bacterium causes tissue degradation, hemolysis, and sepsis in vivo. Genome sequencing of Aliivibrio salmonicida has revealed a mosaic structure of the genome caused by large intrachromosomal rearrangements, gene acquisition, deletion, and duplication of DNA within the chromosomes and between the chromosomes and the plasmids [29].

The genome has many genes that appear to be recently acquired by HGT, and large sections of over 300 coding sequences (CDS) are disrupted by IS elements or contain point mutations causing frame shifts or premature stop codons [29]. The genomic islands (GIs) identified in the bacteria include major virulence-related genes encoding T6SS and Flp-type pilus and genes that appear to provide new functions to the bacteria. The Tad system has been proposed to represent a new subtype of T2SS and is essential for biofilm formation, colonization, and pathogenesis [48].

The genome analysis has unequivocally confirmed that Aliivibrio salmonicida has undergone extensive rearrangement of its genome by losing massive functional genes and acquiring new genes and become host-restricted, allowing the pathogen to adapt to new niches. IS expansion has been related to genome reduction in the evolution and emergence of pathogenicity [49], and accumulation of pseudogenes has been described for several other host restricted pathogens [50, 51].

AEROMONADS

Aeromonas hydrophila and other motile aeromonads are among the most common bacteria in a variety of aquatic environments worldwide, including bottled water, chlorinated water, well water, sewage, and heavily polluted waters, and are frequently associated with severe disease among cultured and feral fishes, amphibians, reptiles, and birds [52]. Aeromonads are also considered serious emerging pathogens of human beings [53]. Determination of the etiology of diseases involving aeromonad infections has been complicated by the genetic, biochemical, and antigenic heterogeneity of members of this group.

The genus Aeromonas has been conveniently divided into a group of nonmotile, psychrophilic species, prominently represented by Aeromonas salmonicida, which is an obligate fish pathogen and a second group of mostly human pathogenic, motile, and mesophilic species including A. hydrophila.

Genome sequencing of A. hydrophila ATCC 7966^T, A. salmonicida subsp. salmonicida A449, A. veronii strain B565, and A. caviae [54–57] has helped in resolving their taxonomic confusion and has brought new insights into the way these bacteria adapt to a myriad of ecological niches, their host adaptive evolution and virulence mechanisms.

Aeromonas salmonicida, the causative agent of furunculosis in salmonid and nonsalmonid fish, is a non-motile, Gram-negative bacterium; furunculosis is an important disease in wild and cultured stocks of fish inflicting heavy losses to aquaculture industry worldwide [58, 59]. A. hydrophila causes a septicemic disease in fish known variously as "motile aeromonas septicemia" (MAS), "hemorrhagic septicemia," "ulcer disease," or "red-sore disease" [60]. The disease caused by this bacterium primarily affects freshwater fish such as catfish, several species of bass, and many species of tropical or ornamental fish. A. veronii is the causative agent of bacterial hemorrhagic septicemia in fish and is becoming a major economic problem in the fish-farming industry [23].

Genome sequencing of the fish pathogen A. salmonicida A449 has confirmed the presence of fully functional genes for a type III secretion system (T3SS) that has been shown to be required for virulence in A. salmonicida[61], and genes for a type VI secretion system (T6SS), which is disrupted by an IS element [55]. The ancestral state of the T3SS in A. salmonicida A449 is ambiguous because of the absence of the genes in A. hydrophilaATCC 7966^T, while other A. hydrophila strains carry T3SS operons on the chromosome [62]. The genome contains a multitude of virulence-related genes including several types of adhesins (e.g., surface layer, flagella, and pili), toxin genes (aerolysin, hemolysin, repeats in toxin (RTX) protein, and cytolytic delta-endotoxin), secreted enzymes (protease, phospholipase, nuclease, amylase, pullulanase, and chitinase), antibiotic resistance genes (tetA, β-lactamase gene, and efflux pumps), and genes involved in iron acquisition and quorum sensing.

Most of the above genes are present in A. hydrophila ATCC 7966^T genome and an expansion of gene families (paralogs) of ABC transporters, two-component signal transduction systems (TCSs), transcriptional regulators, FeS cluster-binding proteins involved in energy transduction at the membrane, and methyl-accepting chemotaxis proteins (MCPs). Interestingly, transposase, resolvase, or insertion sequence element sequences were not discovered in the A. hydrophila ATCC 7966^T genome, whereas these have been identified in A. salmonicida and A. caviae genomes. A. salmonicida possesses 88 copies of 10 different IS elements whereas A. caviae Ae398 has only five different IS elements, and A. hydrophila completely lacks IS elements.

Although A. hydrophila ATCC 7966^T has been demonstrated to be the second most virulent species among Aeromonas [63], a very important virulence determinant, T3SS, which is present in A. salmonicida A449 is strikingly absent in A. hydrophila ATCC 7966^T genome. A. caviae contains many putative virulence genes, including those encoding a type 2 secretion system, an RTX toxin, and polar flagella.

The genome of A. veronii strain B565 contains some putative virulence factors, such as chitinase, RTX protein, adhesion factor, flagella, and mannose-sensitive hemagglutinin (MSHA), all of which are shared with A. hydrophila ATCC 7966^T and A. salmonicida A449. On the other hand, 346 genes including some important putative virulence factors such as hemolysins and the type III secretion protein, which are shared by the latter two species are absent in A. veronii strain B565.

Many unique genes in A. hydrophila ATCC 7966^T and A. salmonicida A449 are virulence genes and often form large clusters, such as the rtx cluster in ATCC 7966^T and the flagellar gene cluster in A449, or are involved in mobile elements such as phages and transposons, highlighting their lateral transfer history [56].

The A. hydrophila ATCC 7966^T and A. salmonicida A449 genomes appear to be very closely related, encoding similar number of proteins with only 9% difference in gene content. However, there are many transposons, phage-related genes, and unique CDS in A. salmonicida A449 genome that are different from A. hydrophilaATCC 7966^T sequences, showing their distinct lineages and adaptive evolution that occurred while segregating into different species of the genus.

In sharp contrast to A. hydrophila ATCC 7966^T genome, the A. salmonicida A449 genome is characterized by the presence of large numbers of several different types of IS elements in multiple copies, with more than 20 genes being interrupted by IS elements. A. hydrophila ATCC 7966^T genome has no IS elements.

There is a higher tendency for genomic reduction in A. salmonicida A449 with the formation of many pseudogenes, and A. hydrophila ATCC 7966^T has only seven pseudogenes. The formation of pseudogenes has resulted in the loss of function of many genes including flagella and type IV pili, transcriptional regulators, genes encoding carbohydrate synthesis, and modification enzymes and genes for basic metabolic pathways, which are some characteristic features of pathogenomic evolution.

Thus, A. salmonicida A449 appears to have evolved much faster than A. hydrophila ATCC 7966^T through genetic rearrangements, genomic reduction, and HGT from common ancestral lineages by acquiring and forming multiple plasmids, prophages, a battery of IS elements, pseudogenes, and several individual genes and operons.

FLAVOBACTERIUM

The genus Flavobacterium includes over 30 species of which Flavobacterium psychrophilum, F. branchiophilum, and F. columnare are important disease agents for salmonids, catfish, and other cultured species [64, 65]. Flavobacteria are significant as they are ubiquitous in the soil, freshwater, and marine environments and are noted for their novel gliding motility and ability to degrade polymeric organic matter such as hydrocarbons [66].

F. psychrophilum is the etiological agent of bacterial coldwater disease (BCWD). It is a serious fish pathogen causing substantial economic losses and rearing difficulties to both commercial and conservation aquaculture.F. psychrophilum infections are found throughout the world. Juvenile rainbow trout and coho salmon are particularly susceptible to BCWD. However, F. psychrophilum infections have been reported in a wide range of hosts, Anguilla japonica, A. anguilla, Cyprinus carpio, Carassius carassius, Tinca tinca, Plecoglossus altivelis,Perca fluviatilis, and Rutilus rutilus [64, 67]. Fry and fingerlings with BCWD often have skin ulcerations on the peduncle, anterior to the dorsal fin, at the anus, or on the lower jaw and mortalities can go up to 70% [68].

F. branchiophilum is the causative organism of bacterial gill disease (BGD) in several parts of the world [69]. This disease is characterized by explosive morbidity and mortality rates attributable to massive bacterial colonization of gill lamellar surfaces and progressive branchial pathology stemming from high rates of lamellar epithelial necrosis [70].

F. columnare (formerly Cytophaga columnaris; Flexibacter columnaris) is the causative agent of columnaris disease of salmonids and other fishes in commercial aquaculture, the ornamental fish industry, and wild fish populations worldwide [71]. Classically, during outbreaks, its morbidity and mortality rates escalate more gradually than for BGD. Additionally, unlike the pattern of necrosis in BGD, fish with columnaris will have severe necrosis of all parts of the gill as the bacterium invades inwardly [72].

The taxonomy of the three species was initially based on phenotypic characteristics and has been revised several times during the years. The latest classification based on G+C content, DNA-ribosomal ribonucleic acid (rRNA) hybridisation, and fatty acid and protein profiles, has confirmed that all the three species now belong to the phylum/division Cytophaga-Flavobacterium-Bacteroides, family Flavobacteriaceae, and genusFlavobacterium [73].

The whole genome sequences of F. psychrophilum and F. branchiophilum have been published [74, 75]. The F. columnare genome sequence is yet to be completed and published [76].

Prominent features of F. psychrophilum infection include the strong adhesion to fish epithelial tissues followed by gliding motility, rapid and mass tissue destruction, and severe muscle tissue ulcerations. Hence, the identification of multiple genes encoding secreted proteases, adhesins, and gliding motility (gld) genes in F. psychrophilum genome indicates their possible involvement in the virulence of the pathogen. However, the gene sequence of a secreted collagenase was disrupted by an insertion sequence of the IS256 family in several strains isolated from rainbow trout [74] indicating the clonal dissemination of strains containing the disrupted gene. The F. psychrophilum seems to have horizontally acquired virulence associated genes from other unrelated bacteria. It has a hemolysin similar to the toxin VAH5, which is a virulence factor in Vibrio anguillarum [77]. It also has a gene encoding a protein that is similar to domains 1–3 of thiol-activated cytolysin family of pore-forming toxins (TACYs), which has been implicated in the pathogenicity of several Gram-positive bacteria [78]. Interestingly, F. psychrophilum lacks the type III and IV secretion systems usually present in Gram-negative pathogens; but, it has genes encoding PorT and PorR proteins, which are involved in transport and anchoring of virulence factors of the bacteria [79, 80]. In addition, the F. psychrophilum genome contains a large repertoire of genes involved in aerobic respiration, psychrotolerance, and stress response.

The sequencing of F. branchiophilum genome has revealed the existence of virulence mechanisms distinctly different from the closest species, F. psychrophilum. The F. branchiophilum genome has the first cholera-like toxin in a nonproteobacteria and an array of adhesins. A comparative analysis of its genome with genomes of other Flavobacterium species revealed a smaller genome size, large differences in chromosome organization, and fewer rRNA and tRNA genes, fitting with its more fastidious growth. In addition, identification of certain virulence factors, genomic islands, and CRISPR (clustered regularly interspaced short palindromic repeats) systems points to the adaptive evolution of F. branchiophilum by horizontal acquisition of genes.

EDWARDSIELLA

The genus Edwardsiella belongs to subgroup 3 of γ-proteobacteria, encompassing a group of Gram-negative enteric bacteria pathogenic to a variety of animals [81]. Two very closely related species, Edwardsiella tarda andE. ictaluri are important fish pathogens. Both are Gram-negative motile rods that are cytochrome oxidase negative and ferment glucose with production of acid and gas. The two species can be differentiated biochemically in that E. tarda produces both indol and hydrogen sulfide, whereas E. ictaluri produces neither. Moreover, the two species do not cross-react serologically. E. tarda has been isolated from many warm water fishes and some coldwater fishes, whereas E. ictaluri has been isolated only from a few species of warm water fishes (Table 2). Additionally, E. tarda causes disease in such other animals as marine mammals, pigs, turtles, alligators, ostriches, skunks, and snakes [81]. It has also occasionally infected humans [82, 83]. In contrast, E. ictaluri is limited to fish, and survivors of epizootics probably become carriers. The geographic range of E. tardais worldwide, whereas that of E. ictaluri is still confined to the catfish growing areas in the United States [84].

E. tarda causes a disease condition in fish called systemic hemorrhagic septicemia with swelling skin lesions as well as ulcer and necrosis in internal organs such as liver, kidney, spleen, and musculature [85]. It has the ability of invading and multiplying in epithelial cells and macrophages in order to subvert the host immune system and to survive in the fish [86].

E. ictaluri is the causative agent of enteric septicemia of catfish (ESC), a major disease affecting the catfish industry. The disease can manifest as an acute form that is characterized by hemorrhagic enteritis and septicemia and a chronic disease that is characterized by meningoencephalitis [87]. Gross external symptoms include hemorrhages on the body, especially around the mouth and fins. Other signs include pale gills, exophthalmia, and small ulcerations on the body [84].

The whole genome sequencing of the two species has recently been completed and published allowing comparative genomic analysis of these very important fish pathogens [88, 89]. The genome sequencing of the two closely related species E. tarda and E. ictaluri has revealed a high level of genomic plasticity with a high content of mobile genetic elements, IS elements, genomic islands, phage-like products, integrases, or recombinases. E. ictaluri displays high biochemical homogeneity with only one serotype, but possess many IS elements in the genome. In addition, highly variable G+C content and a large quantity of variable number of tandem repeats (VNTRs) or direct repeat sequences were identified in the E. tarda genome indicating the rapid genomic evolution undergoing in the species [88]. An interesting feature is the identification of insertion sequence IS Saen1 of Salmonella enterica serovar Enteritidis [90] in both E. tarda EIB202 and E. ictaluri 93–146 genomes. Conversely, the difference in genomic islands among the three species may partially explain their rapid evolutionary changes and diverging lineage from a common ancestor.

The E. tarda genome has a gene cluster sharing high similarities to the pvsABCDE-psuA-pvuA operon, which encodes the proteins for the synthesis and utilization of vibrioferrin, an unusual type of siderophore requiring nonribosomal peptide synthetase (NRPS) independent synthetases (NIS) and usually mediating the iron uptake systems in V. parahaemolyticus and V. alginolyticus [91, 92]. But E. ictaluri genome lacks siderophore biosynthesis genes, even though it possesses heme binding/transport genes.

E. tarda genome is smaller than that of E. ictaluri and other sequenced genomes of Enterobacteriaceae, justifying the hypothesis that E. tarda may not be present as a free living microorganism in natural waters but multiply intracellularly in protozoans and transmitted to fish, reptile, and other animals or humans [81].

The E. tarda and E. ictaluri genomes have a multitude of virulence factors including P pilus, type 1 fimbriae, nonfimbrial adhesins, invasins and hemagglutinins and various secretion pathways including sec-dependent transport system, the components of the main terminal branch of the general secretory pathway (GSP), the signal recognition particle (SRP), and the sec-independent twin arginine transport (Tat), T1SS, TTSS, and T6SS indicating their evolutionary fitness and ability to adapt to a variety of demanding ecological niches and harsh host intracellular environments.

YERSINIA RUCKERI

Yersinia ruckeri, the causal agent of enteric redmouth (ERM) disease, which is a systemic bacterial infection of fishes, but is principally known for its occurrence in rainbow trout, Salmo gairdneri [93]. Y. ruckeri was initially isolated from rainbow trout in the Hagerman Valley, Idaho, USA, in the 1950s [94] and is now widely found in fish populations throughout North America, Australia, South Africa, and Europe [95]. Outbreaks of ERM usually begin with low mortalities which slowly escalate and may result in high losses. The problem may become large-scaled if chronically infected fish are exposed to stressful conditions such as high stocking densities and poor water quality [96]. Y. ruckeri is a nonspore-forming bacterium which does not possess a capsule, but often has a flagellum [97].

Historically, Y. ruckeri is fairly homogenous in biochemical reactions. However, Y. ruckeri strains have recently been grouped into clonal types on the basis of biotype, serotype, and outer membrane protein (OMP) profiles [98]. Strains of serovars I and II [99], equivalent to serotypes O1a and O2b, respectively [100], cause most epizootic outbreaks in cultured salmonids, serovar I being predominant in rainbow trout [101]. Within serovar I, six clonal OMP types have been recognized, but only two are associated with major disease outbreaks: clonal group 5, which includes the so-called Hagerman strain and clonal group 2 [98, 102]. Clonal group 5 comprises the majority of isolates, all of them motile and with a widespread distribution (Europe, North America, and South Africa). Clonal group 2 includes only nonmotile strains isolated in the UK.

More recently, multilocus sequence typing has revealed distinct phylogenetic divergence of Y. ruckeri from the rest of the Yersinia genus raising doubts about its taxonomic position [103]. This view has gained credibility after the genome sequencing of Y. ruckeri, which has a substantially reduced total genome size (3.58 to 3.89 Mb), compared with the 4.6 to 4.8 Mb seen in the genus generally [104]. In addition, Y. ruckeri was found to be the most evolutionarily distant member of the genus with a number of features distinct from other members of the genus.

Several common Yersinia genes were missing in Y. ruckeri. These included genes involved in xylose utilization, urease activity, B12-related metabolism, and the mtnKADCBEU gene cluster that comprises the majority of the methionine salvage pathway [104]. The genomic reduction achieved by losing these and other genes is suggestive of its means of adaptation to an obligatory life style in fish hosts.

RENIBACTERIUM SALMONINARUM

Renibacterium salmoninarum is a small Gram-positive diplobacillus, and the causative agent of bacterial kidney disease (BKD), which is a slowly progressive, systemic infection in salmonid fishes with a protracted course and an insidious nature [105]. The pathogen can be transmitted from fish to fish [106] or from adults to their progeny via eggs [107]. Infected fish may take months to show signs of disease. bacterial kidney disease is one of the most difficult bacterial diseases of fish to treat [108], mainly due to its ability to evade phagocytosis and invade and survive in host cells [109, 110]. R. salmoninarum is very slow growing, and it is extremely difficult to apply genetic manipulation techniques to study its gene functions.

R. salmoninarum, despite being an obligate intracellular pathogen of fish, is phylogenetically closest to the non-pathogenic environmental Arthrobacter species [51]. Based on 16S rRNA phylogenetic analysis, R. salmoninarum has been included in the actinomycetes subdivision and was found related to a subgroup harboring morphologically and chemotaxonomically rather heterogeneous taxa, including Arthrobacter, Micrococcus, Cellulomonas, Jonesia, Promicromonospora, Stomatococcus, and Brevibacterium [111]. In fact,Arthrobacter davidanieli is commercially used as a vaccine (commercially known as Renogen) and can provide significant cross-protection in Atlantic salmon, though not in Pacific salmon [112]. The genome sequencing ofR. salmoninarum ATCC 33209 strain and two Arthrobacter strains, the TC1 and FB24, has revealed many interesting aspects of how this obligates fish pathogen evolved, via genomic reduction and horizontal gene acquisition, from members of the nonpathogenic genus Arthrobacter [51, 113]. A total of 1562 ORF clusters were similar in R. salmoninarum and Arthrobacter spp. demonstrating the genetic basis for the efficiency and cross-protection of the A. davidanieli vaccine.

There is significant genome reduction in R. salmoninarum genome, which is 1.44 Mb smaller than the chromosome of TC1 and 1.55 Mb smaller than the chromosome of FB24. The two Arthrobacter strains have several large plasmids that are not present in the ATCC 33209 strain. In addition, these plasmids do not have high levels of similarity to sequences in the R. salmoninarum chromosome [51].

The presence of many IS elements, pseudogenes, and genomic islands in R. salmoninarum genome coupled with a lack of restriction-modification systems contribute to the extensive disruption of ORFs as a strategy to reduce many pathways in the bacteria. Moreover, the highly homogeneous nature of R. salmoninarum with respect to the overall genomic structure, biochemical properties, and surface antigens [114, 115] points to the evolution of this pathogen towards a strictly intracellular life style.

Several virulence factors including capsular synthesis genes, heme acquisition operons, genes encoding possible hemolysins, and the poorly characterized msa genes identified in the R. salmoninarum genome seems to be horizontally acquired. Arthrobacter spp. lacks most of these gene sequences, thus underlining the differential evolution and adaptation of these two very closely related species to contrasting ecological niches.

STREPTOCOCCUS AND LACTOCOCCUS

Gram-positive cocci belonging to the genera Streptococcus and Lactococcus are increasingly being recognized as important fish pathogens all over the world [116]. There are several different species of Gram-positive cocci, including Streptococcus parauberis, S. iniae, S. agalactiae (syn. Streptococcus difficilis), S. phocae [117, 118],Lactococcus garvieae (syn. Enterococcus seriolicida) [119], L. piscium [120–123], Vagococcus salmoninarum, andCarnobacterium piscicola [124], implicated in infectious diseases of warm water as well as cold water fishes.

Streptococcosis appears to have very few limitations in regard to geographic boundaries or host range, with outbreaks occurring in aquaculture facilities worldwide and in many different cultured species. S. iniae, S.parauberis, S. agalactiae, and L. garvieae are known as the major pathogens of streptococcosis and lactococcosis in Oncorhynchus mykiss, Seriola quinqueradiata, Siganus canaliculatus, and Tilapia spp. [125].Recently, S. iniae and L. garvieae are also recognized as emerging zoonotic pathogens, causing diseases in both fish and human beings [23, 126].

S. iniae is a β-haemolytic, Gram-positive coccus that causes generalized septicaemia and meningoencephalitis in a variety of warm water fishes [127], whereas S. parauberis is an α-hemolytic, Gram-positive coccus, mainly pathogenic in cultured turbot (Scophthalmus maximus) and olive flounder, Paralichthys olivaceus. L. garvieaecauses a hyperacute and haemorrhagic septicemia in fishes particularly during the summer time. General pathological symptoms of streptococcosis and lactococcosis in fishes are hemorrhage, congestion, lethargy, dark pigmentation, erratic swimming, and exophthalmos with clouding of the cornea [117, 128].

Complete genome sequences of different strains of S. parauberis and L. garvieae, important pathogenic species isolated from both fish and human, have been published [129–132].

S. parauberis is recognized as the dominant etiological agent of streptococcosis in fish [117], whereas both S. parauberis and S. uberis are involved the causation of bovine mastitis in dairy cow [133, 134].

S. parauberis is closer to S. uberis than with other Streptococcus spp. and is biochemically and serologically indistinguishable from S. uberis [135]. Both species were earlier considered as type I and II of S. uberis, but later shown to be phylogenetically distinct and renamed the type I as S. uberis and type II as S. parauberis [134].

The S. parauberis strain KCTC11537BP genome size falls in the middle of the 1.8 to 2.3 Mb range of streptococcal genomes sequenced to date and the average G+C content of 35.6% is significantly lower than those of S. pyogenes [132]. About 78% of genes are shared between the genomes of S. parauberis strain KCTC11537BP and S. uberis NC_012004, but they differ significantly at two regions of the genome, demonstrating the genomic basis for their separation into two species.

S. parauberis genome encodes an M-like protein of S. iniae (SiM), which is an important virulence factor in S. iniae [136]. It also encodes hasA and hasB genes that may be involved in capsule production for resistance against phagocytosis. The genome analysis indicates that S. parauberis could possibly possess the ability to regulate the metabolism of more carbohydrates than other Streptococcus species and to synthesize all the aminoacids and regulatory factors required to adapt and survive in a highly hostile host environment.

Complete genome sequences of L. garvieae strain UNIUD074, isolated from diseased rainbow trout in Italy, a virulent strain Lg2 (serotype KG2) and a nonvirulent strain ATCC 49156 (serotype KG+), both isolated from diseased yellowtail in Japan have recently been published [130, 131]. In addition, genome sequence of L. garvieae strain 21881, isolated from a man suffering from septicemia has been published [129].

The strains Lg2 and ATCC 49156 have 99% sequence identity and share 1944 orthologous genes, but are different in 24 Lg2-specific genes that were absent in the ATCC 49156 genome. One of the Lg2-specific genes is a 16.5 kb capsule gene cluster, which confirms the earlier transmission electron microscopic finding that Lg2 is encapsulated, and ATCC 49156 is nonencapsulated [137]. In fact, the capsule gene cluster has the features of a horizontally acquired genomic island conferring virulence to the Lg2 strain but might have been lost from the ATCC 49156 strain while subculturing in the laboratory [131]. Both genomes carried three types of IS elements, prophage sequences, and integrase genes and were found smaller than those of at least five sequencedL. lactis genomes. The Lg2 genome lacks several aminoacid biosynthesis genes, which is a characteristic feature of pathogenic bacteria with reduced genomes. The Lg2 strain contains hemolysins, NADH oxidase and superoxide dismutase (SOD), adhesins and sortase, which are known virulence factors [137–139]. It also encodes a gene for phosphoglucomutase, a virulence factor conferring the resistance to peptide antimicrobials in S. iniae [140].

Although L. garvieae and L. lactis genomes share 75% CDS, about 25% genes are Lg2-specific hypothetical proteins and proteins of unknown functions, which may be involved in the virulence of the Lg2 strain. These findings indicate that L. garvieae and L. lactis have significantly diverged from the common ancestor, and the L. garvieae is evolving into a pathogenic species equipped with virulence features suitable for living in the host environment.

MYCOBACTERIA

Chronic infections in fish caused by different species of mycobacteria have been well recognized [23, 141, 142]. Several slow growing as well as fast growing species of mycobacteria such as Mycobacterium marinum, M. fortuitum, M. chelonae, and M. avium have been isolated from wild and cultured fish suffering from mycobacteriosis in different parts of the world [143–145]. Among them, M. marinum is the most important fish pathogen, frequently isolated from a variety of fish species with granulomas [146]. It is also a known zoonotic pathogen, transmitted to man though fish handling in aquariums and aquaculture tanks, producing superficial and self-limiting lesions called "fish tank or aquarium tank granuloma" involving the cooler parts of the body such as hands, forearms, elbows, and knees [147, 148]. Although strain variation has been reported [149], there is significant intraspecies sequence homogeneity among different M. mrinum strains [150]. However, it is hypothesized that only certain strains of M. marinum have zoonotic potential [151]. Phylogenetic studies have shown that M. marinum is most closely related to M. ulcerans followed by M. tuberculosis [150].Owing to this, M. marinum and M. tuberculosis share many virulence factors and significant pathological features and respond to similar antibiotics [152, 153]. Hence, M. marinum is also an important model organism to study the pathogenesis of tuberculosis [152, 153].

Interestingly, the genome of M. marinum is 50% bigger than that of M. tuberculosis and seems to have acquired a number of genes encoding NRPSs and the huge repertoire of PE, PPE, and ESX systems probably by HGT [154]. Both species might have evolved differently from a common environmental mycobacteria. M. tuberculosis might have adapted to its host intracellular life by extensive genome reduction and M. marinum, by and large retained or obtained genes required for its dual lifestyle and broad-host range.

GENOME SEQUENCING TO FIND NOVEL VACCINE AND DRUG TARGETS IN FISH PATHOGENS

Our understanding of the molecular basis of virulence of certain well-studied fish bacterial pathogens has increased dramatically during the past decade. This has resulted from the application of recombinant DNA technology and cell biology to investigate bacterial infections, and the development of genetic techniques for identifying virulence genes.

More recently, genome sequence information of several bacterial fish pathogens has become available from genome sequencing projects. There is strong reason to believe that this understanding will be exploited to develop new interventions against fish bacterial infections.

The relevance of sequencing projects for drug and vaccine discovery is obvious. During the "pregenomic" era, the vaccine candidate genes were individually identified by tedious gene knockout studies and virulence attenuation. But now, the complete genome sequencing provides information on every virulence gene and all potential vaccine candidates, and the sequence databases will become indispensable for research in fish vaccinology and drug development.

After sequencing, the open reading frames (ORFs) are searched against available databases for sequence similarity with genes of known functions in other organisms. There are several strategies for gene annotation employing the tools of predictive bioinformatics programs combined with analyses of the published literature.

Multiple target vaccine candidate genes can be chosen and deleted simultaneously by various strategies including global transposon mutagenesis and gene replacement techniques [155, 156] to study their effect on virulence and essentiality. A number of important virulence determinants identified in the sequenced genome can be targeted. For example, the sortase enzyme in Gram-positive fish pathogens would be a very attractive universal vaccine and therapeutic drug target, as it mediates covalent anchoring of many surface displayed antigenic and/or virulence related proteins in Gram-positive bacteria [139]. The inactivation or inhibition of the sortase enzyme can simultaneously prevent the surface display of a number of virulence factors, thus effectively attenuating the virulence of the pathogen [110, 157].

The availability of sequences of the complete surface antigenic repertoire of pathogens, including protein and noprotein antigens would facilitate strategies for rational design of vaccines and drugs. In addition, the recent availability of large collections of the "virulogenome" of fish bacterial pathogens will provide enormous virulence sequence information for DNA vaccination studies. The whole complement of IS elements, prophages, and pathogenicity islands that can harbor virulence, and antimicrobial resistance gene clusters can be easily identified in the genomes. The comparison of genomes of different strains of the same bacteria or closely related species can reveal how these strains or species behave differently while infecting fish hosts, thus opening exciting opportunities for functional genomic analysis of infection processes and pathogenesis. However, experimental validation of predicted functions of genes identified from sequencing projects has lagged far behind the speed of annotation, and the major challenge of researchers in the field today is to understand the functional framework of the sequenced genomes.

CONCLUSIONS

There has been a steady increase in the number of species of bacteria implicated in fish diseases. The common fish pathogenic bacterial species belong to the genera Vibrio, Aeromonas, Flavobacterium, Yersinia, Edwardsiella, Streptococcus, lactococcus, Renibacterium, and Mycobacterium [23]. However, there is growing indications that the pathogenic species spectrum as well as the geographic and host range is widening among fish pathogens [158–161], leading to the emergence of new pathogens. Unlike the situation in human and animal medicine, fish diseases pose unique and daunting challenges. Fish are always bathed in a continuous medium of water, and fish disease treatment is essentially a population medicine. In addition, the current treatment methods are largely ineffective, and the biology and genetics of most fish bacterial pathogens are poorly understood, limiting the application of modern science-based pathogen intervention strategies.

Rapid growth and expansion of genome sequencing of human and animal pathogens enabled better understanding of their biology, evolution, and host adaptation strategies, and helped in combating many major diseases. Unfortunately, such developments and progress in the genomics and functional genomics of fish pathogenic bacteria have been very slow. However, recent availability of cost-effective high-throughput sequencing technologies has set the pace of sequencing of more fish pathogenic bacteria. Genome sequencing of a number of important bacterial pathogens of fish has helped us to better understand their biology and genetics. The sequencing projects have unearthed exciting new information on the adaptive evolution of fish pathogens, for example, how the nonpathogenic and ubiquitous soil bacteria such as Arthrobacter sp. has evolved into a strictly obligate fish pathogen, R. salmoninarum, by shedding functional genes through genomic reduction to lead to a very cosy intracellular life style.

On the other hand, phenotypically similar strains of the same species differ in certain set of virulence gene clusters, acquired through HGT and become highly virulent. The capsule gene cluster in the L. garvieae Lg2 strain confers virulence compared to noncapsulated ATCC 49156, which lacks the gene cluster. Nonpathogenic strains acquire genomic islands from distantly related pathogenic species and emerge as new pathogens of fish.

Comparative pathogenomics of closely related bacteria has increased our knowledge of how they vary in their virulence and their ability to adapt to different ecological niches. This is clearly evident in the difference in virulence of various strains of V. anguillarum and V. vulnificus, and among the closely related species of the genus Flavobacterium. As more strain-specific sequence information on bacterial pathogens of fish becomes available, we will have a better understanding of the subtle genomic differences among strains with varying virulence characteristics.

The typical pathogen evolutionary strategy of acquiring, shuffling and shedding genes mediated by IS elements, pseudogenes, prophage sequences, and HGT is also observed in most bacterial pathogens of fish. It is certain that the new genomic information will bring paradigm changes in bacterial pathogenesis and should provide new perspectives to our current thinking on the evolutionary and adaptive strategies of aquatic bacteria and how they colonize and establish in wider ecological niches and new host species. Moreover, the identification of key virulence factors in pathogenic strains should help us design efficient drugs and vaccines to combat major bacterial pathogens of fish.

However, it should be stressed that the genomic information will provide only a snapshot of the microorganism. Highly virulent clones armed with one or more acquired virulence factors can suddenly develop from the existing harmless microorganisms in the face of environmental, antibiotic, and host-induced selective pressures.

More intriguingly, about 40% of the genes in sequenced bacterial genomes constitute new putative genes and hypothetical proteins with mysterious functions and are conserved among several different species of bacteria. Even in Escherichia coli, the most studied of all bacteria, only 54% genes have currently been functionally characterized based on experimental evidence [162]. A close scrutiny of the sequenced genomes of fish pathogens reveals that the above situation is essentially true for these pathogens as well. Although current advances in functional genomics, structural genomics and bioinformatics have contributed immensely to deciphering and extracting useful biological information from the vast genomic data, understanding and assigning functionality to the unique and new gene sequences discovered in the genomes will be the major task of genome biologists in the coming years.

ACKNOWLEDGMENTS

This work was financially supported by various research fundings from the Ministry of Agriculture and Fisheries Wealth, Oman.

REFERENCES

1. J. M. Rothberg and J. H. Leamon, "The development and impact of 454 sequencing," Nature Biotechnology, vol. 26, no. 10, pp. 1117–1124, 2008.
2. J. Zhang, R. Chiodini, A. Badr, and G. Zhang, "The impact of next-generation sequencing on genomics,"Journal of Genetics and Genomics, vol. 38, no. 3, pp. 95–109, 2011.
3. E. Pennisi, "Genome data shake tree of life," Science, vol. 280, no. 5364, pp. 672–674, 1998.
4. W. F. Doolittle, "Phylogenetic classification and the universal tree," Science, vol. 284, no. 5423, pp. 2124–2128, 1999.
5. W. F. Doolittle, "Lateral genomics," Trends in Cell Biology, vol. 9, no. 12, pp. M5–M8, 1999.
6. E. Pennisi, "Is it time to uproot the tree of life?" Science, vol. 284, no. 5418, pp. 1305–1307, 1999.
7. W. F. Doolittle, "Uprooting the tree of life," Scientific American, vol. 282, no. 2, pp. 90–95, 2000.
8. E. V. Koonin, K. S. Makarova, and L. Aravind, "Horizontal gene transfer in prokaryotes: quantification and classification," Annual Review of Microbiology, vol. 55, pp. 709–742, 2001.
9. N. A. Moran, "Microbial minimalism: genome reduction in bacterial pathogens," Cell, vol. 108, no. 5, pp. 583–586, 2002.
10. A. Mira, L. Klasson, and S. G. E. Andersson, "Microbial genome evolution: sources of variability,"Current Opinion in Microbiology, vol. 5, no. 5, pp. 506–512, 2002.
11. L. Holm, "Codon usage and gene expression," Nucleic Acids Research, vol. 14, no. 7, pp. 3075–3087, 1986.
12. L. S. Frost, R. Leplae, A. O. Summers, and A. Toussaint, "Mobile genetic elements: the agents of open source evolution," Nature Reviews Microbiology, vol. 3, no. 9, pp. 722–732, 2005.
13. U. Dobrindt, B. Hochhut, U. Hentschel, and J. Hacker, "Genomic islands in pathogenic and environmental microorganisms," Nature Reviews Microbiology, vol. 2, no. 5, pp. 414–424, 2004.
14. U. Dobrindt and J. Hacker, "How bacterial pathogens were constructed," in Bacterial Virulence: Basic Principles, Models and Global Approaches, P. Sansonetti, Ed., pp. 3–15, Wiley-VCH, GmbH & Co. KGaA, Weinheim, Germany, 2010.
15. M. J. Pallen and B. W. Wren, "Bacterial pathogenomics," Nature, vol. 449, no. 7164, pp. 835–842, 2007.
16. J. G. Lawrence and J. R. Roth, "Genomic flux: genome evolution by gene loss and acquisition," inOrganization of the Prokaryotic Genome, R. L. Charlebois, Ed., pp. 263–289, American Society for Microbiology, Washington, DC, USA, 1999.
17. F. P. Meyer, "Aquaculture disease and health management.," Journal of Animal Science, vol. 69, no. 10, pp. 4201–4208, 1991.
18. C. D. Harvell, K. Kim, J. M. Burkholder et al., "Emerging marine diseases—climate links and anthropogenic factors," Science, vol. 285, no. 5433, pp. 1505–1510, 1999.
19. R. Subasinghe, "Fish health and quarantine; review of the state of the World aquaculture," in FAO Fisheries Circular, no. 886, pp. 45–49, Rome, Italy, Food and Agriculture Organization of the United Nations, 1997.
20. B. Hjeltnes and R. J. Roberts, "Vibriosis," in Bacterial Diseases of Fish, R. J. Roberts, N. R. Bromag, and V. Inglis, Eds., pp. 109–121, Blackwell Scientific, Oxford, UK, 1993.
21. D. V. Lightner and R. M. Redman, "Shrimp diseases and current diagnostic methods," Aquaculture, vol. 164, no. 1–4, pp. 201–220, 1998.
22. O. Bergh, F. Nilsen, and O. B. Samuelsen, "Diseases, prophylaxis and treatment of the Atlantic halibutHippoglossus hippoglossus: a review," Diseases of Aquatic Organisms, vol. 48, no. 1, pp. 57–74, 2001.
23. B. Austin and D. A. Austin, Bacterial Fish Pathogens: Disease in Farmed and Wild Fish, Springer, New York, NY, USA, 3rd edition, 1999.
24. B. Austin and X. H. Zhang, "Vibrio harveyi: a significant pathogen of marine vertebrates and invertebrates," Letters in Applied Microbiology, vol. 43, no. 2, pp. 119–124, 2006.
25. C. Amaro, E. G. Biosca, C. Esteve, et al., "Comparative study of phenotypic and virulence properties inVibrio vulnificus biotypes 1 and 2 obtained from a European eel farm experiencing mortalities," Diseases of Aquatic Organisms, vol. 13, pp. 29–35, 1992.
26. E. G. Biosca, H. Llorens, E. Garay, and C. Amaro, "Presence of a capsule in Vibrio vulnificus biotype 2 and its relationship to virulence for eels," Infection and Immunity, vol. 61, no. 5, pp. 1611–1618, 1993.
27. E. Benediktsdóttir, L. Verdonck, C. Spröer, S. Helgason, and J. Swings, "Characterization of Vibrio viscosus and Vibrio wodanis isolated at different geographical locations: a proposal for reclassification ofVibrio viscosus as Moritella viscosa comb. nov.," International Journal of Systematic and Evolutionary Microbiology, vol. 50, no. 2, pp. 479–488, 2000.
28. L. A. Actis, M. E. Tolmasky, and J. H. Crosa, "Vibriosis," in Fish Diseases and Disorders, Viral, Bacterial, and Fungal Infections, P. T. K. Woo and D. W. Bruno, Eds., vol. 3, pp. 570–605, CABI International, Oxfordshire, UK, 2nd edition, 2011.
29. E. Hjerde, M. Lorentzen, M. T. G. Holden et al., "The genome sequence of the fish pathogen Aliivibrio salmonicida strain LFI1238 shows extensive evidence of gene decay," BMC Genomics, vol. 9, article 616, 2008.
30. P. A. Gulig, V. D. Crécy-Lagard, A. C. Wright, B. Walts, M. Telonis-Scott, and L. M. McIntyre, "SOLiD sequencing of four Vibrio vulnificus genomes enables comparative genomic analysis and identification of candidate clade-specific virulence genes," BMC

Genomics, vol. 11, no. 1, article 512, 2010.

31. H. Naka, G. M. Dias, C. C. Thompson, C. Dubay, F. L. Thompson, and J. H. Crosa, "Complete genome sequence of the marine fish pathogen Vibrio anguillarum harboring the pJM1 virulence plasmid and genomic comparison with other virulent strains of V. anguillarum and V. ordalii," Infection and Immunity, vol. 79, no. 7, pp. 2889–2900, 2011.

32. J. L. Larsen, K. Pedersen, and I. Dalsgaard, "Vibrio anguillarum serovars associated with vibriosis in fish,"Journal of Fish Diseases, vol. 17, no. 3, pp. 259–267, 1994.

33. U. B. S. Sorensen and J. L. Larsen, "Serotyping of Vibrio anguillarum," Applied and Environmental Microbiology, vol. 51, no. 3, pp. 593–597, 1986.

34. A. E. Toranzo and J. L. Barja, "A review of the taxonomy and seroepizootiology of Vibrio anguillarum, with special reference to aquaculture in the NorthWest Spain," Diseases of Aquatic Organisms, vol. 9, pp. 73–82, 1990.

35. J. H. Crosa, "A plasmid associated with virulence in the marine fish pathogen Vibrio anguillarumspecifies an iron-sequestering system," Nature, vol. 284, no. 5756, pp. 566–568, 1980.

36. A. E. Toranzo, J. L. Barja, S. A. Potter, et al., "Molecular factors associated with virulence of marine vibrios isolated from striped bass in Chesapeake Bay," Infection and Immunity, vol. 39, no. 3, pp. 1220–1227, 1983.

37. M. H. Schiewe, T. J. Trust, and J. H. Crosa, "Vibrio ordalii sp. nov.: a causative agent of vibriosis in fish,"Current Microbiology, vol. 6, no. 6, pp. 343–348, 1981.

38. D. P. Ransom, C. N. Lannan, J. S. Rohovec, and J. L. Fryer, "Comparison of histopathology caused byVibrio anguillarum and Vibrio ordalii in three species of Pacific salmon," Journal of Fish Diseases, vol. 7, pp. 107–115, 1984.

39. K. Makino, K. Oshima, K. Kurokawa et al., "Genome sequence of Vibrio parahaemolyticus: a pathogenic mechanism distinct from that of V. cholerae," The Lancet, vol. 361, no. 9359, pp. 743–749, 2003.

40. N. Okada, S. Matsuda, J. Matsuyama et al., "Presence of genes for type III secretion system 2 in Vibrio mimicus strains," BMC Microbiology, vol. 10, article 302, 2010.

41. A. Alam, K. A. Miller, M. Chaand, J. S. Butler, and M. Dziejman, "Identification of Vibrio cholerae type III secretion system effector proteins," Infection and Immunity, vol. 79, no. 4, pp. 1728–1740, 2011.

42. C. R. Osorio, S. Juiz-Rio, and M. L. Lemos, "A siderophore biosynthesis gene cluster from the fish pathogen Photobacterium damselae subsp. piscicida is structurally and functionally related to the Yersiniahigh-pathogenicity island," Microbiology, vol. 152, no. 11, pp. 3327–3341, 2006.

43. E. S. Yip, B. T. Grublesky, E. A. Hussa, and K. L. Visick, "A novel, conserved cluster of genes promotes symbiotic colonization and σ-dependent biofilm formation by Vibrio fischeri," Molecular Microbiology, vol. 57, no. 5, pp. 1485–1498, 2005.

44. H. S. Kim, S. J. Park, and K. H. Lee, "Role of NtrC-regulated exopolysaccharides in the biofilm formation and pathogenic interaction of Vibrio vulnificus," Molecular Microbiology, vol. 74, no. 2, pp. 436–453, 2009.

45. E. Sanjuan, B. Fouz, J. D. Oliver, and C. Amaro, "Evaluation of genotypic and phenotypic methods to distinguish clinical from environmental Vibrio vulnificus strains," Applied and Environmental Microbiology, vol. 75, no. 6, pp. 1604–1613, 2009.

46. N. Bisharat, D. I. Cohen, M. C. Maiden, D. W. Crook, T. Peto, and R. M. Harding, "The evolution of genetic structure in the marine pathogen, Vibrio vulnificus," Infection, Genetics and Evolution, vol. 7, no. 6, pp. 685–693, 2007.

47. M. B. Schrøder, S. Espelid, and T. Ø. Jørgensen, "Two serotypes of Vibrio salmonicida isolated from diseased cod (Gadus morhua L.); virulence, immunological studies and vaccination experiments," Fish and Shellfish Immunology, vol. 2, no. 3, pp. 211–221, 1992.

48. M. Tomich, P. J. Planet, and D. H. Figurski, "The tad locus: postcards from the widespread colonization island," Nature Reviews Microbiology, vol. 5, no. 5, pp. 363–375, 2007.

49. P. Siguier, J. Filee, and M. Chandler, "Insertion sequences in prokaryotic genomes," Current Opinion in Microbiology, vol. 9, no. 5, pp. 526–531, 2006.

50. J. Parkhill, G. Dougan, K. D. James et al., "Complete genome sequence of a multiple drug resistantSalmonella enterica serovar Typhi CT18," Nature, vol. 413, no. 6858, pp. 848–852, 2001.

51. G. D. Wiens, D. D. Rockey, Z. Wu et al., "Genome sequence of the fish pathogen Renibacterium salmoninarum suggests reductive evolution away from an environmental Arthrobacter ancestor," Journal of Bacteriology, vol. 190, no. 21, pp. 6970–6982, 2008.

52. A. Martin-Carnahan and S. W. Joseph, "Aeromonadaceae," in Bergey›s Manual of Systematic Bacteriology, G. M. Garrity, Ed., vol. 2, pp. 556–580, Springer, New York, NY, USA, 2nd edition, 2005.

53. M. J. Figueras, "Clinical relevance of Aeromonas sM503," Reviews in Medical Microbiology, vol. 16, no. 4, pp. 145–153, 2005.

54. R. Seshadri, S. W. Joseph, A. K. Chopra et al., "Genome sequence of Aeromonas hydrophila ATCC 7966T: jack of all trades," Journal of Bacteriology, vol. 188, no. 23, pp. 8272–8282, 2006.

55. M. E. Reith, R. K. Singh, B. Curtis et al., "The genome of Aeromonas salmonicida subsp. salmonicidaA449: Insights into the evolution of a fish pathogen," BMC Genomics, vol. 9, article 427, 2008.

56. Y. Li, Y. Liu, Z. Zhou et al., "Complete genome sequence of Aeromonas veronii strain B565," Journal of Bacteriology, vol. 193, no. 13, pp. 3389–3390, 2011.

57. S. A. Beatson, M. D. G. de Luna, N. L. Bachmann et al., "Genome sequence of the emerging pathogenAeromonas caviae," Journal of Bacteriology, vol. 193, no. 5, pp. 1286–1287, 2011.

58. E. M. Bernoth, "Furunculosis: the history of the

diseases and of disease research," in Furunculosis, Multidisciplinary Fish Disease Research, E. M. Bernoth, A. E. Ellis, P. J. Midtlyng, G. Olivier, and P. Smith, Eds., pp. 1–20, Academic Press, London, UK, 1997.

59. M. Hiney and G. Olivier, "Furunculosis (Aeromonas salmonicida)," in Fish Diseases and Disorders III: Viral, Bacterial and Fungal Infections, P. T. K. Woo and D. W. Bruno, Eds., pp. 341–425, CAB Publishing, Oxford, UK, 1999.

60. C. Paniagua, O. Rivero, J. Anguita, and G. Naharro, "Pathogenicity factors and virulence for rainbow trout (Salmo gairdneri) of motile Aeromonas spp. isolated from a river," Journal of Clinical Microbiology, vol. 28, no. 2, pp. 350–355, 1990.

61. S. E. Burr, D. Pugovkin, T. Wahli, H. Segner, and J. Frey, "Attenuated virulence of an Aeromonas salmonicida subsp. salmonicida type III secretion mutant in a rainbow trout model," Microbiology, vol. 151, no. 6, pp. 2111–2118, 2005.

62. J. Sha, L. Pillai, A. A. Fadl, C. L. Galindo, T. E. Erova, and A. K. Chopra, "The type III secretion system and cytotoxic enterotoxin alter the virulence of Aeromonas hydrophila," Infection and Immunity, vol. 73, no. 10, pp. 6446–6457, 2005.

63. J. M. Janda and R. P. Kokka, "The pathogenicity of Aeromonas strains relative to genospecies and phenospecies identification," FEMS Microbiology Letters, vol. 90, no. 1, pp. 29–33, 1991.

64. J. Madetoja, I. Dalsgaard, and T. Wiklund, "Occurrence of Flavobacterium psychrophilum in fish-farming environments," Diseases of Aquatic Organisms, vol. 52, no. 2, pp. 109–118, 2002.

65. A. Nematollahi, A. Decostere, F. Pasmans, and F. Haesebrouck, "Flavobacterium psychrophilum infections in salmonid fish," Journal of Fish Diseases, vol. 26, no. 10, pp. 563–574, 2003.

66. J. G. Leahy and R. R. Colwell, "Microbial degradation of hydrocarbons in the environment,"Microbiological Reviews, vol. 54, no. 3, pp. 305–315, 1990.

67. J. Lehmann, D. Mock, F. J. Stuerenberg, and J. F. Bernardet, "First isolation of Cytophaga psychrophilafrom a systemic disease in eel and cyprinids," Diseases of Aquatic Organisms, vol. 10, pp. 217–220, 1991.

68. R. A. Holt, Cytophaga psychrophila, the causative agent of bacterial cold-water disease in salmonid fish, Ph.D. thesis, Oregon State University, Corvallis, Ore, USA, 1987.

69. G. J. Heo, K. Kasai, and H. Wakabayashi, "Occurrence of Flavobacterium branchiophila associated with bacterial gill disease at a trout hatchery," Fish Pathology, vol. 25, pp. 99–105, 1990.

70. V. E. Ostland, J. S. Lumsden, D. D. MacPhee, and H. W. Ferguson, "Characteristics of Flavobacterium branchiophilum, the cause of salmonid bacterial gill disease in Ontario," Journal of Aquatic Animal Health, vol. 6, no. 1, pp. 13–26, 1994.

71. J. F. Bernardet, "Flexibacter columnaris: first description in France and comparison with bacterial strains from other origins," Diseases of Aquatic Organisms, vol. 6, pp. 37–44, 1989.

72. D. J. Speare and H. W. Ferguson, "Clinical and pathological features of common gill diseases of cultured salmonids in Ontario," Canadian Veterinary Journal, vol. 30, pp. 882–887, 1989.

73. J. F. Bernardet, P. Segers, M. Vancanneyt, F. Berthe, K. Kersters, and P. Vandamme, "Cutting a gordian knot: emended classification and description of the genus Flavobacterium, emended description of the family Flavobacteriaceae, and proposal of Flavobacterium hydatis nom. nov. (basonym, Cytophaga aquatilis Strohl and Tait 1978)," International Journal of Systematic Bacteriology, vol. 46, no. 1, pp. 128–148, 1996.

74. E. Duchaud, M. Boussaha, V. Loux et al., "Complete genome sequence of the fish pathogenFlavobacterium psychrophilum," Nature Biotechnology, vol. 25, no. 7, pp. 763–769, 2007.

75. M. Touchon, P. Barbier, J. F. Bernardet, et al., "Complete genome sequence of the fish pathogenFlavobacterium branchiophilum," Applied and Environmentl Microbiology, vol. 77, no. 21, pp. 7656–7662, 2011.

76. M. L. Lawrence, A. Karsi, H. C. Tekedar, et al., "Comparative genomics of Flavobacterium columnareisolates from two genetic divisions and with different pathogenic potential for channel catfish (Ictalurus punctatus)," in Proceedings of the Conference on FLAVOBACTERIUM, Paris, France, September 2009.

77. C. Rodkhum, I. Hirono, J. H. Crosa, and T. Aoki, "Four novel hemolysin genes of Vibrio anguillarum and their virulence to rainbow trout," Microbial Pathogenesis, vol. 39, no. 4, pp. 109–119, 2005.

78. M. Palmer, "The family of thiol-activated, cholesterol-binding cytolysins," Toxicon, vol. 39, no. 11, pp. 1681–1689, 2001.

79. M. Shoji, D. B. Ratnayake, Y. Shi et al., "Construction and characterization of a nonpigmented mutant ofPorphyromonas gingivalis: Cell surface polysaccharide as an anchorage for gingipanis," Microbiology, vol. 148, no. 4, pp. 1183–1191, 2002.

80. K. Sato, E. Sakai, P. D. Veith et al., "Identification of a new membrane-associated protein that influences transport/maturation of gingipains and adhesins of Porphyromonas gingivalis," The Journal of Biological Chemistry, vol. 280, no. 10, pp. 8668–8677, 2005.

81. S. L. Abbott and J. M. Janda, "The genus Edwardsiella," Prokaryotes, vol. 6, pp. 72–89, 2006.

82. J. E. Clarridge, D. M. Musher, V. Fanstein, and R. J. Wallace Jr., "Extra-intestinal human infection caused by Edwardsiella tarda," Journal of Clinical Microbiology, vol. 11, no. 5, pp. 511–514, 1980.

83. P. Nagel, A. Serritella, and T. J. Layden, "Edwardsiella tarda gastroenteritis associated with a pet turtle,"Gastroenterology, vol. 82, no. 6, pp. 1436–1437, 1982.

84. W. A. Rogers, "Edwardsiellosis in fishes," in Antigens of Fish Pathogens. Les Antigenes des Microorganisms

Pathogenes des Poissons, D. P. Anderson, M. Dorson, and Ph. Dubourget, Eds., pp. 153–159, Fondation Marcel Merieux, Lyon, France, 1983.

85. F. P. Meyer and G. L. Bullock, "Edwardsiella tarda, a new pathogen of channel catfish (Ictalurus punctatus)," Applied Microbiology, vol. 25, no. 1, pp. 155–156, 1973.

86. S. H. M. Ling, X. H. Wang, L. Xie, T. M. Lim, and K. Y. Leung, "Use of green fluorescent protein (GFP) to study the invasion pathways of Edwardsiella tarda in in vivo and in vitro fish models," Microbiology, vol. 146, no. 1, pp. 7–19, 2000.

87. J. C. Newton, L. G. Wolfe, J. M. Grizzle, and J. A. Plumb, "Pathology of experimental enteric septicaemia in channel catfish, Ictalurus punctatus (Rafinesque), following immersion-exposure to Edwardsiella ictaluri," Journal of Fish Diseases, vol. 12, no. 4, pp. 335–347, 1989.

88. Q. Wang, M. Yang, J. Xiao et al., "Genome sequence of the versatile fish pathogen Edwardsiella tardaprovides insights into its adaptation to broad host ranges and intracellular niches," PLoS One, vol. 4, no. 10, Article ID e7646, 2009.

89. M. L. Williams, A. F. Gillaspy, D. W. Dyer, et al., "Genome sequence of Edwardsiella ictaluri 93–146, a strain associated with a natural channel catfish outbreak of enteric septicemia of catfish," Journal of Bacteriology, vol. 194, no. 3, pp. 740–741, 2012.

90. S. R. Partridge and R. M. Hall, "The IS1111 family members IS4321 and IS5075 have sub-terminal inverted repeats and target the terminal inverted repeats of Tn21 family transposons," Journal of Bacteriology, vol. 185, no. 21, pp. 6371–6384, 2003.

91. T. Tanabe, T. Funahashi, H. Nakao, S. I. Miyoshi, S. Shinoda, and S. Yamamoto, "Identification and characterization of genes required for biosynthesis and transport of the siderophore vibrioferrin in Vibrio parahaemolyticus," Journal of Bacteriology, vol. 185, no. 23, pp. 6938–6949, 2003.

92. Q. Wang, Q. Liu, Y. Ma, L. Zhou, and Y. Zhang, "Isolation, sequencing and characterization of cluster genes involved in the biosynthesis and utilization of the siderophore of marine fish pathogen Vibrio alginolyticus," Archives of Microbiology, vol. 188, no. 4, pp. 433–439, 2007.

93. M. D. Furones, C. J. Rodgers, and C. B. Munn, "Yersinia ruckeri, the causal agent of enteric redmouth disease (ERM) in fish," Annual Review of Fish Diseases, vol. 3, no. C, pp. 105–125, 1993.

94. R. R. Rucker, "Redmouth disease of rainbow trout (Salmo gairdneri)," Bulletin de l›Office International des Epizooties, vol. 65, no. 5, pp. 825–830, 1966.

95. E. Tobback, A. Decostere, K. Hermans, F. Haesebrouck, and K. Chiers, "Yersinia ruckeri infections in salmonid fish," Journal of Fish Diseases, vol. 30, no. 5, pp. 257–268, 2007.

96. M. T. Horne and A. C. Barnes, "Enteric redmouth disease (Y. ruckeri)," in Fish Diseases and Disorders: Viral, Bacterial and Fungal Infections, P. T. K. Woo and D. W. Bruno, Eds., vol. 3, pp. 455–477, CABI Publishing, Oxfordshire, UK, 1999.

97. A. J. Ross, R. R. Rucker, and W. H. Ewing, "Description of a bacterium associated with redmouth disease of rainbow trout (Salmo gairdneri)," Canadian Journal of Microbiology, vol. 12, no. 4, pp. 763–770, 1966.

98. R. L. Davies, "Virulence and serum-resistance in different clonal groups and serotypes of Yersinia ruckeri," Veterinary Microbiology, vol. 29, no. 3-4, pp. 289–297, 1991.

99. R. M. V. Stevenson and D. W. Airdrie, "Serological variation among Yersinia ruckeri strains," Journal of Fish Diseases, vol. 7, pp. 247–254, 1984.

100. J. L. Romalde, B. Magarinos, J. L. Barja, and A. E. Toranzo, "Antigenic and molecular characterization ofYersinia ruckeri proposal for a new intraspecies classification," Systematic and Applied Microbiology, vol. 16, no. 3, pp. 411–419, 1993.

101. R. M. Stevenson, "Immunization with bacterial antigens: yersiniosis," Developments in Biological Standardization, vol. 90, pp. 117–124, 1997.

102. R. L. Davies, "Yersinia ruckeri produces four iron-regulated proteins but does not produce detectable siderophores," Journal of Fish Diseases, vol. 14, pp. 563–570, 1991.

103. M. Kotetishvili, A. Kreger, G. Wauters, J. G. Morris, A. Sulakvelidze, and O. C. Stine, "Multilocus sequence typing for studying genetic relationships among Yersinia species," Journal of Clinical Microbiology, vol. 43, no. 6, pp. 2674–2684, 2005.

104. P. E. Chen, C. Cook, A. C. Stewart et al., "Genomic characterization of the Yersinia genus," Genome Biology, vol. 11, no. 1, article R1, 2010.

105. J. E. Sanders and J. L. Fryer, "Renibacterium salmoninarum gen. nov., sp. nov., the causative agent of bacterial kidney disease in salmonid fishes," International Journal of Systematic Bacteriology, vol. 30, no. 2, pp. 496–502, 1980.

106. D. L. Mitchum and L. E. Sherman, "Transmission of bacterial kidney disease from wild to stocked hatchery trout," Canadian Journal of Fisheries and Aquatic Sciences, vol. 38, pp. 547–551, 1981.

107. G. L. Bullock, Bacterial Kidney Disease of Salmonid Fishes Caused by Renibacterium salmoninarum, no. 60, U.S. Fish and Wildlife Service, Fish Disease Leaflet, Washington, DC, USA, 1980.

108. K. E. Wolf and C. E. Dunbar, "Test of 34 therapeutic agents for control of kidney disease in trout,"Transactions of American Fisheries Society, vol. 88, pp. 117–124, 1959.

109. S. K. Gutenberger, J. R. Duimstra, J. S. Rohovec, and J. L. Fryer, "Intracellular survival of Renibacterium salmoninarum in trout mononuclear phagocytes," Disease of Aquatic Organisms, vol. 28, no. 2, pp. 93–106, 1997.

110. P. S. Sudheesh, S. Crane, K. D. Cain, and M. S. Strom, "Sortase inhibitor phenyl vinyl sulfone inhibitsRenibacterium salmoninarum adherence and invasion of host cells," Diseases of Aquatic Organisms,

vol. 78, no. 2, pp. 115–127, 2007.

111. E. Stackebrandt, U. Wehmeyer, H. Nader, and F. Fiedler, "Phylogenetic relationship of the fish pathogenicRenibacterium salmoninarum to Arthrobacter, Micrococcus and related taxa," FEMS Microbiology Letters, vol. 50, no. 2-3, pp. 117–120, 1988.

112. K. Salonius, C. Siderakis, A. M. MacKinnon, and S. G. Griffiths, "Use of Arthrobacter davidanieli as a live vaccine against Renibacterium salmoninarum and Piscirickettsia salmonis in salmonids," Developments in Biologicals, vol. 121, pp. 189–197, 2005.

113. E. F. Mongodin, N. Shapir, S. C. Daugherty et al., "Secrets of soil survival revealed by the genome sequence of Arthrobacter aurescens TC1," PLoS Genetics, vol. 2, no. 12, article e214, 2006.

114. R. G. Getchell, J. S. Rohovec, and J. L. Fryer, "Comparison of Renibacterium salmoninarum isolates by antigenic analysis," Fish Pathology, vol. 20, pp. 149–159, 1985.

115. F. Fiedler and R. Draxl, "Biochemical and immunochemical properties of the cell surface ofRenibacterium salmoninarum," Journal of Bacteriology, vol. 168, no. 2, pp. 799–804, 1986.

116. T. Kitao, "Streptococcal infections," in Bacterial Disease of Fish, V. Inglis, R. J. Roberts, and N. R. Bromage, Eds., pp. 196–210, Blackwell Scientific, Oxford, UK, 1993.

117. A. Domenech, J. F. Fernandez-Garayzabal, C. Pasqual, et al., "Streptococcosis in cultured turbot,Scophthalmus maximus (L.), associated with Streptococcus parauberis," Journal of Fish Diseases, vol. 19, pp. 33–38, 1996.

118. J. L. Romalde, C. Ravelo, I. Valdés et al., "Streptococcus phocae, an emerging pathogen for salmonid culture," Veterinary Microbiology, vol. 130, no. 1-2, pp. 198–207, 2008.

119. D. Vendrell, J. L. Balcázar, I. Ruiz-Zarzuela, I. de Blas, O. Gironés, and J. L. Múzquiz, "Lactococcus garvieae in fish: a review," Comparative Immunology, Microbiology and Infectious Diseases, vol. 29, no. 4, pp. 177–198, 2006.

120. M. D. Collins, J. A. E. Farrow, B. A. Phillips, and O. Kandler, "Streptococcus garvieae sp. nov. andStreptococcus plantarum sp. nov.," Journal of General Microbiology, vol. 129, no. 11, pp. 3427–3431, 1983.

121. A. M. Williams, J. L. Fryer, and M. D. Collins, "Lactococcus piscium sp. nov. a new Lactococcus species from salmonid fish," FEMS Microbiology Letters, vol. 68, no. 1-2, pp. 109–113, 1990.

122. A. Domenech, J. Prieta, J. F. Fernandez-Garayzabal, et al., "Phenotypic and phylogenetic evidence for a close relationship between Lactococcus garvieae and Enterococcus seriolicida," Microbiologia SEM, vol. 9, no. 1, pp. 63–68, 1993.

123. A. Eldar, C. Ghittino, L. Asanta et al., "Enterococcus seriolicida is a junior synonym of Lactococcus garvieae, a causative agent of septicemia and meningoencephalitis in fish," Current Microbiology, vol. 32, no. 2, pp. 85–88, 1996.

124. S. Wallbanks, A. J. Martinez-Murcia, J. L. Fryer, B. A. Phillips, and M. D. Collins, "16S rRNA sequence determination for members of the genus Carnobacterium and related lactic acid bacteria and description of Vagococcus salmoninarum sp. nov.," International Journal of Systematic Bacteriology, vol. 40, no. 3, pp. 224–230, 1990.

125. E. S. Bromage and L. Owens, "Infection of barramundi Lates calcarifer with Streptococcus iniae: effects of different routes of exposure," Diseases of Aquatic Organisms, vol. 52, no. 3, pp. 199–205, 2002.

126. J. A. Elliott, M. D. Collins, N. E. Pigott, and R. R. Facklam, "Differentiation of Lactococcus lactis andLactococcus garvieae from humans by comparison of whole-cell protein patterns," Journal of Clinical Microbiology, vol. 29, no. 12, pp. 2731–2734, 1991.

127. R. P. Perera, S. K. Johnson, M. D. Collins, and D. H. Lewis, "Streptococcus iniae associated with mortality of Tilapia nilotica X T.aurea hybrids," Journal of Aquatic Animal Health, vol. 6, no. 4, pp. 335–340, 1994.

128. R. Kusuda, K. Kawai, F. Salati, C. R. Banner, and J. L. Fryer, "Enterococcus seriolicida sp. nov., a fish pathogen," International Journal of Systematic Bacteriology, vol. 41, no. 3, pp. 406–409, 1991.

129. M. Aguado-Urda, G. H. López-Campos, M. M. Blanco et al., "Genome sequence of Lactococcus garvieae21881, isolated in a case of human septicemia," Journal of Bacteriology, vol. 193, no. 15, pp. 4033–4034, 2011.

130. P. Reimundo, M. Pignatelli, L. D. Alcaraz, G. D›Auria, A. Moya, and J. A. Guijarro, "Genome sequence ofLactococcus garvieae UNIUD074, isolated in Italy from a lactococcosis outbreak," Journal of Bacteriology, vol. 193, no. 14, pp. 3684–3685, 2011.

131. H. Morita, H. Toh, K. Oshima et al., "Complete genome sequence and comparative analysis of the fish pathogen Lactococcus garvieae," PLoS One, vol. 6, no. 8, Article ID e23184, 2011.

132. S. W. Nho, J. I. Hikima, I. S. Cha et al., "Complete genome sequence and immunoproteomic analyses of the bacterial fish pathogen Streptococcus parauberis," Journal of Bacteriology, vol. 193, no. 13, pp. 3356–3366, 2011.

133. J. L. Watts, "Etiological agents of bovine mastitis," Veterinary Microbiology, vol. 16, no. 1, pp. 41–66, 1988.

134. A. M. Williams and M. D. Collins, "Molecular taxonomic studies on Streptococcus uberis types I and II. Description of Streptococcus parauberis sp. nov.," Journal of Applied Bacteriology, vol. 68, no. 5, pp. 485–490, 1990.

135. R. W. Bentley, J. A. Leigh, and M. D. Collins, "Intrageneric structure of Streptococcus based on comparative analysis of small-subunit rRNA sequences," International Journal of Systematic Bacteriology, vol. 41, no. 4, pp. 487–494, 1991.

136. J. B. Locke, R. K. Aziz, M. R. Vicknair, V. Nizet, and J. T. Buchanan, "Streptococcus iniae M-like protein contributes to virulence in fish and is a target for live attenuated vaccine development," PLoS One, vol. 3, no. 7, Article ID e2824, 2008.

137. M. Kawanishi, T. Yoshida, M. Kijima et al., "Characterization of Lactococcus garvieae isolated from radish and broccoli sprouts that exhibited a KG+ phenotype, lack of virulence and absence of a capsule," Letters in Applied Microbiology, vol. 44, no. 5, pp. 481–487, 2007.
138. T. J. Mitchell, "The pathogenesis of streptococcal infections: from tooth decay to meningitis," Nature Reviews. Microbiology, vol. 1, no. 3, pp. 219–230, 2003.
139. L. A. Marraffini, A. C. Dedent, and O. Schneewind, "Sortases and the art of anchoring proteins to the envelopes of gram-positive bacteria," Microbiology and Molecular Biology Reviews, vol. 70, no. 1, pp. 192–221, 2006.
140. J. T. Buchanan, J. A. Stannard, X. Lauth et al., "Streptococcus iniaephosphoglucomutase is a virulence factor and a target for vaccine development," Infection and Immunity, vol. 73, no. 10, pp. 6935–6944, 2005.
141. G. N. Frerichs, "Acid-fast fish pathogens," in Bacterial Diseases of Fish, V. Inglis, R. J. Roberts, and N. R. Bromage, Eds., pp. 217–233, Blackwell Scientific, Oxford, UK, 1993.
142. E. J. Noga, Fish disease diagnosis and treatment, Mosby, St. Louis, Mo, USA, 1996.
143. L. G. Wayne, G. P. Kubica, et al., "Genus Mycobacterium Lehmann and Neumann 1896, 363AL," inBergey›s Manual of Systematic Bacteriology, P. H. A. Sneath, N. S. Mair, M. E. Sharpe, and J. G. Holt, Eds., vol. 2, pp. 1436–1457, Williams & Wilkins, Baltimore, Md, USA, 1986.
144. C. O. Thoen and T. A. Schliesser, "Mycobacterial infections in cold-blooded animals," in The Mycobacteria, G. P. Kubica and L. G. Wayne, Eds., part B, p. 1471, 1st edition, 1984.
145. G. E. Sanders and L. E. Swaim, "Atypical piscine Mycobacteriosis in Japanese medaka (Oryzias latipes),"Comparative Medicine, vol. 51, no. 2, pp. 171–175, 2001.
146. M. K. Stoskopf, Fish Medicine, WB Saunders, Philadelphia, Pa, USA, 1993.
147. D. T. Smith and H. P. Willett, "Other Mycobacterium species," in Zinsser Microbiology, K. Joklik, H. P. Willet, and D. B. Amos, Eds., pp. 674–698, Appleton-Century-Crofts, New York, NY, USA, 17th edition, 1980.
148. D. Huminer, S. D. Pitlik, C. Block, et al., "Aquarium-borne Mycobacterium marinum skin infection. Report of a case and review of the literature," Archives of Dermatology, vol. 122, no. 6, pp. 698–703, 1986.
149. M. Ucko, A. Colorni, H. Kvitt, A. Diamant, A. Zlotkin, and W. R. Knibb, "Strain variation inMycobacterium marinum fish isolates," Applied and Environmental Microbiology, vol. 68, no. 11, pp. 5281–5287, 2002.
150. M. J. Yip, J. L. Porter, J. A. M. Fyfe et al., "Evolution of Mycobacterium ulcerans and other mycolactone-producing mycobacteria from a common Mycobacterium marinum progenitor," Journal of Bacteriology, vol. 189, no. 5, pp. 2021–2029, 2007.
151. M. Ucko and A. Colorni, "Mycobacterium marinum infections in fish and humans in Israel," Journal of Clinical Microbiology, vol. 43, no. 2, pp. 892–895, 2005.
152. H. E. Volkman, H. Clay, D. Beery, J. C. W. Chang, D. R. Sherman, and L. Ramakrishnan, "Tuberculous granuloma formation is enhanced by a Mycobacterium virulence determinant," PLoS Biology, vol. 2, no. 11, article e367, 2004.
153. L. E. Swaim, L. E. Connolly, H. E. Volkman, O. Humbert, D. E. Born, and L. Ramakrishnan, "Mycobacterium marinum infection of adult zebrafish causes caseating granulomatous tuberculosis and is moderated by adaptive immunity," Infection and Immunity, vol. 74, no. 11, pp. 6108–6117, 2006.
154. T. P. Stinear, T. Seemann, P. F. Harrison et al., "Insights from the complete genome sequence ofMycobacterium marinum on the evolution of Mycobacterium tuberculosis," Genome Research, vol. 18, no. 5, pp. 729–741, 2008.
155. J. McFadden, "Recombination in mycobacteria," Molecular Microbiology, vol. 21, no. 2, pp. 205–211, 1996.
156. N. R. Salama, B. Shepherd, and S. Falkow, "Global transposon mutagenesis and essential gene analysis ofHelicobacter pylori," Journal of Bacteriology, vol. 186, no. 23, pp. 7926–7935, 2004.
157. P. Cossart and R. Jonquières, "Sortase, a universal target for therapeutic agents against Gram-positive bacteria?" Proceedings of the National Academy of Sciences of the United States of America, vol. 97, no. 10, pp. 5013–5015, 2000.
158. M. Maher, R. Palmer, F. Gannon, and T. Smith, "Relationship of a novel bacterial fish pathogen toStreptobacillus monilifomris and the fusobacteria group, based on 16S ribosomal RNA analysis,"Systematic and Applied Microbiology, vol. 18, no. 1, pp. 79–84, 1995.
159. J. L. Fryer and M. J. Mauel, "The rickettsia: an emerging group of pathogens in fish," Emerging Infectious Diseases, vol. 3, no. 2, pp. 137–144, 1997.
160. T. H. Birkbeck, L. A. Laidler, A. N. Grant, and D. I. Cox, "Pasteurella skyensis sp. nov., isolated from Atlantic salmon (Salmo salar L.)," International Journal of Systematic and Evolutionary Microbiology, vol. 52, no. 3, pp. 699–704, 2002.
161. C. Michel, J. F. Bernardet, P. Daniel, S. Chilmonczyk, M. Urdaci, and P. de Kinkelin, "Clinical and aetiological aspects of a summer enteritic syndrome associated with the sporulating segmented filamentous bacterium "Candidatus Arthromitus" in farmed rainbow trout, Oncorhynchus mykiss(Walbaum)," Journal of Fish Diseases, vol. 25, no. 9, pp. 533–543, 2002.
162. M. Riley, T. Abe, M. B. Arnaud et al., "Escherichia coli K-12: a cooperatively developed annotation snapshot—2005," Nucleic Acids Research, vol. 34, no. 1, pp. 1–9, 2006.

CHAPTER 8

NANODELIVERY SYSTEMS AS NEW TOOLS FOR IMMUNOSTIMULANT OR VACCINE ADMINISTRATION: TARGETING THE FISH IMMUNE SYSTEM

Jie Ji, Debora Torrealba, Àngels Ruyra, and Nerea Roher

Keywords: Nanoparticles, fish, immunostimulation, alginate, liposome, chitosan, PLGA, carbon nanotubes

INTRODUCTION

The use of vaccines has been essential in aquaculture and has been under development for over 70 years since the first successful fish vaccine was formulated [1]. Vaccines stimulate the immune system to mount a defence against a pathogen and as such to protect the host from infection by this pathogen. While they are extremely important to control infectious diseases in farmed fish, there are still some hurdles affecting the development of effective vaccines against viruses, parasites, and intracellular pathogens. One of these bottlenecks is the vaccine administration system [2,3]. Different approaches have been employed in aquaculture to improve the vaccine efficacy and to explore alternative routes of immunization. Traditional adjuvants such as mineral oils have been routinely used for vaccine injection, the most common examples are Freund's complete adjuvant (FCA), Freund's incomplete adjuvant (FIA) and more recently Montanide. Although adjuvants are very effective in potentiating the immune response against the pathogen, they present different side effects. There are three main methods for vaccine administration in fish: orally, by immersion or by injection. Vaccination by injection is the most reliable and effective system for vaccine delivery in fish and the protection is by far the most effective and long lasting [4,5]. However, the injectable vaccines are usually prepared with oil/water adjuvant formulations (FCA or FIA) which result in adverse effects such as the appearance of granulomas [2,6] adhesions between organ and peritoneal wall [7], injection site lesions [8,9], reduced appetite and growth [10], or deformations of the skeleton [11]. Added to this, the anaesthesia, handling, and injection may cause occasional mortality. Importantly, there are also logistical challenges to inject fish of less than 20 g in large numbers, but these fish need vaccination the most because they are the most susceptible to disease [12]. The disadvantages of vaccination by immersion are: the large amounts of vaccine required [13], the difficulty to measure the efficiency of the uptake [14] and the degradation of the compound in the water. Like for immersion, oral vaccines offer the advantages of being stress free and easy to administer to large numbers of fish but it is also difficult to determine the dose of antigen received. Also depending whether fish are gastric or agastric the intact antigen has to pass through the digestive system to reach the second segment of the hindgut where antigens are absorbed [15,16]. In this context, the use of nanodelivery systems has been proposed as an alternative strategy to address not only the above mentioned problems, but also to enhance the efficacy since some of these delivery systems may act also as a potent adjuvant, which is extremely important for anti-viral vaccines. Therefore, searching for new delivery systems is required to improve the administration and the efficacy of vaccines and immunostimulants.

Delivery systems are those materials used for the administration of pharmaceuticals in a controlled manner aimed to achieve a therapeutic effect. These systems provide: cell or tissue targeted delivery of active compounds, improved bioavailability, improved solubility of hydrophobic drugs, sustained release and protection of the therapeutic agent from degradation [17]. Nanoencapsulation involves forming drug loaded particles with diameters ranging from 1 to 1000 nm, although other stricter definitions refer only to structures in the 1–100 nm range (*US National Nanotechnology Initiative, What*

Citation: Jie Ji, Debora Torrealba, Àngels Ruyra, and Nerea Roher. Nanodelivery systems as new tools for immunostimulant or vaccine administration: targeting the fish immune system, Biology, 2015; 4(4), 664-696. ISSN 2079-7737. DOI:10.3390/biology4040664.

is nanotechnology?). This size property enables the nanoscale devices to readily interact with biomolecules, such as enzymes and receptors, both on the surface and inside the cells.

Since 1960, when the first liposomes for drug delivery were described, a variety of other organic and inorganic biomaterials were developed for drug delivery [17]. In 1980 more complex drug delivery systems capable of responding to pH changes to trigger drug release or the first examples of cell specific targeting of liposomes were described [18]. Nowadays, nanoparticles can be easily tuned to have unique physical characteristics in size, shape, surface chemistry, or targeted surface ligand/receptor. The benefits of nanoparticles as delivery tools are the reduction of the doses, tissue specific targeting, reduction of the toxic or secondary effects of the drug and increase in the delivery efficiency [19]. The encapsulated molecules will generally have completely different properties (e.g., solubility or circulating half-life) compared to the non-encapsulated ones. Thus, it is very important to understand and control the *in vivo* behaviour on cells or tissues of these bioactive compounds once encapsulated, to know their efficacy and side effects. As mentioned, the size of the nanoparticle is not only important for the interaction with biomolecules but also because it will influence its biodistribution*in vivo*. In mammals it has been extensively studied that particles of less than 5 nm are cleared from the circulation through extravasation or renal clearance, whereas bigger nanoparticles (up to 15 μm) accumulate in the spleen, liver, and bone marrow [20]. The particle size also influences the preferred mechanism of cellular internalization, such as phagocytosis, macropinocytosis, caveolae-mediated-endocytosis, or others. Of note, the fate of the internalized material will be different in each case [21,22].

In teleosts, there are few reports on how the nanoparticles spread throughout the organism through the circulatory system, gills, gut, spleen, liver, or brain depending on the administration route [23,24]. Current findings indicate that particle shape and rigidity are also key factors for the kinetics and fate of the nanoparticles, mainly affecting the endocytosis. The vast majority of nanoparticles have a spherical shape, however similar volumes with different shapes are internalized at different rates [25,26]. Increased nanoparticle rigidity is related to enhanced phagocytosis by macrophages [27]. Finally, the nanoparticle surface charge critically affects how they interact with serum proteins and cell membranes. Highly charged particles fix more complement proteins [28], a process that can only be inhibited by addition of a hydrophilic coating. The surface charge will also determine the interaction with cell membranes. In general, neutral and anionic nanoparticles will be less internalized than positively charged ones [22,29]. Different studies using the same nanoparticle with different surface charges have shown that those with cationic groups were internalized more efficiently, [30,31] mostly due to the high affinity for the negatively charged proteoglycans present on the surface of cells [32]. The use of nanoparticles does also have some limitations. For example, their small size and large surface area can lead to particle aggregation and result in limited drug loading and burst release, making physical handling of nanoparticles difficult in liquid and dry forms [33]. Another issue that should be addressed in the future is the safety, for both human and animals, not only of the delivery system itself but also of the degradation products of the nanoparticles. These biosafety issues should be carefully addressed to avoid environmental contamination that can provoke detrimental effects on animal and human health.

In this review, we summarized different nano- and micro-sized delivery systems that have been described as delivery tools for fish vaccination or immunostimulation. Calcium phosphate nanoparticles, carbon nanotubes, chitosan nanoparticles, liposomes, poly-lactic-glycolic acid nanoparticles, or alginate micro-particles are described in detail below.

NANODELIVERY SYSTEMS

Alginate

Alginate is found naturally in brown algae, such as *Laminaria hyperborea*, *Laminaria digitata*, *Macrocystes pyrifera*, and *Lessonia nigrescens*. It can also be found as a polysaccharide in some bacteria such as *Azotobacter vinelandii* and *Pseudomonas*[34]. Alginate is a generic name used to define a complex molecule made of repeated units of the unbranched polyanionic polysaccharides α-l-guluronic acid (G) and β-d-mannuronic acid (M). Alginate is built by combination of G-G, G-M, and/or M-M blocks. These blocks can be found in different G/M composition and chain arrangements, which gives them its differential physico-chemical properties [35,36]. The mechanical and the physical stability of alginate mainly depend on the G content, the greater the G content, the more rigid and brittle the matrix [37]. Alginate-microparticles (alginate-MPs) are eroded at neutral and basic pH allowing the release of the cargo by diffusion, while at low pH values they are extremely stable [38]. This stability at low pH makes alginate-MPs suitable for oral administration, since in the fish stomach (pH between 2 and 4) the release will be low while the release in the foregut or hindgut at neutral-basic pH (pH 7 and 8.3, respectively) will be high [39,40]. Notably, alginate is mucoadhesive allowing the adhesion to the epithelial mucus (e.g., intestinal mucosa) and making it very attractive for oral administration. Other important features of alginate-MPs are the high biocompatibility and the low cost of production.

Alginate-MPs can be produced by classical techniques such as air atomization, emulsification, gelation, and complexation with counterion polymers, or by new methods, such as spray-drying, electrohydrodynamic atomization, impinging aerosols, and inkjet/drying process, that enable a better control of the size [37]. For application in fish, alginate-MPs are generally produced by emulsification [39,40,41,42] that is one of the fastest methods for nanoparticle preparation and is readily scalable [43], and to a lesser extent by other methodologies such as the orifice-ionic gelation and the spray method [44,45] (Table 1).

In mammals, alginate nanoparticles have been used for the delivery of different drugs [46,47,48,49], but to date there are no alginate nano-sized particles routinely used for delivery of active compounds in fish. Nevertheless, micro-sized alginate particles are one of the most common delivery systems assayed in fish with promising results for viral diseases.

Encapsulation of Bacterial Antigens in Alginate Microparticles

To date, the main bacterial antigens encapsulated into alginate-MPs have been formalin-killed bacteria (FKB) from different species (Table 1). FKB have been widely used as antigens for fish vaccination in some diseases, mainly those caused by Gram-negative bacteria. In general, FKB vaccines provide excellent levels of protection by itself or in combination with an adjuvant (e.g., FCA) [40,50]. In general, oral administration of FKB encapsulated in alginate-MPs does not work very well alone, and only when combining the alginate-MPs with the FKB vaccine they obtained a longer lasting protection. The oral administration of alginate-FKB from *Lactococcus garviaeae* in rainbow trout (*Oncorhynchus mykiss*) provided low levels of protection against *L. garviaeae* infection (35% Relative percent survival (RPS) at 30 days) compared with the naked FKB vaccine administrated intraperitoneally (100% RPS at 30 days and 40% at 90 days) (Table 1). These results were improved when fish were immunized orally a second time with the alginate-vaccine three months later (61% RPS at 180 days) [44]. Altun and coworkers observed similar result with this alginate-construct administrated orally in rainbow trout (Table 1). It did not provide better protection against *L. garviaeae* infection (53% RPS at 30 days and 38% RPS at 60 days) than the naked vaccine (95% RPS at 30 days and 82% RPS at 60 days). The protection was again increased when fish was immunized a second time with alginate-FKB-LG at day 61 (67% RPS at 90 days and 62% RPS at 120 days) or with a first administration of naked vaccine and then a second administration of the alginate-construct at day 61 (86% RPS at 90 days and 81% RPS at 120 days) [45].

Leal *et al.* [40] evaluated the alginate-MPs formulated with FKB from *Flavobacterium columnare* in nile tilapia (*Oreochromis niloticus*) (Table 1). Alginate-vaccine and naked vaccine administrated orally did not provide protection against *F. columnare*challenge (0% of RPS at 21 days in both cases) and did not stimulate the production of specific antibodies against *F. columnare* in immunized fish [40].

Encapsulation of Viral DNA in Alginate Microparticles

For viral diseases, alginate-MPs have been used to encapsulate DNA vaccines made with plasmids coding for viral proteins. The alginate-MPs loaded with DNA vaccines are smaller (≤ 10 μm) than the alginate-MPs loaded with bacterial antigens (10–30 μm) [44] and this seems to favor the targeting of different organs, such as spleen, kidney, liver, pyloric caeca, heart, intestine, or gills [41,42]. Alginate-MPs containing the plasmid coding for the major capsid protein (MCP) of Lymphocystis Disease Virus (LCDV) increased the titer of specific antibodies against LCDV in olive flounder (*Paralichthys olivaceus*) serum after oral administration (Table 1). The results showed a progressive increase until week 11 while the naked DNA vaccine did not stimulate any increase in the antibody titer. The naked DNA vaccine might thus be hydrolyzed in the gastrointestinal tract while the alginate-MPs can reach the tissues [41].

Alginate-MPs with a plasmid coding for VP2, one of the major structural proteins of Infectious Pancreatic Necrosis Virus (IPNV) stimulated the production of specific neutralizing antibodies in *O. mykiss* until eight weeks after oral administration (Table 1). In infection experiments with this virus, alginate-MPs orally administrated to *O. mykiss* and *Salmo trutta* increased the protection levels nearly to 80% RPS at 15 and 30 days post-vaccination [42]. These levels of protection were comparable with a commercial subunit vaccine (e.g., Microtek) administrated by intraperitoneal injection [51].

Table 1. Microparticles used as delivery systems in fish.

Microparticle	Size	Production Technique and Composition	Encapsulated Molecule	Administration	Species	Fish Size	**RPS**	**Reference**
Alginate	30 μm	Spray method, sodium alginate, 0.5% (*w/v*)	FKB from*Lactococcus garviaeae*	Oral	*Onchorhynchus mykiss*	22 g	35% E and 100% N at 30 DPV; 5% E and 40% N at 90 DPV; 61% first V with N and second with E at 180 DPV	[44]
	n.d.	Orifice-ionic gelation, Sodium alginate, 4% (*w/v*)	FKB From*Lactococcus garviaeae*	Oral	*Onchorhynchus mykiss*	20 g	53% E and 95% N at 30 DPV; 38% E and 82% at 60 DPV; 67% first V with N and second with E at 90 DPV; 62% first and second V with E at 120 DPV	[45]
	n.d.	Emulsification, sodium alginate, 3.5% (*w/v*)	FKB from*Flavobacterium columnare*	Oral	*Oreochromis niloticus*	15.7 g	0% E and 0% N at 21 DPC	[40]
	≤ 10 μm	Emulsification, sodium alginate, 3% (*w/v*)	Plasmid DNA: MCP from LCDV	Oral	*Paralichthys olivaceus*	40–60 g	n.d.	[41]

	10 µm	Emulsification, sodium alginate, 3% (*w/v*)	Plasmid DNA: VP2 from IPNV	Oral	*Salmo trutta*	1.5 g/3 cm	At 15 DPV: 78% E and 0% empty plasmid at 30 DPC, At 30 DPV: 79% and 0% empty plasmid at 30 DPC (*)	[42]
	10 µm	Emulsification, sodium alginate, 3% (*w/v*)	Plasmid DNA: VP2 from IPNV	Oral	*Onchorhynchus mykiss*	1 g/3.5 cm	At 15 DPV: 80% E and 5% empty plasmid at 30 DPC; At 30 DPV: 67% and 0% empty plasmid at 30 DPC (*)	[42]
Chitosan	≤ 10 µm	Emulsification, 3% chitosan (*m/v*)	Plasmid DNA: MCP from LCDV	Oral	*Paralichthys olivaceus*	50–100 g and 13–15 cm	n.d.	[52]
	< 5 µm	Spray drying, 240 mg of PVMMA and 250 mg of chitosan Seacure 210 HCl	Surface antigens (Ag) from*Philasterides dicentrarchi*	i.p. injection	*Scophthalmus maximus*	50 g	68% E, 58% Ag in FCA and 43% FCA at 20 DPC	[53]
	4.28 ± 0.4 µm	Spray drying, 240 mg of GantrezAN119 and 250 mg of chitosan Seacure 210 HCl	Surface antigens (Ag) from*Philasterides dicentrarchi*	*In vitro*, anterior kidney leukocytes	*Scophthalmus maximus*	n.d	n.d.	[54]
	1.101 ± 0.0103 µm	TPP ionic gelation, 5 mg/mL chitosan in sodium alginate solution at concentration of 10 mg/mL	FKB from*Aeromonas hydrophila*	Oral	*Labeo rohita*	Juveniles	13% alginate and chitosan E, 13% chitosan E, 16% alginate and chitosan, 0% N at 15 DPC (*)	[55]
PLGA	1.12 µm	D.E., PLGA 50:50, MW: 30–70 kDa	OMP from*Aeromona hydrophila*	Parenteral	*Labeo rohita*	30–40 g and 250–300 g	n.d.	[56]
	< 10 µm	D.E., L:G = 75:25, MW:50 kDa	Plasmid DNA: MCP from LCDV	Oral	*Paralichthys olivaceus*	500–1000 g	n.d.	[57]
	1 µm	Emulsion, PLGA 50:50	γ-globulins from human blood	Oral	*Onchorhynchus mykiss*	100–200 g	n.d.	[58]
	n.d.	D.E., PLGA 50:50	i-antigen from*Uromena marinum*	i.p. injection	*Epinephelus bruneus*	31.4 ± 2.3 g	78% E and 66% N at 30 DPC (*)	[59]
PLGA/ Liposome	5–10 µm	Film dispersion method, PS, PC, and Chol (molar ratio 1:10:5)	FKB from*Aeromonas hydrophila*	Oral	*Cyprinus carpio*	30 g	64% E at 12 DPC	[60]
	n.d.	D.E., PLGA 50:50	ODN1668	i.p. injection	*Epinephelus bruneus*	36.7 ± 2.8 g	78% PLGA E, 83% Liposome E, 83% PLGA/ Liposome E and 78% N at 30 DPC (*)	[61]

Chol: Cholesterol; D.E.: double emulsion; DPC: days post-challenge; DPV: days post-vaccination; E.: encapsulated antigen; FCA: Freund's complete adjuvant; FKB: formalin killed bacteria; GantrezAN119: methyl vinyl ether-co-maleic anhydride; i.p.: intraperitoneal; IPNV: Infectious pancreatic necrosis virus; MCP: major capsid protein; N: naked antigen; n.d.: not described; LCDV: Lymphocystis disease virus; ODN1668: oligodeoxynucleotide 1668; Omp: outer membrane protein; PC: Phosphatidylcholine; PLGA: poly(lactic-co-glycolic acid); PS: Phosphatidylserine; PVMMA: Poly (methyl vinyl ether)-co-(maleic anhydre); RPS: Relative percent survival; V: vaccination; VP2: Viral protein 2; (*): calculated RPS.

Carbon Nanotubes

Carbon nanotubes (CNTs) were discovered in 1991 by Iijima [62]. CNTs are allotropes of carbon with a cylindrical nanostructure and this network of carbon atoms can reach several micrometers in length with a nanosized diameter. CNTs can be produced at large scale by three methods: discharge, laser ablation, and chemical vapor deposition. During the production process with all these methods impurities are formed, thus requiring an additional purification step [63]. Pure CNTs are not soluble in aqueous solutions because they have highly hydrophobic surfaces and an additional functionalization step is needed. There are two main types of carbon nanotubes, single-walled, and multi-walled. Single-walled CNTs are flexible but require catalytic synthesis making its bulk production difficult and leading to poor levels of purity. Multi-walled CNTs are formed by several concentric layers and thus are more rigid. They can be produced without catalyst, which allows bulk synthesis and high purity [64]. CNTs are chemically stable, relatively inert, non-immunogenic, and non-toxic. Additionally, CNTs have a large surface area available and are able to absorb or to be conjugated to a wide variety of antigens, presenting high stability *in vivo* [65,66]. In mammals, CNTs are being investigated as a delivery system for genes, peptides, oligonucleotides, antimicrobial agents, and cytotoxic drugs [67,68,69,70]. In fish, the study of CNTs as delivery systems has recently started, focusing on its functionalization with chemical groups and proteins and on the encapsulation of DNA vaccines [71,72,73] (Table 2).

Functionalization of CNTs

As mentioned above, functionalization is required to solubilize the CNTs and to make them biocompatible. This process can be divided in two different approaches, depending on the covalent/non-covalent nature of the linked antigens [64]. The covalent attachment of different chemical groups (e.g., sulfonate) and proteins (e.g., bovine serum albumin) has been used to design nanoparticles for fish [71,72,73]. Different studies warn about the potential for these manufactured nanomaterials to contaminate the aquatic environment. To evaluate immunotoxicity, functionalized single-walled and multi-walled CTNs with chemical groups, such as sulfonate, sulfonic acid, and polyethylene glycol were tested for toxicity in head kidney macrophages isolated from *O. mykiss* (Table 2). The CNTs formulations did not decrease the cell viability after 24 h treatment [72]. None of these formulations stimulated the expression of interferon alpha (IFNα) gene, however CNTs with and without functionalization stimulated interleukin 1 beta (IL-1β) gene expression in trout macrophages indicating that they can be pro-inflammatory if they gain entry to the body. Multi-walled CNTs containing anionic groups (sulfonate groups) caused the highest IL-1β stimulation, while single-walled CNTs containing neutral groups (polyethylene glycol groups) caused the least reaction. The functionalized CNTs were also more potent in stimulating gene expression than the non-functionalized counterparts [72].

The functionalized CNTs thus produce a stimulation of the immune system by themselves without any loaded antigen [71,72], although there is no information about the levels of protection that they may provide in a challenge. Fluorescent multi-walled CNTs functionalized with bovine serum albumin (BSA) were tested in zebrafish (*Danio rerio*) embryos by microinjection into the circulation at 72 h post fertilization (Table 2). These CNTs distributed all along the blood circulation and then moved to the muscle, brain ventricle and notochord, being finally cleared out at 96 h after injection. The immune response of the embryos was studied by *in situ* hybridization of Matrix Metalloproteinase 9 (MMP9). At early stages, the injected embryos showed an increase in MMP9 expression levels and changes in the expression pattern. These results suggest that embryos may generate an innate immune response when being injected with CNTs at the 1-cell stage. The injected zebrafish embryos had normal primordial germ cells and were able to produce a new generation at the adult stage. However, the larvae of the second generation showed lower survival rates as compared with the untreated group, suggesting a negative effect on the reproduction potential [71].

Encapsulation of Viral DNA in CNTs

To date only one work has evaluated CNTs as a DNA delivery system in fish, but with promising results. Single-walled CNTs were loaded with a plasmid encoding the VP7 protein of Grass Carp Reovirus (GCRV). The plasmid expression after intramuscular injection in grass carp (*Ctenopharyngodon idella*) was detected at high levels in muscle at 28 days post-injection. At the level of the humoral response, specific VP7 antibody production was detected during eight weeks with a peak titer at four weeks post-vaccination. Other immune parameters such as respiratory burst, serum lysozyme activity, complement activity, or superoxide dismutase activity were also stimulated. Importantly, in a challenge against GCRV, the treated fish showed good protection levels even at low plasmid doses (1 μg: 73% RPS, 5 μg: 91% RPS and 10 μg: 100% RPS) when compared with the naked plasmid (1 μg: 9% RPS, 5 μg: 27% RPS and 10 μg: 44% RPS) [73] (Table 2).

Chitosan

Chitin is a natural, biodegradable, biocompatible, and nontoxic biopolymer derived from the shells of crustaceans, insects, and some microorganisms. It can be converted to chitosan, a linear polysaccharide compound of β-(1-4)-linked d-glucosamine and N-acetyl-d-glucosamine obtained from the *N*-deacetylated derivative of chitin by enzymatic or chemical processes. Chemical methods are used extensively for chitosan preparation for commercial purposes because of their low cost and scalability, but have a high energetic cost and produce a concentrated alkaline waste solution. In contrast, enzymatic methods offer the possibility of a controlled process, resulting in the production of well-defined chitosan [74]. Chitosan nanoparticles are prepared by ionic gelation [75], followed by freeze-drying (or spray-drying) to recover these particles.

The solubilization of chitosan occurs by protonation of the -NH_2 group on the C-2 position of the d-glucosamine repeat unit, whereby the polysaccharide is converted to a polyelectrolyte in acidic media. Being soluble in aqueous solutions, chitosan and its derivatives are largely used in medical and pharmaceutical applications like artificial matrices for tissue engineering, targeted drug delivery, drug transport, protein delivery or gene transfer [76,77,78,79]. They can be functionalized to display antimicrobial activity against many bacteria, filamentous fungi and yeasts [80], hemostatic potential

[81], and antioxidant activity [82]. The beneficial applications of chitosan on fish have been demonstrated in different studies in which chitosan nanoparticles were administrated through diet. Diets supplemented with chitosan for rainbow trout [83], olive flounder [84], koi [85], kelp grouper [86], turbot [87], gibel carp [88], mrigal carp [89], and Asian seabass [90] have proved that the chitosan could enhance growth, the innate immunity, disease, and stress resistance, improve haematological parameters and improve water quality. From fish immunological perspective, chitosan nanoparticles have been used for the delivery of vitamin C [91], RNA [92], or DNA [93,94,95] due to their positive charge and solubility in aqueous solution. In addition, chitosan protects encapsulated active compounds from the harsh conditions in the gastrointestinal tract and enhances their absorption [96]. Therefore, chitosans can be used for delivering immunostimulants or vaccines to fish in aquaculture.

Encapsulation of Compounds in Chitosan Nanoparticles

The DNA that encodes for the 38 kDa protein of the external membrane (OMP38) of *Vibrio anguillarum* was encapsulated in chitosan and administered with food to Asian sea bass (*Lates calcarifer*). It induced a significant antibody immune response and was able to give moderate levels of protection (RPS 46%) against experimental challenge with *V. anguillarum*[94] (Table 2). Another pDNA vaccine constructed with the outer membrane protein K of *Vibrio parahaemolyticus* was encapsulated in chitosan nanoparticles and mixed with dry fish food powder and used to feed blackhead seabream (*Acanthopagrus schlegelii*). The outer membrane protein K gene and protein were expressed in muscle, liver, kidney, and mid-intestine of the vaccinated animals. Furthermore, blackhead seabream were protected from *V. parahaemolyticus* challenge with 72.3% RPS after 21 days post-vaccination [95] (Table 2). Ramos and coworkers also clearly showed that chitosan is an excellent DNA delivery system through oral administration, either by feeding with plasmid DNA-chitosan incorporated into the food, or by direct intrabuccal delivery [93] (Table 2).

In a different setup, dietary RNA (*i.e.*, nucleotides derived from yeast) was loaded into chitosan nanoparticles at a chitosan/RNA ratio of 2:1 and were fed during 60 days to fingerlings of *Labeo rohita*. The body composition in terms of protein and lipid content was not affected by RNA-loaded chitosan nanoparticles (chitosan-NPs) while the growth, performance, immunity, and survival following a bacterial challenge (*Aeromonas hydrophila*) were significantly increased compared to only chitosan or bare RNA. Unaffected glucose and serum uric acid levels, and decreased transaminases and dehydrogenases, coupled with improved performance, indicated an enhanced energetic efficiency for anabolic processes and the safety of RNA-loaded chitosan-NPs as a nutraceutical [92] (Table 2).

Finally, chitosan-NPs are very suitable to encapsulate Vitamin C. In the gastrointestinal tract of rainbow trout (*O. mykiss*), the release of Vitamin C was regulated by the chitosan encapsulation up to 48 h. The innate immunity indices (lysozyme and complement proteins) were considerably increased in the treated rainbow trout and even the non-specific defense mechanisms were stimulated as a result of the synergistic effects caused by Vitamin C and the chitosan nanoparticle itself [91] (Table 2). Vitamin C was also administered in this way to post-metamorphic larvae of *Solea senegalensis* and rotifers (*Brachionus plicatilis*). The NPs were stable in seawater and *in vitro* assays with a zebrafish liver cell-line showed a statistically significant increase in total antioxidant capacity. In addition, the nanoparticles were able to penetrate through the intestinal epithelium in *S. senegalensis* larvae and could be used as an enriching additive for rotifers [97] (Table 2).

Encapsulation of Compounds in Chitosan Microparticles

Not only chitosan-NPs but also the larger chitosan microparticles (chitosan-MPs) are intensively studied. A plasmid containing the major capsid protein (MCP) gene of Lymphocystis Disease Virus (LCDV) was encapsulated in chitosan-MPs using an emulsion-based methodology. Oral administration led to an increase in the immune response in Japanese flounder (*Paralichythys olivaceus*) compared to injection immunization with naked plasmid DNA [52] (Table 1). The surface antigens (Ag) of the parasite *Philasterides dicentrarchi* were encapsulated and covalently linked to a polymeric microparticle formulation composed of two biodegradable polymers (chitosan and Gantrez). Poly (methyl vinyl ether)-co-(maleic anhydride) (Gantrez AN119) is a polymer belonging to the vinyl ether group that it is widely used for pharmaceutical purposes and has also been used to prepare ligand-nanoparticle conjugates for eliciting immune responses [98]. These chitosan and gantrez MPs encapsulated vaccine induced higher level of antibody than that induced by the same vaccine emulsified in FCA [53] (Table 1). These MPs could also significantly stimulate the phagocytic activity of leukocytes and the levels of the pro-inflammatory cytokine tumor necrosis factor alpha (TNFα) and also increased the production of reactive oxygen and nitrogen species in the anterior kidney of turbot (*Scophthalmus maximus*) [54] (Table 1). Finally, alginate-coated chitosan-MPs were evaluated through oral dietary administration. A potent humoral and innate immune response was elicited but it was not sufficient to induce protection against *Aeromonas hydrophila* infection under these conditions [55] (Table 1).

Liposomes

Liposomes are spherical, closed structures, composed of phospholipid bilayers, which enclose part of the surrounding solvent into their interior [99]. They are self-sealing and have the capacity to incorporate both hydrophilic and lipophilic drugs. Since the early 1980s, liposomes have been extensively studied as a drug carrier transport to target cells or tissues [100,101,102,103]. The drug delivery properties of liposomes are largely determined by factors such as the lipid composition, the particle size, the net charge and the loaded compound [104]. The liposome charge needs to be considered when administering molecules to fish, since fish gills contain a high level of mucin. In rainbow trout fry, a mechanism of acute toxicity after liposome treatment was suggested to be an interaction between the cationic liposomes and anionic components of gill mucin [105]. However, no toxicity was reported in zebrafish after immersion administration of nanoliposomes [23].

There is a wide variety of techniques that can be used to produce liposomal formulations, such as the Bangham method, detergent depletion method or extrusion [106]. All methods for producing liposomes require lipids to be combined by some means with an aqueous phase [107]. The extrusion technique is the most common method to prepare liposomes because it allows a better control of the size and the polydispersity index [108,109]. Extrusion is a process in which micrometric liposomes are structurally modified to large unilamellar vesicles or nanoliposomes depending on the pore-size of the filters used [108,110,111]. Compared to micro-liposomes, nanoliposomes provide more surface area and have the potential to increase solubility, enhance bioavailability, improve controlled release, and enable precision targeting of the encapsulated material to a greater extent [112].

Encapsulation of Bacterial Antigens in Liposomes

FKB vaccines composed of liposomes entrapping *Vibrio harveyi* were tested in *Epinephelus bruneus*. In *in vivo* infection assays, the cumulative mortality was 10%, 15%, and 65% lower in this immunized group compared to treatment with *V. harveyi* alone, liposome alone and non-immunized groups, respectively [113] (Table 2). In another study in carp (*Cyprinus carpio*), the oral administration of liposomes containing *Aeromonas salmonicida* antigen was investigated. The survival of carp after the challenge was 83% when they were immunised with *A. salmonicida* antigen-containing liposomes, whereas non-immunized carp showed 66% survival. Furthermore, the development of skin ulcers was significantly inhibited in carp immunized with liposomes containing *A. salmonicida* antigen [114].

Lipopolysaccharide (LPS) from *A. salmonicida* was also incorporated into liposomes in order to enhance the immune response in rainbow trout (*O. mykiss*). LPS incorporated into multilamellar vesicles or large unilamellar vesicles prolonged the period of serum anti-LPS antibody levels to 6–14 weeks comparing to free-LPS (2–4 weeks) when administered intraperitoneally [115] (Table 2). Also in rainbow trout, the immune efficacy of vaccine containing liposome particles with vaccine alone against furunculosis was compared [116]. Results indicated that the protection level was significantly enhanced when the vaccine also contained liposomes (Table 2). In addition, vaccinated fish appeared to be significantly larger than control fish.

Lastly, the *A. hydrophila* antigens entrapped in liposomes were developed for oral administration to immunize common carp (*C. carpio)*. The levels of antibodies in the serum rose at two and three weeks post-vaccination and the vaccination protected the fish after injection with live *A. hydrophila* at 22 days post-vaccination [60] (Table 2).

Encapsulation of Viral Antigens in Liposomes

Formalin-inactivated koi herpesvirus entrapped within liposomes was used for oral vaccination of common carp (*C. carpio*). Specific antibody titer was significantly increased and challenge experiments revealed that orally vaccinated fish were protected from infection with two different isolates of koi herpesvirus (NKC03 and IKC03) showing high RPS (75% and 65%, respectively) [117] (Table 2).

Distinct from classical vaccines, immunostimulant-loaded liposomes are also developed to protect fish against bacterial and viral infections. For example, polyinosinic-polycytidylic acid (Poly I:C) is a synthetic analog of double-stranded RNA and is a typical molecular pattern associated with viral infections. When combined with LPS, it is a strong stimulus to the innate immune system. Liposomes encapsulating both Poly I:C and LPS elicited a pro-inflammatory and anti-viral response in zebrafish hepatocytes and trout macrophages. When administrated *in vivo* they accumulated in immune tissue and specifically in macrophages. Of interest, they protected zebrafish against otherwise lethal bacterial (*Pseudomonas aeruginosa*PAO1) and viral (Spring Viraemia of Carp Virus) infections regardless of whether they were administered by injection or by immersion. No stimulation of innate immunity was observed in the treatment with empty liposomes or with the free immunostimulants [23,118] (Table 2).

Table 2. Nanoparticles used as delivery system in fish.

Nanoparticle	Size	Production Technique and Composition	Encapsulated Molecule	Administration	Species	Fish Size	RPS	Reference
Calcium phosphate	224.98 ± 14.62 nm	n.d.	S-layer protein from *Aeromonas hydrophila*	i.p. injection	*Labeo rohita*	100–150 g	97% E, 13% N and 94% E with FIA at 15 DPC (*)	[119]
Carbon nanotubes	d: 10-20 nm; l: 1–2 μm	n.d., SWCNTs and MWCNTs	Sulfonate group, polyethyleglycol and sulfonic acid	*In vitro*, head kidney monocytes	*Oncorhynchus mykiss*	0.5–1 kg	n.d.	[72]
	d: 19.9 ± 8.25 nm; l: 0.8 ± 0.5 μm	n.d., MWCNTs	BSA	Microinjection	*Danio rerio*	embryos/ larvae	n.d.	[71]

	n.d.	n.d., SWCNTs	Plasmid DNA: VP7 from grass carp reovirus	i.p. injection	*Ctenopharyngodon idella*	25–30 g	73% E (1 µg), 91% E (5 µg) 100% E (10 µg), 9% N (1 µg), 27% N (5 µg) and N (10 µg) at 15 DPC	[73]
Chitosan	n.d.	0.02% chitosan in sodium acetate buffer	Plasmid DNA: OMP38	Oral	*Lates calcarifer*	Juveniles	46% E at 14 DPC	[94]
	218.9 nm	TPP ionic gelation, 2 mg/mL chitosan in 3% (*v/v*) acetate	Plasmid DNA: OMPK	Oral	*Acanthopagrus schlegelii*	15–16 cm	72.3% E and 0% N 14 DPC	[95]
	n.d.	Complex coacervation, 0.02% (*w/v*) powdered chitosan	Plasmid DNA: βgalactosidase	Oral	*Oreochromis niloticus*	5–10 cm and 33–40 g	n.d.	[93]
	287.1 ± 1.49 nm	Complex coacervation, chitosan to RNA ratio: 1:1, 2:1, and 3:1	Bare RNA	Oral	*Labeo rohita*	2.7–3.1 g	83% E (2:1) and 33% N at 15 DPC (*)	[92]
	185.4 ± 2.1 nm	TPP ionic gelation, chitosan in 1% (*w/v*) acetic acid solution	Vitamin C	Oral	*Onchorhynchus mykiss*	Adult	n.d.	[91]
	253–258 nm	Ionotropic gelation, chitosan at concentration of 2.4 mg/mL in acetic acid solution (0.4% *v/v*)	Vitamin C	Oral	*Solea senegalensis*	Larvae	n.d.	[97]
Liposomes	n.d.	Film dispersion method. DPPC, DPPS, Chol (molar ratio 1:10:5)	FKB *Vibrio harveyi*	i.p.injection	*Epinephelus bruneus*	29.5 ± 2.1 g	75% E, 65% N and 60% liposome at 30 DPC	[113]
	n.d.	Film dispersion method. DPPC (0.5 µmol), DPPS (0.5 µmol) and Chol (1 µmol)	*Aeromonas salmonicida* total extract	Oral	*Cyprinus carpio*	350 g	54% E at 30 DPC (*)	[114]
	200 nm	Extrusion method. PC:Chol: PG or PC:Chol:SA in a 6:3:1 molar ratio	LPS from*Aeromonas salmonicida*	i.p.injection	*Onchorhynchus mykiss*	40 and 80 g	n.d.	[115]
	n.d.	Film dispersion method. 600 mg of phosphatidylcholine in 25 mL chloroform	FKB *Aeromonas salmonicida*, inactivated toxin and LPS	Immersion	*Salmo gairdneri*	Fry	70% E and 59% N at 126 DPC (*)	[116]
	n.d.	Film dispersion method. PS, PC, and Chol (molar ratio 1:10:5)	Koi herpesvirus whole extract	Oral	*Cyprinus carpio*	30 g	74% E (NKC03) and 65% E (IKC03) at 23 DPC	[117]

	125 nm	Extrusion method. DOPA, DLPC, Chol, Cholesteryl and Chol-PEG600	LPS and Poly I:C	*In vitro,* zebrafish hepatocytes and head kidney macrophages	*Danio rerio* and*Oncho-rhynchus mykiss*	Zebrafish hepatocytes, trout macrophages	n.d.	[118]
	125 nm	Extrusion method. DOPA, DLPC, Chol, Cholesteryl and Chol-PEG600	LPS and Poly I:C	Injection and immersion	*Danio rerio*	Adult	33% E, 21% N and 20% liposome at 15 DPC	[23]
	n.d.	High-pressure homogenization. 6% (*wt/v*) cinnamaldehyde, 10% (*v/v*) lecithin and 0.5% (*v/v*) α-tocopherol	Cinnamaldehyde	Immersion	*Danio rerio*	Adult	58% E at 11 DPC (*Vibrio Vulnificus*),35% E at 8 DPC (*Aeromonas hydrophila*) and 31% E at 8 DPC (*Streptococcus agalactiae*) (*)	[120]
Liposome	n.d.	Lipid film hydration, lipid:peptide ratio of 1:50	Melittin	*In vitro,* EPC cell line	*Pimephales promelas*	EPC cell line	n.d.	[121]
	n.d	Film dispersion method, DPPC (0.5 µmol), DPPS (0.5 µmol) and Chol (1 µmol), or DPPC (3.5 µmol) and Chol (1 µmol)	BSA	Oral	*Cyprinus carpio*	350 g	n.d.	[122]
PLGA	125–225 nm	D.E., PLGA: 50:50 (40–75 kDa); PLA (85–160 kDa)	OMP from*Aeromonas hydrophila*	i.p. injection	*Labaeo rohita*	50 ± 10 g	75% PLA, 55% PLGA and 38 % N at 42 DPV	[123]
	320–500 nm	D.E., n.d.	Plasmid DNA: Firefly luciferase gene	i.m. injection	*Salmo salar*	30 g	n.d.	[124]
	< 500 nm	D.E., n.d.	Plasmid DNA: MCP from LCDV	Oral	*Paralich thys olivaceus*	50–100 g	n.d.	[125]
	n.d.	D.E., 5% of PLGA/methylene chloride and 5% of PVA/water soluble	Plasmid DNA: protein-G from IHNV	Oral	*Onchorhynchus mykiss*	5 g	11% E low dose, 22% E high dose and 82% N at 180 DPC; 0% E low dose, 19% E high dose and 55% N at 300 DPC	[126]
	300–400 nm	D.E., PLGA : 50:50 (5–15 kDa; 40–75 kDa); 75:25 (66–107 kDa); PLA (24–47 kDa)	Hemocyanin from *Limulus polyphemus*	i.p. injection	*Salmo salar*	29 ± 3.1 g	n.d.	[127]
	300–400 nm	D.E., PLGA: 50:50 (5–15 kDa; 40–75 kDa); 75:25 (66–107 kDa); PLA (24–47 kDa)	β-glucan	i.p. injection	*Salmo salar*	29 ± 3.1 g	n.d.	[127]

	300–400 nm	D.E., PLGA: 50:50 (5–15 kDa; 40–75 kDa); 75:25 (66–107 kDa); PLA (24–47 kDa)	β-glucan	i.p. injection	*Salmo salar*	29 ± 3.1 g	n.d.	[127]
	< 1000 nm	D.E., n.d.	γ-globulins from human blood	i.p. injection	*Salmo salar*	30 g	n.d.	[128]
	< 1000 nm	D.E., n.d.	β-glucan	i.p. injection	*Salmo salar*	30 g	n.d.	[128]
SLN	141–335 nm	n.d.	6-Coumarin	*In vitro*, SAF-1 cell line and HK leukocytes	*Sparus aurata*	100 g	n.d.	[129]

BSA: Bovine serum albumin; Chol: Cholesterol; D.E.: double emulsion; DLPC: 1,2-didodecanoyl-sn-glycero-3-phosphocholine; DOPA: 1,2-dioleoyl-sn-glycero-3-phosphoric acid monosodium salt; DPC: days post-challenge; DPPC: Dipalmitoylphosphatidylcholine; DPPS: Dipalmitoylphosphatidylserine; E: encapsulated antigen; FIA: Freund`s incomplete adjuvant; FKB: formalin killed bacteria; i.m.: intramuscular injection; i.p.: intraperitoneal; IHNV: Infectious haematopoetic necrosis virus; LCDV: Lymphocystis disease virus; LPS: Lipopolysaccharide; MCP: major capsid protein; MWCNTs: Multi-Walled Carbon Nanotubes; N: naked antigen ; n.d.: not described; NKC03 and IKC03: two koi herpesvirus isolates; OMP: outer membrane protein; OMP38: outer membrane protein of *Vibrio anguillarum*; OMPK: Outer membrane protein K; PC: Phosphatidylcholine; PG: Phosphatidylglycerol; PLA: Polylactic acid; PLGA: poly(lactic-co-glycolic acid); Poly I:C: Polyinosinic:polycytidylic acid; PS: Phosphatidylserine; PVA: polyvinyl alcohol ; RPS: Relative percent survival; SA: Stearylamine; SWCNTs: Single-Walled Carbon Nanotubes; βgal: β-galactosidase; VP7: Viral protein 7; (*): calculated RPS.

Encapsulation of Other Compounds in Liposomes

Cinnamaldehyde, a natural compound extracted from cinnamon, was encapsulated in liposomes. These liposomes displayed antimicrobial activity *in vitro* against aquatic pathogens such as *Streptococcus agalactiae*, *Aeromonas hydrophila*, and*Vibrio vulnificus*, as well as the antibiotic resistants *Vibrio parahaemolyticus* and *Vibrio alginolyticus*. The *in vivo* results using an immersion treatment demonstrated an increased survival rate and bacterial growth inhibition in zebrafish infected with *S. agalactiae*, *A. hydrophila* and *V. vulnificus* [120] (Table 2).

Also melittin, an antimicrobial peptide, was loaded into liposomes with covalently attached antibodies directed against Viral Haemorrhagic Septicemia Rhabdovirus (VHSV) glycoprotein G (Table 2). These melittin-immunoliposomes were capable of inhibiting the VHSV infectivity by 95% via direct inactivation of the virus. To our knowledge, this is the first report on fish pathogen targeted liposomes. However, the characterization of this formulation was not described nor the size or the charge of this formulation [121].

Finally, humoral immune responses were analyzed in a study of oral administration of liposome-entrapped BSA in carp (*C. carpio*). The BSA-containing liposomes were stable in carp bile and induced significant antibody responses against BSA in serum as well as in intestinal mucus and bile. BSA-specific antibody secreting lymphocytes were detected in the spleen and head kidney of immunized fish. In contrast, no serum antibody responses were observed when fish were orally immunized with BSA-containing unstable liposomes or BSA alone [122] (Table 2).

Poly (Lactic-co-Glycolic Acid) (PLGA)

PLGA, Poly (Lactic-co-Glycolic Acid) is a biodegradable polymer and probably the most extensively investigated carrier for drug delivery in mammals [130,131]. PLGA is a copolymer synthesized by two different monomers, lactic acid and glycolic acid. The forms of PLGA depend on the monomer ratio used during the polymerization process. PLGA nanoparticles (PLGA-NPs) are degraded by hydrolysis and the degradation time depends on the monomer ratio and on the molecular weight of the polymers [132]. The PLGA-NPs were approved for human use by the Food and Drug Administration (FDA, USA) and the European Medicines Agency (EMA), because they are highly biodegradable and biocompatible [133]. When the polymer is hydrolyzed the glycolic and lactic acid monomers are released and are eventually removed from the body through the citric acid cycle [134]. For this reason, there is minimal toxicity associated with the use of PLGA as a nanodelivery system since these components are present in different metabolic pathways. The most common PLGA-NPs preparation method used in fish is the double emulsion method [123,124,125,126,127,128] which is based on the dissolution of an appropriate amount of polymer (PLGA) in an organic solvent (oil phase) such as dichloromethane, chloroform or ethylacetate [135]. Hydrophobic drugs can be added directly to the oil phase, whereas hydrophilic drugs must be first emulsified with the polymer solution prior to the formation of particles [136]. Then, the solution is emulsified by the addition of an aqueous solution containing a surfactant or an emulsifying agent (e.g., polyvinyl alcohol). By reducing the pressure or by continuous stirring the organic solvent evaporates and this results in the formation of solid nanoparticles. The encapsulation efficiency and the particle size can be controlled by the solvent choice and the stirring rate [135]. The internalization of PLGA-NPs by cells involves different uptake mechanisms. In mammals, the PLGA-NPs are able to avoid the endo-lysosomal system and are maintained in the cytoplasm. The uptake mechanisms of PLGA-NPs in teleosts is poorly understood and few published work have addressed this issue [118]. Despite this, it is known that nanoparticles of less than 500 nm in size are able to enter the bloodstream and subsequently they are cleared by phagocytes in the head kidney, spleen, and/or liver [137]. PLGA-NPs are highly versatile loading bioactive compounds. For applications in fish, γ-globulins from human blood, β-glucan [127,128], DNA vaccines [124,125,126] or the bacterial outer membrane complex [123] have been encapsulated (Table 2).

Encapsulation of Bacterial Antigens in PLGA-NPs and -MPs

PLGA-NPs were compared with polylactic acid-NPs (PLA-NPs), which also have good mechanical strength, to encapsulate the outer membrane complex from *A. hydrophyla* [138,139]. This complex consists mainly of lipopolysaccharide, phospholipids, and a group of outer membrane proteins. The encapsulation efficiency was higher in PLGA-NPs compared with PLA-NPs (59% and 44%, respectively) but the release *in vitro* was slower from PLA-NPs than from PLGA-NPs (50% at 24 h and 4 h, respectively). This might be explained by the higher hydrophilic nature of the PLGA- compared to PLA-NPs. The non-specific and specific immunity were stimulated in *L. rohita* by both PLGA- and PLA-NPs and this at higher levels than the naked antigen (Table 2). Finally, in a challenge against *A. hydrophyla*, the PLA-vaccine provided higher levels of protection compared with the PLGA-vaccine (75% RPS and 55% RPS, respectively) and with the naked antigen (38% RPS) [123]. In addition, the encapsulation of the same antigen in PLGA-microparticles (PLGA-MPs) was evaluated [56]. These PLGA-MPs were studied in *L. rohita* administrated parenterally (Table 2). Encapsulation efficiency of PLGA-MPs was lower compared to PLGA-NPs (25% and 50%, respectively). Both the microparticles and the nanoparticles significantly stimulated non-specific (myeloperoxidase, respiratory burst activity, haemagglutination, *etc.*) and specific immune response parameters at similar levels at 21 and 42 days after vaccination. Finally, in a challenge study PLGA-NPs provided protection (55% RPS) against *A. hydrophila* infection, while no data was reported for PLGA-MPs [123].

Oligodeoxynucleotides (ODNs) are short single-stranded synthetic DNA molecules that contain unmethylated CpG motifs. These motifs are highly abundant in bacterial DNA and extremely rare in vertebrates, and they are classified as a type of *Pathogen Associated Molecular Pattern* (PAMP). In mammals, they are recognized by Toll-like receptor 9 leading to strong immunostimulatory effects and also fish are able respond to CpG binding to TLR9 [140]. This antigen was encapsulated in PLGA/liposome-MPs and used to stimulate the immune system of *E. bruneus* using intraperitoneal injection [61] (Table 2). Superoxide dismutase, respiratory burst, and complement activity were mainly stimulated by the PLGA/Liposome microparticles. In contrast, the adaptive immune response and the specific *V. alginolyticus* serum antibody levels were significantly higher with the PLGA-MPs. Finally, the treatment with the PLGA-, Liposome- and PLGA/Liposome-MPs encapsulating ODNs provided good protection levels (78%, 83%, and 83% of RPS, respectively) against a *V. alginolyticus* infection [61] but they are not significantly different from the naked ODN (78% RPS).

The alginate-MPs mentioned in section 2.1.1 were compared with PLGA-MPs as vehicles for the delivery of FKB from *L. garvieae* (Table 2). The PLGA-NPs provided similar protection levels than alginate-FKB-LG (about 63% RPS at 30 days), but lower when compared with the conventional FKB vaccine intraperitoneally injected (95% RPS) [45].

Hølvold *et al.* [124] encapsulated in PLGA-NPs a plasmid containing the firefly luciferase gene under the control of the CMV-IEP promoter. Despite the fact that this formulation does not contain a specific bacterial antigen, the plasmid itself is from bacterial origin and acts as an immunostimulant (bacterial CpG). The PLGA-NPs showed a fast release of the plasmid (80% after 1 h), induced a significant increase in IL-1β and IFNα gene expression in muscle at the injection site in comparison with naked plasmid and stimulated TNFα expression in head kidney. The PLGA-NPs labeled with [^{125}I]-fluorescein were detected until day 70 in trunk kidney, muscle and organ package (liver, heart, gastrointestinal tract and interstitial adipose tissue) [124]. The performance of these NPs was here also compared to MPs. The PLGA-MPs showed a lower release of antigen than the PLGA-NPs (49% at 1 h and an accumulative release of 69% at day 70; 81% at 1 h and an accumulative release of 96% at day 70, respectively). Additionally, PLGA-MPs had a higher retention than PLGA-NPs at the injection site, contributing to the onset of severe histopathological inflammation. This suggests that nanoparticles are more suited to avoid potential tissue damage. Both PLGA-MPs and PLGA-NPs showed better performance than naked plasmid DNA for the induction of pro-inflammatory and antiviral immune responses [124].

Encapsulation of Viral Antigens in PLGA-NPs and -MPs

PLGA-NPs have been mainly used to encapsulate DNA vaccines aiming to protect against viral diseases. Lymphocystis Disease Virus (LCDV) infection is not lethal, but infected fish are more susceptible to secondary microbial infection [141]. The progression of the disease correlates with an increase in the presence of nodules. A plasmid coding for the major capsid protein (MCP) of LCDV was encapsulated in PLGA-NPs and PLGA-MPs and tested in *P. olivaceus* [125]. The encapsulation efficiency was 64% and full release (100%) was achieved after 60 h at pH 2.0 and after 90 h at pH 9.0. MCP gene expression was detected in gills, intestine, spleen, and kidney from 10 to 90 days after oral administration. Specific serum antibody titers against LCDV reached a maximum at 30 days post-administration. Importantly, in a challenge against LCDV, the presence of nodules was significantly lower in PLGA-vaccinated fish compared to naked DNA vaccinated fish (17% *versus*100%, respectively) [125]. Of note, Tian and coworkers showed that the encapsulation efficiency in PLGA microparticles was more stable than in the nanoparticle system (78%–88% and 64%–96%, respectively) [57,125]. PLGA-NPs and PLGA-MPs displayed similar performance except that the nanoparticles showed higher release characteristics. The study concluded that PLGA-MPs were also effective oral carriers for the transfer of plasmid DNA [57].

Other PLGA-NPs containing a DNA vaccine against Infectious Haematopoetic Necrosis Virus (IHNV) were used to vaccinate *O. mykiss* (Table 2). In this case, the release of plasmid DNA was not clearly pH dependent, nor were there significant differences between the number of fish expressing the plasmid gene compared to the naked plasmid treatment. This PLGA-vaccine was also not able to confer protection against IHNV [126].

Encapsulation of Other Antigens into PLGA-NPs and -MPs

As mentioned above, PLGA-NPs allow maximal versatility in encapsulating molecules of different nature. Other representative examples of this are immunostimulants such as γ-globulins from human blood, β-glucan from *L. hyperborea*and hemocyanin from *Limulus polyphemus* (Table 2). Three different loaded PLGA-NPs (β-glucan, hemocyanin and both combined) were administrated by intraperitoneal injection in *S. salar* (Table 2). The gene expression profile showed that even PLGA-NPs alone induced a mild inflammatory response in *S. salar* having potential as an adjuvant in salmon vaccine [127]. In a subsequent study, the same group assessed different formulations of PLGA-NPs and -MPs at different monomer ratios. The release of the antigen was similar (around 10%) for all formulations, however the nanoparticles co-encapsulating γ-globulins and β-glucan induced the highest specific antibody response [128]. In another study, PLGA-MPs loaded with γ-globulins were also investigated in *S. salar* by oral administration. The encapsulation in PLGA-MPs allowed its stability in the stomach for longer periods of time, slowing down the passage into the intestine and increasing the levels of intact antigen reaching the blood stream. Also, the PLGA-MPs stimulated the antibody titer in serum but not in cutaneous mucus, gut mucus, or in bile [58].

Parasitic protozoa have developed sophisticated evasion mechanisms to evade the host's innate immune defenses and currently, there are no anti-parasitic vaccines commercially available for farmed fish. A unique study aimed to design a specific delivery system for parasite disease prevention [59]. Formalin-killed parasite (i-antigen) from *Uronema marimun,* an opportunistic pathogen infecting flounder (*P. olivaceus*) and grouper (*E. bruneus*), was encapsulated into PLGA-MPs. The PLGA-MPs were administrated to *E. bruneus* by intraperitoneal injection and different innate immune response parameters such as respiratory burst activity, serum lysozyme activity, or complement activity were evaluated. All of them were stimulated by PLGA-MPs and were sustained from one to four weeks, whereas the treatment with the free i-antigen was detected only at week four and at lower levels. The specific i-antigen antibody levels were stimulated both by the free i-antigen and the PLGA-i-antigen, but again at higher levels by the PLGA-MPs [59]. Interestingly, the levels of protection from the loaded PLGA-MPs against *U. marinum* infection were notably high (only 20% of cumulative mortality after 30 days), but only slightly different from the empty PLGA-MPs or the free i-antigen [59].

Other Nanodelivery Systems

Although the amount of research done in mammals and fish is not comparable, there exists a large effort to discover new nanodelivery systems in teleost to cover the different needs for the prevention of diseases in aquaculture. Here we mention two additional approaches aiming to develop new nanomaterials for *in vivo* delivery in fish: calcium phosphate nanoparticles (CaP-NPs) [119] and solid lipid nanoparticles (SL-NPs) [129].

Calcium phosphate is a natural, inorganic, and biocompatible material. CaP-NPs are synthesized using different methods such as mechanochemical synthesis, combustion preparation, wet chemistry techniques, and others [142]. CaP-NPs can be produced in different morphologies, such as spheres, plate-like crystals, needles, or blades [142,143], however, the size and the stability of CaP-NPs are very difficult to control [143]. CaP-NPs are a potential nanodelivery system due to their high bioactivity, biocompatibility, biodegradability and strong adsorption ability under physiological conditions. In mammals, they have been used as nanodelivery system for drugs, vectors, antibacterial agents, or as a vaccine adjuvant [142]. In *L. rohita* CaP-NPs loaded with the S-layer from *A. hydrophila* was assessed by intraperitonal injection (Table 2). The non-specific immune responses (superoxide dismutase, myeloperoxidase, respiratory burst, *etc.*) and the specific immune response (antibody titers) were stimulated and detected at 21, 42, and 63 days post-vaccination. When fish were challenged with *A. hydrophila*, loaded CaP-NPs were able to provide good levels of protection 15 days post-vaccination (97% RPS) with a significant difference in comparison with non-encapsulated S-layer (13% RPS), but with similar level (94% of RPS) when compared with CaP-NPs with only Freund`s incomplete adjuvant [119].

Solid lipid nanoparticles are produced in solution using solid lipidic materials with surfactants that confer stability and co-surfactants that confer specific ligand properties [144]. SL-NPs can be prepared by different techniques, such as high-pressure homogenization, high-shear mixing, ultrasound, or solvent emulsification/evaporation methods [143,145]. Additionally, SL-NPs present a range of characteristic advantages, such as biocompatibility, non-toxicity, high bioavailability, high-antigen loading ability, controlled release, physical stability, and protection of encapsulated antigens. Finally, SL-NPs can be easily scaled-up for industrial purposes [145,146]. A preliminary *in vitro* characterization of SL-NPs was performed in fish [129] (Table 2). The loaded SL-NPs had a mean diameter of 235–335 nm depending on the amount of cargo, with a net surface charge between −12.5 and −16.5 mV. These nanoparticles were loaded with a fluorescent molecule (6-Coumarin) and were tested for uptake and toxicity in a cell line (SAF-1) and in leukocyte primary cell culture of *S. aurata*head kidney. Release of 6-Coumarin from SL-NPs was around 1% over the course of 48 h at 22 °C and both the cell line and the primary leukocytes were able to internalize these SL-NPs without affecting the cell viability. SL-NPs internalization was dose- and time-dependent. The uptake in SAF-1 cells decreased over time indicating that the SL-NPs in SAF-1 cells are likely processed in the endolysosomal compartment, while the fluorescent signal was stable over the time in primary leukocytes [129].

DISCUSSION

Nowadays there is a large variety of materials that can be used as delivery systems for vaccine/immunostimulant administration in fish. This diversity provides a wide range of options to respond to the high number of different farmed species and the challenge to achieve a good health status in the presence of different potentially harmful microorganisms. However, the study of nanoparticles for aquacultural use is still in its early stage. Research shows variable efficiency of protection depending on the nature of the nanomaterials, the method to produce the nanoparticles, the antigens encapsulated or the fish species assessed. Often there is a lack of information about the manufacturing process as well as the physico-chemical characteristics of the nanoparticles and the properties of the antigens after the encapsulation process, hampering a correct analysis and comparison between delivery systems. For instance, not all publications mention the size of the particles, the efficiency of encapsulation or the release of the antigens in *in vitro* conditions (see Table 1 and Table 2). In many cases, the starting point is the "recycling" of particles for mammalian use that are applied directly in fish, without considering the evident differences between mammals and fish. The researcher working on fish health should make a strong effort to design or adapt nanoparticles in order to reach optimal compatibility with the fish characteristics. However, this is difficult because the fish immune system has several differences with the mammalian immune system regarding cell types, cell biology, tissues involved in immune response, *etc.* Also, the fish immune system is not well known in many aspects such as how the adaptive system memory works, which cell types are involved or the role of mucosal immunity.

In general, the use of microparticles is more frequent than the use of nanoparticles even when the surface area/volume ratio is much more advantageous in nanoparticles. An explanation of this is that some materials are not easy to manipulate in the nanosize range or there are not protocols to nanosize such materials. An example is alginate that has been mainly used to produce microparticles and only recently has it started to be used to produce sub-200 nm particles [147]. Again, the characterization of the manufactured nanoparticles should be detailed in the publications in order to compare between the administration routes, the adjuvant properties, the potential degradation of the loaded compound, the efficacy of the system protecting against infection, and the targeted cells or tissues. Different fish species have different responses to vaccination [148] and this fact should be considered when choosing an encapsulation system because they may not be transferable from one species to another. Additionally, the encapsulated antigens modify the physico-chemical characteristics of the nanodelivery system so that the results of the assays on stability, size, surface charge, and organ biodistribution cannot be extrapolated from one molecule to another using the same encapsulating particle. Similarly, the characteristics of the antigen can be changed when it is encapsulated, and thus the functional structure, stability, and immunogenicity of the antigen need to be verified. For example, the size and the surface charge are extremely important for interaction with cells and should be characterized in the loaded system because they can change easily [118]. Of note, in some cases the encapsulation did not provide any protection [40] or did not improve the protection with respect to conventional immunoprotective therapies [55]. Overall, the administration of nanoparticles by intraperitoneal injection achieves good protection levels against infections while the oral administration is at this moment less efficient. One of the exceptions is the system developed with alginate or chitosan to encapsulate DNA vaccines [41,42]. DNA vaccines are still under development and only one commercial vaccine has been licensed in Canada. They are the most promising tools to fight viral infections and thus, the development of novel encapsulation systems to improve their administration and the efficiency is very important. Several new nanomaterials such as carbon nanotubes or solid lipid NP are still in the early steps of development but have shown promising results. For example, CNTs have been very effective for encapsulating a DNA vaccine and to confer protection against infection even at low DNA concentrations [73]. It is important to mention that in some studies, the adjuvant effect of the nanodelivery system is almost as potent as the loaded antigen itself. The adjuvant effect of the system itself has been extensively reported in mammals (e.g., liposomes) and it is also clearly observed in some fish species [124] but not in others [23].

Also an important point that should be taken into consideration is the final cost of the encapsulation system for industrial production. Some of the systems developed under research conditions are expensive and may not be affordable for the fish farmers. Finally, all of the nanodelivery systems included in this review have been characterized as non-toxic for cells (*in vitro* viability assays) and similarly *in vivo*. However, the toxicity of nanomaterials and, more importantly, the toxicity of the nanomaterial degradation products that could be detected in water should be addressed carefully [149,150].

CONCLUSIONS

Altogether, nano-encapsulation is a very promising strategy with a potential to substantially improve the development of effective vaccines for farmed fish. The research on the delivery of viral vaccines using nanoparticles will be the more important milestone in fish vaccinology. In this context, more traditional biomaterials such as alginate and chitosan have shown good results but new materials such as CNTs or solid lipid NP could improve the delivery of DNA vaccines. More research is still needed to specifically design encapsulation systems adapted to the fish immune system and to decipher the basis of the fish immune system.

ACKNOWLEDGMENTS

Jie Ji thanks the Chinese Science Council (CSC, China) for the PhD fellowship, Debora Torrealba thanks CONICYT-Chile for a PhD fellowship and Nerea Roher is supported by the Ramón y Cajal program (RYC-2010-06210) and by AGL2012-33877 (MINECO, Spain). We would like to thank Alexandros Nikolaou for the English editing of the manuscript.

AUTHOR CONTRIBUTIONS

Jie Ji mainly wrote the chitosan and liposome sections and the Table 1, Debora Torrealba mainly wrote the alginate, PLGA and CNTs sections and Table 2, Àngels Ruyra and Nerea Roher wrote the introduction and discussion and Nerea Roher revised and corrected the whole manuscript.

REFERENCES

1. Duff, D. The oral immunization of trout against *Bacterium salmonicida*. *J. Immunol.* 1942, *44*, 87–94.

2. Hastein, T.; Gudding, R.; Evensen, O. Bacterial vaccines for fish—An update of the current situation worldwide. *Dev. Biol.* 2005, *121*, 55–74.

3. Sommerset, I.; Krossoy, B.B.E.; Frost, P. Vaccines for fish in aquaculture. *Expert Rev. Vaccines* 2005, *4*, 89–101.

4. Mitchell, H. Choosing a furunculosis vaccine: Points to consider. *Bull. Aquac. Assoc. Can.* 1995, *95*, 30–37.

5. Kibenge, F.S.B.; Godoy, M.G.; Fast, M.; Workenhe, S.; Kibenge, M.J.T. Countermeasures against viral diseases of farmed fish. *Antivir. Res.* 2012, *95*, 257–281.

6. Koppang, E.O.; Haugarvoll, E.; Hordvik, I.; Aune, L.; Poppe, T.T. Vaccine-associated granulomatous inflammation and melanin accumulation in Atlantic salmon, *Salmo salar* L., white muscle. *J. Fish Dis.* 2005, *28*, 13–22.

7. Midtlyng, P.J. A field study on intraperitoneal vaccination of Atlantic salmon (*Salmo salar* L) against furunculosis. *Fish Shellfish Immunol.* 1996, *6*, 553–565.

8. Anderson, D.P. Immunostimulants, adjuvants, and vaccine carriers in fish: Applications to aquaculture. *Ann. Rev. Fish Dis.* 1992, *2*, 281–307.

9. Lund, R.A.; Midtlyng, P.J.; Hansen, L.P. Post-vaccination intra-abdominal adhesions as a marker to identify Atlantic salmon, *Salmo salar* L., escaped from commercial fish farms. *Aquaculture* 1997, *154*, 27–37.

10. Sorum, U.; Damsgard, B. Effects of anaesthetisation and vaccination on feed intake and growth in Atlantic salmon (*Salmo salar* L.). *Aquaculture* 2004, *232*, 333–341.

11. Berg, A.; Rødseth, O.M.; Tangerås, A.; Hansen, T. Time of vaccination influences development of adhesions, growth and spinal deformities in Atlantic salmon *Salmo salar*. *Dis. Aquat. Organ* 2006, *69*, 239–248.

12. Horne, M.T. Technical aspects of the administration of vaccines. *Dev. Boil. Stand.* 1996, *90*, 79–89.

13. Nakanishi, T.; Ototake, M. Antigen uptake and immune responses after immersion vaccination. *Dev. Boil. Stand.* 1996,*90*, 59–68.

14. Kai, Y.H.; Chi, S.C. Efficacies of inactivated vaccines against betanodavirus in grouper larvae (*Epinephelus coioides*) by bath immunization. *Vaccine* 2008, *26*, 1450–1457.

15. Quentel, C.; Vigneulle, M. Antigen uptake and immune responses after oral vaccination. *Dev. Boil. Stand.* 1996, *90*, 69–78.

16. Plant, K.P.; LaPatra, S.E. Advances in fish vaccine delivery. *Dev. Comp. Immunol.* 2011, *35*, 1256–1262.

17. Shi, J.; Votruba, A.R.; Farokhzad, O.C.; Langer, R. Nanotechnology in drug delivery and tissue engineering: From discovery to applications. *Nano Lett.* 2010, *10*, 3223–3230.

18. Yatvin, M.B.; Kreutz, W.; Horwitz, B.A.; Shinitzky, M. pH-sensitive liposomes: possible clinical implications. *Science*1980, *210*, 1253–1255.

19. Toyokawa, H.; Nakao, A.; Bailey, R.J.; Nalesnik, M.A.; Kaizu, T.; Lemoine, J.L.; Ikeda, A.; Tomiyama, K.; Papworth, G.D.; Huang, L. Relative contribution of direct and indirect allorecognition in developing tolerance after liver transplantation. *Liver Transplantat.* 2008, *14*, 346–357.

20. Li, S.D.; Huang, L. Pharmacokinetics and biodistribution of nanoparticles. *Mol. Pharm.* 2008, *5*, 496–504.

21. Hillaireau, H.; Couvreur, P. Nanocarriers' entry into the cell: Relevance to drug delivery. *Cell. Mol. Life Sci.* 2009, *66*, 2873–2896.

22. Canton, I.; Battaglia, G. Endocytosis at the nanoscale. *Chem. Soc. Rev.* 2012, *41*, 2718–2739.

23. Ruyra, A.; Cano-Sarabia, M.; Garcia-Valtanen, P.; Yero, D.; Gibert, I.; Mackenzie, S.A.; Estepa, A.; Maspoch, D.; Roher, N. Targeting and stimulation of the zebrafish (*Danio rerio*) innate immune system with LPS/dsRNA-loaded nanoliposomes. *Vaccine* 2014, *32*, 3955–3962.

24. Handy, R.D.; Von der Kammer, F.; Lead, J.R.; Hassellov, M.; Owen, R.; Crane, M. The ecotoxicology and chemistry of manufactured nanoparticles. *Ecotoxicology* 2008, *17*, 287–314.

25. Doshi, N.; Mitragotri, S. Designer biomaterials for nanomedicine. *Adv. Funct. Mater.* 2009, *19*, 3843–3854.

26. Gratton, S.E.; Ropp, P.A.; Pohlhaus, P.D.; Luft, J.C.; Madden, V.J.; Napier, M.E.; DeSimone, J.M. The effect of particle design on cellular internalization pathways. *Proc. Natl. Acad. Sci.* 2008, *105*, 11613–11618.

27. Champion, J.A.; Mitragotri, S. Role of target geometry in phagocytosis. *Proc. Natl. Acad. Sci.* 2006, *103*, 4930–4934.

28. Devine, D.V.; Wong, K.; Serrano, K.; Chonn, A.; Cullis, P.R. Liposome-complement interactions in rat serum—Implications for liposome survival studies. *BBA Biomembr.* 1994, *1191*, 43–51.

29. Verma, A.; Stellacci, F. Effect of surface properties on nanoparticle-cell interactions. *Small* 2010, *6*, 12–21.

30. Cho, E.C.; Xie, J.W.; Wurm, P.A.; Xia, Y.N. Understanding the role of surface charges in cellular adsorption *versus*internalization by selectively removing gold nanoparticles on the cell surface with a I-2/KI etchant. *Nano Lett.* 2009, *9*, 1080–1084.

31. Miller, C.R.; Bondurant, B.; McLean, S.D.; McGovern, K.A.; O'Brien, D.F. Liposome-cell interactions *in vitro*: Effect of liposome surface charge on the binding and endocytosis of conventional and sterically stabilized liposomes.*Biochemistry* 1998, *37*, 12875–12883.

32. Mislick, K.A.; Baldeschwieler, J.D. Evidence for the role of proteoglycans in cation-mediated gene transfer. *Proc. Natl. Acad. Sci.* 1996, *93*, 12349–12354.

33. Mohanraj, V.J.; Chen, Y. Nanoparticles—A review. *Trop. J. Pharm. Res.* 2007, *5*, 561–573.

34. Gombotz, W.R.; Wee, S.F. Protein release from alginate matrices. *Adv. Drug Deliv. Rev.* 2012, *31*, 267–285.

35. Østgaard, K.; Knutsen, S.H.; Dyrset, N.; Aasen, I.M. Production and characterization of guluronate lyase from*Klebsiella pneumoniae* for applications in seaweed biotechnology. *Enzyme Microbial. Technol.* 1993, *15*, 756–763.

36. De Vos, P.; Lazarjani, H.A.; Poncelet, D.; Faas, M.M.

Advanced drug delivery reviews. *Adv. Drug Deliv. Rev.* 2014, *67–68*, 15–34.

37. Sosnik, A. Alginate particles as platform for drug delivery by the oral route: State-of-the-Art. *ISRN Pharm.* 2014, *2014*, 1–17.
38. Tanaka, H.; Matsumura, M. Diffusion characteristics of substrates in Ca-alginate gel beads. *Biotechnol. Bioeng.* 1984, *26*, 53–58.
39. Rodrigues, A.P.; Hirsch, D.; Figueiredo, H.C.P.; Logato, P.V.R.; Moraes, Â.M. Production and characterisation of alginate microparticles incorporating Aeromonas hydrophila designed for fish oral vaccination. *Proc. Biochem.* 2006,*41*, 638–643.
40. Leal, C.A.G.; Carvalho-Castro, G.A.; Sacchetin, P.S.C.; Lopes, C.O.; Moraes, Â.M.; Figueiredo, H.C.P. Oral and parenteral vaccines against *Flavobacterium columnare*: evaluation of humoral immune response by ELISA and *in vivo*efficiency in Nile tilapia (*Oreochromis niloticus*). *Aquacult. Int.* 2010, *18*, 657–666.
41. Tian, J.; Sun, X.; Chen, X. Formation and oral administration of alginate microspheres loaded with pDNA coding for lymphocystis disease virus (LCDV) to Japanese flounder. *Fish Shellfish Immunol.* 2008, *24*, 592–599.
42. Ana, I.; Saint-Jean, S.R.; Pérez-Prieto, S.I. Immunogenic and protective effects of an oral DNA vaccine against infectious pancreatic necrosis virus in fish. *Fish Shellfish Immunol.* 2010, *28*, 562–570.
43. Pinto Reis, C.; Neufeld, R.J.; Ribeiro, A.J.; Veiga, F. Nanoencapsulation I. Methods for preparation of drug-loaded polymeric nanoparticles. *Nanomedicine* 2006, *2*, 8–21.
44. Romalde, J.L.; Luzardo-Alvárez, A.; Ravelo, C.; Toranzo, A.E.; Blanco-Méndez, J. Oral immunization using alginate microparticles as a useful strategy for booster vaccination against fish lactoccocosis. *Aquaculture* 2004, *236*, 119–129.
45. Altun, S.; Kubilay, S.; Ekici, S.; Didinen, B.; Diler, O. Oral vaccination against lactococcosis in rainbow trout (*Oncorhynchus mykiss*) using sodium alginate and poly (lactide-co-glycolide) carrier. *Kafkas Univ. Vet. Fak. Derg.* 2010,*16*, 211–217.
46. Sarmento, B.; Ferreira, D.; Veiga, F.; Ribeiro, A. Characterization of insulin-loaded alginate nanoparticles produced by ionotropic pre-gelation through DSC and FTIR studies. *Carbohydr. Polym.* 2006, *66*, 1–7.
47. Sarmento, B.; Ribeiro, A.J.; Veiga, F.; Ferreira, D.C.; Neufeld, R.J. Insulin-loaded nanoparticles are prepared by alginate ionotropic pre-gelation followed by chitosan polyelectrolyte complexation. *J. Nanosci. Nanotech.* 2007, *7*, 2833–2841.
48. Ahmad, Z.; Khuller, G.K. Alginate-based sustained release drug delivery systems for tuberculosis. *Expert Opin. Drug Deliv.* 2008, *5*, 1323–1334.
49. Fonte, P.; Araújo, F.; Silva, C.; Pereira, C.; Reis, S.; Santos, H.A.; Sarmento, B. Fonte. Polymer-based nanoparticles for oral insulin delivery: Revisited approaches. *Biotechnol. Adv.* 2015, *33*, 1342–1354.
50. Grabowski, L.D.; LaPatra, S.E.; Cain, K.D. Systemic and mucosal antibody response in tilapia, *Oreochromis niloticus*(L.), following immunization with *Flavobacterium columnare. J. Fish Dis.* 2004, *27*, 573–581.
51. Dhar, A.K.; Manna, S.K.; Thomas Allnutt, F.C. Viral vaccines for farmed finfish. *Virus Dis.* 2014, *25*, 1–17.
52. Tian, J.; Yu, J.; Sun, X. Chitosan microspheres as candidate plasmid vaccine carrier for oral immunisation of Japanese flounder (*Paralichthys olivaceus*). *Vet. Immunol. Immunopathol.* 2008, *126*, 220–229.
53. León-Rodríguez, L.; Luzardo-Álvarez, A.; Blanco-Méndez, J.; Lamas, J.; Leiro, J. A vaccine based on biodegradable microspheres induces protective immunity against scuticociliatosis without producing side effects in turbot. *Fish Shellfish Immunol.* 2012, *33*, 21–27.
54. León-Rodríguez, L.; Luzardo-Álvarez, A.; Blanco-Méndez, J.; Lamas, J.; Leiro, J. Biodegradable microparticles covalently linked to surface antigens of the scuticociliate parasite *P. dicentrarchi* promote innate immune responses *in vitro. Fish Shellfish Immunol.* 2013, *34*, 236–243.
55. Behera, T.; Swain, P. Antigen encapsulated alginate-coated chitosan microspheres stimulate both innate and adaptive immune responses in fish through oral immunization. *Aquacult. Int.* 2014, *22*, 673–688.
56. Behera, T.; Nanda, P.K.; Mohanty, C.; Mohapatra, D.; Swain, P.; Das, B.K.; Routray, P.; Mishra, B.K.; Sahoo, S.K. Parenteral immunization of fish, *Labeo rohita* with Poly d,l-lactide-co-glycolic acid (PLGA) encapsulated antigen microparticles promotes innate and adaptive immune responses. *Fish Shellfish Immunol.* 2010, *28*, 320–325.
57. Tian, J.; Sun, X.; Chen, X.; Yu, J.; Qu, L.; Wang, L. The formulation and immunisation of oral poly (dl-lactide-co-glycolide) microcapsules containing a plasmid vaccine against lymphocystis disease virus in Japanese flounder (*Paralichthys olivaceus*). *Int. Immunopharmacol.* 2008, *8*, 900–908.
58. Lavelle,E.C.;Jenkins,P.G.;Harris,J.E.Oral immunization of rainbow trout with antigen microencapsulated in poly (dl-lactide-co-glycolide) microparticles. *Vaccine* 1997, *15*, 1070–1078.
59. Harikrishnan, R.; Balasundaram, C.; Heo, M.S. Poly d,l-lactide-co-glycolic acid (PLGA)-encapsulated vaccine on immune system in *Epinephelus bruneus* against *Uronema marinum. Exp. Parasitol.* 2012, *131*, 325–332.
60. Yasumoto, S.; Yoshimura, T.; Miyazaki, T. Oral immunization of common carp with a liposome vaccine containing*Aeromonas hydrophila* antigen. *Fish Pathol.* 2006, *41*, 45–49.
61. Harikrishnan, R.; Balasundaram, C.; Heo, M.-S. Poly d,l-lactide-co-glycolic acid-liposome encapsulated ODN on innate immunity in *Epinephelus bruneus* against *Vibrio alginolyticus. Vet. Immunol. Immunopathol.* 2012, *147*, 77–85.
62. Iijima, S. Helical microtubules of graphitic carbon. *Nature* 1991, *354*, 56–58.

63. Pagona, G.; Tagmatrachis, N. Carbon nanotubes: Materials for medicinal chemistry and biotechnological applications.*Curr. Med. Chem.* 2006, *13*, 1789–1798.

64. He, H.; Pham-Huy, L.A.; Dramou, P.; Xiao, D.; Zuo, P.; Pham-Huy, C. Carbon nanotubes: Applications in pharmacy and medicine. *Bio. Med. Res. Int.* 2013, *2013*, 1–12.

65. Dumortier, H.; Lacotte, S.; Pastorin, G.; Marega, R.; Wu, W.; Bonifazi, D.; Briand, J.P.; Prato, M.; Muller, S.; Bianco, A. Functionalized carbon nanotubes are non-cytotoxic and preserve the functionality of primary immune cells. *Nano Lett.*2006, *6*, 1522–1528.

66. Kostarelos, K.; Bianco, A.; Prato, M. Promises, facts and challenges for carbon nanotubes in imaging and therapeutics.*Nature Nanotechnol.* 2009, *4*, 627–633.

67. Pantarotto, D.; Partidos, C.D.; Hoebeke, J.; Brown, F.; Kramer, E.; Briand, J.P.; Muller, S.; Prato, M.; Bianco, A. Immunization with peptide-functionalized carbon nanotubes enhances virus—Specific neutralizing antibody responses. *Chem. Biol.* 2003, *10*, 961–966.

68. Pantarotto, D.; Singh, R.; McCarthy, D.; Erhardt, M.; Briand, J.P.; Prato, M.; Kostarelos, K.; Bianco, A. Functionalized carbon nanotubes for plasmid DNA gene delivery. *Angew. Chem.* 2004, *43*, 5242–5246.

69. Chakravarty, P.; Marches, R. Thermal ablation of tumor cells with antibody-functionalized single-walled carbon nanotubes. *Proc. Natl. Acad. Sci.* 2008, *105*, 8697–8702.

70. Yacoby, I.; Benhar, I. Antibacterial nanomedicine. *Nanomedicine* 2008, *3*, 329–341.

71. Cheng, J.; Chan, C.M.; Veca, L.M.; Poon, W.L.; Chan, P.K.; Qu, L.; Sun, Y.P.; Cheng, S.H. Acute and long-term effects after single loading of functionalized multi-walled carbon nanotubes into zebrafish (*Danio rerio*). *Toxicol. Appl. Pharmacol.* 2009, *235*, 216–225.

72. Klaper, R.; Arndt, D.; Setyowati, K.; Chen, J.; Goetz, F. Functionalization impacts the effects of carbon nanotubes on the immune system of rainbow trout, *Oncorhynchus mykiss*. *Aqua. Toxicol.* 2010, *100*, 211–217.

73. Zhu, B.; Liu, G.L.; Gong, Y.X.; Ling, F.; Wang, G.X. Protective immunity of grass carp immunized with DNA vaccine encoding the vp7 gene of grass carp reovirus using carbon nanotubes as a carrier molecule. *Fish Shellfish Immunol.*2015, *42*, 325–334.

74. Younes, I.; Rinaudo, M. Chitin and chitosan preparation from marine sources. Structure, properties and applications.*Mar. Drugs* 2015, *13*, 1133–1174.

75. Calvo, P.; Remunan-Lopez, C.; Vila-Jato, J.L.; Alonso, M.J. Development of positively charged colloidal drug carriers: Chitosan-coated polyester nanocapsules and submicron-emulsions. *Colloid. Polym. Sci.* 1997, *275*, 46–53.

76. Pan, Y.; Li, Y.J.; Zhao, H.Y.; Zheng, J.M.; Xu, H.; Wei, G.; Hao, J.S. Bioadhesive polysaccharide in protein delivery system: chitosan nanoparticles improve the intestinal absorption of insulin *in vivo*. *Int. J. Pharm.* 2002, *249*, 139–147.

77. Lavertu, M.; Methot, S.; Tran-Khanh, N.; Buschmann, M.D. High efficiency gene transfer using chitosan/DNA nanoparticles with specific combinations of molecular weight and degree of deacetylation. *Biomaterials* 2006, *27*, 4815–4824.

78. Wei, H.; Zhang, X.; Cheng, C.; Cheng, S.X.; Zhuo, R.X. Self-assembled, thermosensitive micelles of a star block copolymer based on PMMA and PNIPAAm for controlled drug delivery. *Biomaterials* 2007, *28*, 99–107.

79. Kim, I.Y.; Seo, S.J.; Moon, H.S.; Yoo, M.K.; Park, I.Y.; Kim, B.C.; Cho, C.S. Chitosan and its derivatives for tissue engineering applications. *Biotechnol. Adv.* 2008, *26*, 1–21.

80. Kong, M.; Chen, X.G.; Xing, K.; Park, H.J. Antimicrobial properties of chitosan and mode of action: A state of the art review. *Int. J. Food Microbial.* 2010, *144*, 51–63.

81. Rao, S.B.; Sharma, C.P. Use of chitosan as a biomaterial: studies on its safety and hemostatic potential. *J. Biomed. Mat. Res.* 1997, *34*, 21–28.

82. Chiang, M.T.; Yao, H.T.; Chen, H.C. Effect of dietary chitosans with different viscosity on plasma lipids and lipid peroxidation in rats fed on a diet enriched with cholesterol. *Biosci. Biotechnol. Biochem.* 2000, *64*, 965–971.

83. Meshkini, S.; Tafy, A.A.; Tukmechi, A.; Farhang-Pajuh, F. Effects of chitosan on hematological parameters and stress resistance in rainbow trout (*Oncorhynchus mykiss*). *Vet. Res. Forum* 2012, *3*, 49–54.

84. Cha, S.H.; Lee, J.S.; Song, C.B.; Lee, K.J.; Jeon, Y.J. Effects of chitosan-coated diet on improving water quality and innate immunity in the olive flounder, *Paralichthys olivaceus*. *Aquaculture* 2008, *278*, 110–118.

85. Lin, S.; Pan, Y.; Luo, L.; Luo, L. Effects of dietary beta-1, 3-glucan, chitosan or raffinose on the growth, innate immunity and resistance of koi (*Cyprinus carpio koi*). *Fish Shellfish Immunol.* 2011, *31*, 788–794.

86. Harikrishnan, R.; Kim, J.S.; Balasundaram, C.; Heo, M.S. Dietary supplementation with chitin and chitosan on haematology and innate immune response in *Epinephelus bruneus* against *Philasterides dicentrarchi*. *Exp. Parasitol.* 2012,*131*, 116–124.

87. Cui, L.Q.; Xu, W.; Ai, Q.H.; Wang, D.F.; Mai, K.S. Effects of dietary chitosan oligosaccharide complex with rare earth on growth performance and innate immune response of turbot, *Scophthalmus maximus* L. *Aquac. Res.* 2013, *44*, 683–690.

88. Chen, Y.; Zhu, X.; Yang, Y.; Han, D.; Jin, J.; Xie, S. Effect of dietary chitosan on growth performance, haematology, immune response, intestine morphology, intestine microbiota and disease resistance in gibel carp (*Carassius auratus gibelio*). *Aquacult. Nutr.* 2014, *20*, 532–546.

89. Mari, L.S.S.; Jagruthi, C.; Anbazahan, S.M.; Yogeshwari, G.; Thirumurugan, R.; Arockiaraj, J.; Mariappan, P.; Balasundaram, C.; Harikrishnan, R. Protective effect of chitin and chitosan enriched diets on immunity and disease resistance in *Cirrhina mrigala* against *Aphanomyces invadans*. *Fish Shellfish Immunol.* 2014, *39*, 378–385.

90. Ranjan, R.; Prasad, K.P.; Vani, T.; Kumar, R. Effect of dietary chitosan on haematology, innate immunity and disease resistance of Asian seabass *Lates calcarifer* (Bloch). *Aquac. Res.* 2014, *45*, 983–993.
91. Alishahi, A.; Mirvaghefi, A.; Tehrani, M.R.; Farahmand, H.; Koshio, S.; Dorkoosh, F.A.; Elsabee, M.Z. Chitosan nanoparticle to carry vitamin C through the gastrointestinal tract and induce the non-specific immunity system of rainbow trout (*Oncorhynchus mykiss*). *Carbohydr. Polym.* 2011, *86*, 142–146.
92. Ferosekhan, S.; Gupta, S.; Singh, A.R.; Rather, M.A.; Kumari, R.; Kothari, D.C.; Pal, A.K.; Jadhao, S.B. RNA-loaded chitosan nanoparticles for enhanced growth, immunostimulation and disease resistance in fish. *Curr. Nanosci.* 2014,*10*, 453–464.
93. Ramos, E.A.; Relucio, J.L.V.; Torres-Villanueva, C.A.T. Gene expression in tilapia following oral delivery of chitosan-encapsulated plasmid DNA incorporated into fish feeds. *Marine biotechnol.* 2005, *7*, 89–94.
94. Kumar, S.R.; Ahmed, V.P.I.; Parameswaran, V.; Sudhakaran, R.; Babu, V.S.; Hameed, A.S.S. Potential use of chitosan nanoparticles for oral delivery of DNA vaccine in Asian sea bass (*Lates calcarifer*) to protect from *Vibrio (Listonella) anguillarum*. *Fish Shellfish Immunol.* 2008, *25*, 47–56.
95. Li, L.; Lin, S.L.; Deng, L.; Liu, Z.G. Potential use of chitosan nanoparticles for oral delivery of DNA vaccine in black seabream *Acanthopagrus schlegelii* Bleeker to protect from *Vibrio parahaemolyticus*. *J. Fish Dis.* 2013, *36*, 987–995.
96. Aranaz, I.; Mengíbar, M.; Harris, R.; Paños, I.; Miralles, B.; Acosta, N.; Galed, G.; Heras, Á. Functional characterization of chitin and chitosan. *Curr. Chem. Biol.* 2009, *3*, 203–230.
97. Jimenez-Fernandez, E.; Ruyra, A.; Roher, N.; Zuasti, E.; Infante, C.; Fernandez-Diaz, C. Nanoparticles as a novel delivery system for vitamin C administration in aquaculture. *Aquaculture* 2014, *432*, 426–433.
98. Arbos, P.; Wirth, M.; Arangoa, M.A.; Gaborb, F.; Irache, J.M. AN as a new polymer for the preparation of ligand-nanoparticle conjugates. *J. Control. Release* 2002, *83*, 321–330.
99. Trotta, M.; Gasco, M.R.; Morel, S. Release of drugs from oil-water microemulsions. *J. Control. Release* 1989, *10*, 237–243.
100. Mezei, M.; Gulasekharam, V. Liposomes-a selective drug delivery system for the topical route of administration I. Lotion dosage form. *Life Sci.* 1980, *26*, 1473–1477.
101. Lieb, L.M.; Ramachandran, C.; Egbaria, K.; Weiner, N. Topical delivery enhancement with multilamellar liposomes into pilosebaceous units: I. *In vitro* evaluation using fluorescent techniques with the hamster ear model. *J. Investing. Dermatol.* 1992, *99*, 108–113.
102. Honzak, L.; Šentjurc, M. Development of liposome encapsulated clindamycin for treatment of acne vulgaris. *Pflügers Archiv.* 2000, *440*, 44–45.
103. Verma, D.D.; Verma, S.; Blume, G.; Fahr, A. Particle size of liposomes influences dermal delivery of substances into skin. *Int. J. Pharm.* 2003, *258*, 141–151.
104. Knudsen, N.O.; Jorgensen, L.; Hansen, J.; Vermehren, C.; Frokjaer, S.; Foged, C. Targeting of liposome-associated calcipotriol to the skin: effect of liposomal membrane fluidity and skin barrier integrity. *Int. J. Pharm.* 2011, *416*, 478–485.
105. Romoren, K.; Thu, B.J.; Smistad, G.; Evensen, O. Immersion delivery of plasmid DNA I. A study of the potentials of a liposomal delivery system in rainbow trout (*Oncorhynchus mykiss*) fry. *J. Control. Release* 2002, *85*, 203–213.
106. Maherani, B.; Arab-Tehrany, E.; Mozafari, M.R.; Gaiani, C.; Linder, M. Liposomes: A review of manufacturing techniques and targeting strategies. *Curr. Nanosci.* 2011, *7*, 436–452.
107. Meure, L.A.; Foster, N.R.; Dehghani, F. Conventional and dense gas techniques for the production of liposomes: A review. *AAPS Pharm. Sci. Tech.* 2008, *9*, 798–809.
108. Berger, N.; Sachse, A.; Bender, J.; Schubert, R.; Brandl, M. Filter extrusion of liposomes using different devices: comparison of liposome size, encapsulation efficiency, and process characteristics. *Int. J. Pharm.* 2001, *223*, 55–68.
109. Maherani, B.; Arab-Tehrany, E.; Kheirolomoom, A.; Reshetov, V.; Stebe, M.J.; Linder, M. Optimization and characterization of liposome formulation by mixture design. *The Analyst.* 2012, *137*, 773–786.
110. Hope, M.J.; Bally, M.B.; Webb, G.; Cullis, P.R. Production of large unilamellar vesicles by a rapid extrusion procedure—Characterization of size distribution, trapped volume and ability to maintain a membrane-potential. *BBA Biomembr.* 1985, *812*, 55–65.
111. Macdonald, R.C.; Macdonald, R.I.; Menco, B.P.M.; Takeshita, K.; Subbarao, N.K.; Hu, L.R. Small-volume extrusion apparatus for preparation of large, unilamellar vesicles. *BBA Biomembr.* 1991, *1061*, 297–303.
112. Mozafari, M.R. *Nanocarrier Technologies: Frontiers of Nanotherapy*, 1st ed.; Springer Netherland: Dordrecht, The Netherlands, 2006.
113. Harikrishnan, R.; Kim, J.S.; Balasundaram, C.; Heo, M.S. Vaccination effect of liposomes entrapped whole cell bacterial vaccine on immune response and disease protection in *Epinephelus bruneus* against *Vibrio harveyi*. *Aquaculture*2012, *342*, 69–74.
114. Irie, T.; Watarai, S.; Iwasaki, T.; Kodama, H. Protection against experimental *Aeromonas salmonicida* infection in carp by oral immunisation with bacterial antigen entrapped liposomes. *Fish Shellfish Immunol.* 2005, *18*, 235–242.
115. Nakhla, A.N.; Szalai, A.J.; Banoub, J.H.; Keough, K.M.W. Serum anti-LPS antibody production by rainbow trout (*Oncorhynchus mykiss*) in response to the administration of free and liposomally-incorporated LPS from Aeromonas salmonicida. *Fish Shellfish Immunol.* 1997, *7*, 387–401.
116. Rodgers, S. Immersion vaccination for control of fish furunculosis. *Dis. Aqua. Org.* 1990, *8*, 69–72.
117. Yasumoto, S.; Kuzuya, Y.; Yasuda, M.; Yoshimura, T.;

Miyazaki, T. Oral immunization of common carp with a liposome vaccine fusing koi herpesvirus antigen. *Fish Pathol.* 2006, *41*, 141–145.

118. Ruyra, A.; Cano-Sarabia, M.; Mackenzie, S.A.; Maspoch, D.; Roher, N. A novel liposome-based nanocarrier loaded with an LPS-dsRNA cocktail for fish innate immune system stimulation. *PLoS ONE* 2013, *8*, e76338.

119. Behera, T.; Swain, P. Antigen adsorbed calcium phosphate nanoparticles stimulate both innate and adaptive immune response in fish, *Labeo rohita* H. *Cell. Immunol.* 2011, *271*, 350–359.

120. Faikoh, E.N.; Hong, Y.H.; Hu, S.Y. Liposome-encapsulated cinnamaldehyde enhances zebrafish (*Danio rerio*) immunity and survival when challenged with *Vibrio vulnificus* and *Streptococcus agalactiae*. *Fish Shellfish Immunol.* 2014, *38*, 15–24.

121. Falco, A.; Barrajón-Catalán, E.; Menéndez-Gutiérrez, M.P.; Coll, J.; Micol, V.; Estepa, A. Melittin-loaded immunoliposomes against viral surface proteins, a new approach to antiviral therapy. *Antivir. Res.* 2013, *97*, 218–221.

122. Irie, T.; Watarai, S.; Kodama, H. Humoral immune response of carp (*Cyprinus carpio*) induced by oral immunization with liposome-entrapped antigen. *Dev. Comp. Immunol.* 2003, *27*, 413–421.

123. Rauta, P.R.; Nayak, B. Parenteral immunization of PLA/PLGA nanoparticle encapsulating outer membrane protein (Omp) from *Aeromonas hydrophila*: Evaluation of immunostimulatory action in *Labeo rohita* (rohu). *Fish Shellfish Immunol.* 2015, *44*, 287–294.

124. Hølvold, L.B.; Fredriksen, B.N.; Bøgwald, J.; Dalmo, R.A. Transgene and immune gene expression following intramuscular injection of Atlantic salmon (*Salmo salar* L.) with DNA-releasing PLGA nano- and microparticles. *Fish Shellfish Immunol.* 2013, *35*, 890–899.

125. Tian, J.; Yu, J. Poly (lactic-co-glycolic acid) nanoparticles as candidate DNA vaccine carrier for oral immunization of Japanese flounder (*Paralichthys olivaceus*) against lymphocystis disease virus. *Fish Shellfish Immunol.* 2011, *30*, 109–117.

126. Adomako, M.; St-Hilaire, S.; Zheng, Y.; Eley, J.; Marcum, R.D.; Sealey, W.; Donahower, B.C.; LaPatra, S.; Sheridan, P.P. Oral DNA vaccination of rainbow trout, *Oncorhynchus mykiss* (Walbaum), against infectious haematopoietic necrosis virus using PLGA [Poly(d,l-Lactic-Co-Glycolic Acid)] nanoparticles. *J. Fish Dis.* 2012, *35*, 203–214.

127. Fredriksen, B.N.; Sævareid, K.; McAuley, L.; Lane, M.E.; Bogwald, J.; Dalmo, R.A. Early immune responses in Atlantic salmon (*Salmo salar* L.) after immunization with PLGA nanoparticles loaded with a model antigen and β-glucan. *Vaccine* 2011, *29*, 8338–8349.

128. Fredriksen, B.N.; Grip, J. PLGA/PLA micro- and nanoparticle formulations serve as antigen depots and induce elevated humoral responses after immunization of Atlantic salmon (*Salmo salar* L.). *Vaccine* 2012, *30*, 656–667.

129. Trapani, A.; Mandracchia, D.; Di Franco, C.; Cordero, H.; Morcillo, P.; Comparelli, R.; Cuesta, A.; Esteban, M.A. *In vitro* characterization of 6-Coumarin loaded solid lipid nanoparticles and their uptake by immunocompetent fish cells. *Coll. Surf.* 2015, *127*, 79–88.

130. Jain, R.A. The manufacturing techniques of various drug loaded biodegradable poly (lactide-co-glycolide) (PLGA) devices. *Biomaterials* 2000, *21*, 2475–2490.

131. Danhier, F.; Ansorena, E.; Silva, J.M.; Coco, R.; Le Breton, A.; Préat, V. PLGA-based nanoparticles: An overview of biomedical applications. *J. Control. Release* 2012, *161*, 505–522.

132. Vert, M.; Mauduit, J.; Li, S. Biodegradation of PLA/GA polymers: increasing complexity. *Biomaterials* 1994, *15*, 1209–1213.

133. O'Hagan, D.T.; Jeffery, H.; Roberts, M.; McGee, J.P. Controlled release microparticles for vaccine development. *Vaccine*1991, *10*, 768–771.

134. Panyam, J.; Labhasetwar, V. Biodegradable nanoparticles for drug and gene delivery to cells and tissue. *Adv. Drug Deliv. Rev.* 2003, *55*, 329–347.

135. Kapoor, D.N.; Bhatia, A.; Kaur, R.; Sharma, R.; Kaur, G.; Dhawan, S. PLGA: A unique polymer for drug delivery. *Ther. Deliv.* 2015, *6*, 41–58.

136. McCall, R.L.; Sirianni, R.W. PLGA Nanoparticles Formed by Single- or Double-emulsion with Vitamin E-TPGS. *JoVE*2013, *82*, e51015.

137. Newman, K.D.; Elamanchili, P.; Kwon, G.S.; Samuel, J. Uptake of poly(d,l-lactic-co-glycolic acid) microspheres by antigen-presenting cells *in vivo*. *J. Biomed. Mat. Res.* 2002, *60*, 480–486.

138. Lassalle, V.; Ferreira, M.L. PLA nano- and microparticles for drug delivery: An overview of the methods of preparation. *Macromol. Biosci.* 2007, *7*, 767–783.

139. Xiao, R.Z.; Zeng, Z.; Zhou, G.L.; Wang, J.J.; Li, F.Z.; Wang, A.M. Recent advances in PEG–PLA block copolymer nanoparticles. *Int. J. Nanomed.* 2010, *5*, 1057–1065.

140. Iliev, D.B.; Skjæveland, I.; Jørgensen, J.B. CpG oligonucleotides bind TLR9 and RRM-Containing proteins in Atlantic Salmon (*Salmo salar*). *BMC Immunol.* 2013.

141. Alonso, M.C.; Cano, I.; Garcia-Rosado, E.; Castro, D.; Lamas, J.; Barja, J.L.; Borrego, J.J.; Bergmann, S.M. Isolation of lymphocystis disease virus from sole, *Solea senegalensis* Kaup, and blackspot sea bream, *Pagellus bogaraveo* (Brunnich).*J. Fish Dis.* 2005, *28*, 221–228.

142. Loomba, L.; Sekhon, B.S. Calcium Phosphate Nanoparticles and their Biomedical Potential. *J. Nanomater. Mol. Nanotechnol.* 2015, *4*, 1–12.

143. Guo, S.; Huang, L. Nanoparticles containing insoluble drug for cancer therapy. *Biotechnol. Adv.* 2014, *32*, 778–788.

144. De Jesus, M.B.; Zuhorn, I.S. Solid lipid nanoparticles as nucleic acid delivery system: Properties and molecular mechanisms. *J. Control. Release* 2015, *201*, 1–13.

145. Harde, H.; Das, M.; Jain, S. Solid lipid nanoparticles: An oral bioavailability enhancer vehicle. *Expert Opin. Drug Deliv.*2011, *8*, 1407–1424.

146. Parhi, R.; Suresh, P. Preparation and characterization of solid lipid nanoparticles—A review. *Curr. Drug Discov. Technol.* 2012, *9*, 2–16.

147. Machado, A.H.E.; Lundberg, D.; Ribeiro, A.J.; Veiga, F.J.; Miguel, M.G.; Lindman, B.; Olsson, U. Encapsulation of DNA in macroscopic and nanosized calcium alginate gel particles. *Langmuir* 2013, *29*, 15926–15935.

148. Brudeseth, B.E.; Wiulsrød, R.; Fredriksen, B.N.; Lindmo, K.; Løkling, K.-E.; Bordevik, M.; Steine, N.; Klevan, A.; Gravningen, K. Status and future perspectives of vaccines for industrialised fin-fish farming. *Fish Shellfish Immunol.*2013, *6*, 1759–1768.

149. Jovanović, B.; Palić, D. Immunotoxicology of non-functionalized engineered nanoparticles in aquatic organisms with special emphasis on fish—Review of current knowledge, gap identification, and call for further research. *Aqua. Toxicol.* 2012, *118*, 141–151.

150. Jovanović, B.; Whitley, E.M.; Kimura, K.; Crumpton, A.; Palić, D. Titanium dioxide nanoparticles enhance mortality of fish exposed to bacterial pathogens. *Environ. Pollut.* 2015, *203*, 153–164.

CHAPTER 9

PROBIOTIC SUPPLEMENTATION IN DIET AND VACCINATION OF HYBRID SURUBIM (PSEUDOPLATYSTOMA RETICULATUM♀ X P. CORRUSCANS♂)

Gabriella do Vale Pereira, Gabriel Fernandes Alves Jesus, Felipe do Nascimento Vieira, Scheila Anelise Pereira, Thiago Tetsuo Ushizima, José Luiz Pedreira Mouriño, and Maurício Laterça Martins

Keywords: Fish, vaccine, lactic acid bacteria, Aeromonas hydrophila

INTRODUCTION

In surubins (***Pseudoplatystoma corruscans*** Spix and Agassiz 1829 and ***P. reticulatum*** Eigenmann and Eigenmann 1889) culture farms outbreaks of bacterial diseases are frequent during the winter because of the temperature change during the day. The clinical signs of septicemia caused by ***Aeromonas hydrophila*** were described previously by CAMPOS (2004) and further, strains of ***A. hydrophila*** were isolated and characterized by SILVA et al. (2012) as the main surubim hybrid pathogen. Aiming bacterial control, many producers use antibiotics in fish farming, however the use of these antibiotics leads to the selection of resistant bacteria (SASAN et al., 2011). To avoid antibiotic using, prophylactic methods such as vaccine and probiotics have been developed and have been shown to be promising tools for fighting fish bacterial diseases (MOURIÑO et al., 2012).

Probiotic bacteria are widely used in aquaculture in different species (MOURIÑO et al., 2012). They are well known as promote fish health and enhance survival (JATOBÁ et al., 2011), improve immunological parameters (JATOBÁ et al., 2012) and protect fish against possible pathogenic bacteria infection (BALCÁZAR et al., 2007). Among the different methods of vaccine preparation, the inactivated vaccines by heat or formalin are mostly used (MAGNADOTTIR, 2010). These vaccines can contain only inactivated cells (bacterin) or inactivated extracellular products as well (toxoids) which are secreted by the bacteria (DA COSTA et al., 2008). The intraperitoneal injection is one of vaccine's administration routes that provide the best results, although it is required high labour costs. This method of administration can be improved by using a booster (second vaccine application) to enhance vaccine durability, antibody production, and immune memory cells (CHENG et al., 2010). The oral vaccine is a good way to provide the booster, because it is practical, low cost, and viable in culture farms (SILVA et al., 2009).

Evaluating the influence of probiotic feeding on the intraperitoneal vaccination effectiveness in fish is important in order to assess possible interactions between vaccine and probiotics in improving fish health. Thus, the present study aimed to evaluate dietary supplementation with the probiotic ***W. cibaria*** on the efficacy of the hybrid surubim immunization against hemorrhagic septicemia caused by ***A. hydrophila*** based on immunological and hematological parameters.

MATERIAL AND METHODS

A total of 192 healthy hybrid surubim (***P. corruscans*** x ***P. reticulatum***) were obtained from the Piraí farm located in Mato Grosso do Sul state, Brazil. The pathogenic hemolytic strain ***A. hydrophila*** (CPQBA 228-08 DRM) was isolated during mortality in Mato Grosso do Sul state by SILVA et al (2012) and the probiotic strain ***W. cibaria***(CPQBA 001-10 DRM 02) was isolated from healthy and asymptomatic hybrid sorubins (MOURIÑO et al., 2012).

For vaccine preparation, the ***A. hydrophila*** strain was grown in the same condictions and following the same methodology used by SILVA et al. (2012). The vaccination was by intraperitoneal rote (i.p.) injection utilizing an insulin syringe (1mL) with a needle (13x0.45mm Injex(r)). For oral booster, the

Citation: Gabriella do Vale Pereira, Gabriel Fernandes Alves Jesus, Felipe do Nascimento Vieira, Scheila Anelise Pereira, Thiago Tetsuo Ushizima, José Luiz Pedreira Mouriño, and Maurício Laterça Martins. Probiotic supplementation in diet and vaccination of hybrid surubim (Pseudoplatystoma reticulatum♀ x P. corruscans♂), Ciência Rural, Santa Maria, 2016; vol.46. no.2, p.348-353. http://dx.doi.org/10.1590/0103-8478cr20150543. ISSN 0103-8478.

same vaccine formulation and concentration ($2x10^8$ CFU mL^{-1}) was used as described above. Thus, 100mL of vaccine was sprayed using a sterile plastic spray bottle for each kilogram of commercial diet or the diet containing probiotic following methodology previously described by (PEREIRA et al., 2015).

The probiotic bacteria ***W. cibaria*** was grown in vials tubes containing MRS broth (Difco(r)), and incubated at 35°C for 48h. After that, 100mL of the probiotic at concentration of $1x10^8$ CFU mL^{-1} was sprayed (using a sterile plastic spray bottle) on each kilogram of the commercial diet. The diet was dried at 30°C for 12h. This process was repeated every 15 days to achieve a high probiotic concentration in the diet.

To ascertain the ***W. cibaria*** concentration in the diet, 1g of the diet with probiotic was macerated in a sterile mortar with 1mL of saline sterile solution (NaCl 0.65%) and then serially diluted nine times in test tubes at 1:10 factor. The dilutions 10^{-5} to 10^{-9} were plated in petri plates containing MRS agar. The plates were incubated at 35°C for 48h. This process was repeated every time the diet with probiotic was prepared. The probiotic concentration in diet was 5.53 x 10^6 CFU g^{-1}.

Fish with an initial weight of 44.35±3.07g (mean ± standard deviation), were distributed in 24 tanks with 100L capacity (8 fish per tank) and fed commercial diet (Supra Aqualine) four times a day, approximately 3% of fish biomass. The water quality parameters were monitored daily. When necessary, the water was exchanged from 20% to 60% of the system. The tanks were maintained in a closed-water recirculating system with mechanical and biological filters under constant heat at 28°C and UV sterilization and photoperiod was 24h of darkness according (PIAIA et al., 1999).

The treatments were as follows: (1) fish fed a supplemented probiotic diet; (2) vaccinated fish; (3) vaccinated fish fed a supplemented probiotic diet; and (4) untreated fish (control). There were six repeatitions per each treatment. After the acclimation period, the fish from the groups 1 and 3 received the commercial diet with the probiotic. The others two groups received the same commercial feed without probiotic. The feed was maintained for 41 days, sufficient time for probiotic ***W. cibaria*** colonization (MOURIÑO et al., 2012). After 15 days from the beginning of the experiment, the fish from groups 2 and 3 were vaccinated against ***A. hydrophila*** with 0.01mL of vaccine per gram of fish. Fifteen days after the first vaccination, the oral booster began and lasted four consecutive days. The diet containing vaccine was also provided four times a day (3% of the biomass). The vaccination protocol and sampling was carried out according PEREIRA et al. (2015).

A week after the end of the oral booster, three fish from each experimental unit were sampled. They were anaesthetized with benzocaine (0.1g L^{-1}), and the blood was collected by puncture of the caudal vessel in 3mL syringes (21G) containing 10% anticoagulant (EDTA), according to the ethic committee of Universidade Federal de Santa Catarina. A small amount of blood was used for the haematological analysis, and the other part was separated for phagocytosis analysis. A pool of blood was made from the remainder of the blood from each experimental unit to obtain the plasma by centrifugation.

The collected blood was used to make the blood smears stained with Giemsa/ May Grunwald (ROSENFELD, 1947) for differential leukocyte counts, as well as to obtain the total leukocytes and thrombocytes number. An aliquot was used to determine the haematocrit (GOLDENFARB et al., 1971) and the remainder stored in glass vials on ice to quantify the total number of erythrocytes (RBC) by Neubauer chamber. The total numbers of thrombocytes and leukocytes (WBC), as well as differential leukocytes was obtained from blood smears by the indirect method (ISHIKAWA et al., 2008).

The vials tubes containing the blood pool (from three fish) from each experimental unit were centrifuged at 1400g for 10 min to obtain blood plasma and were stored at -20°C. The protein in the blood plasma was measured using the Total Protein kit according to the manufacturer's instructions (Lab Test(r), Lagoa Santa, MG, Brazil). The total immunoglobulin concentration was measured after infection (AMAR et al., 2000).

The lysozyme activity of the blood plasma was determined using the methodology adapted by SANKARAN AND GURNANI (1972), The initial and final absorbance of the samples were measured in a microplate reader (Expert Plus Asys(r)) at 492nm and the rate of reduction in absorbance of the samples was converted to lysozyme concentration (µg mL^{-1}).

The title serum agglutination as well as the blood serum antimicrobial activity were prepared according methodology described by SILVA et al. (2009). To perform the agglutination activity the bacteria ***Aeromonas hydrophila*** (CPQBA 228 08) was used. The antimicrobial activity was performed using the same strain of ***A. hydrophila*** as (Gram negative) mentioned before, further the strain of ***Enterococcus durans*** (Gram positive ATCC 19492).

To determine the percentage of leukocyte phagocytosis, 0.5mL of blood and 0.25mL of a suspension of $1x10^6$CFU mL^{-1} formalin-inactivated ***A. hydrophila*** were added in centrifuge tubes, which were kept at 28°C for 30 min and homogenized every 10 min. After hatching, the blood was seeded in blood smears, in duplicate, and the slides were stained with Giemsa/ May Grunwald (ROSENFELD, 1947). The number of phagocytic leukocytes was counted by percentage to the total number of leukocytes in the smear analysis (MARTINS et al., 2004).

Data were submitted for normality and the Bartlett test was performed to check the homogeneity of variance. The agglutination results were converted to log2. Data that did not have homogeneous variances were transformed into log10 (x+1) and subsequently subjected to a 2x2 factorial analysis of the variance supplemented by the Tukey test for mean separation. All analyses were subjected to a significance level of 5%.

RESULTS AND DISCUSSION

The water quality parameters including, dissolved oxygen (8.06±0.56mg L^{-1}), total ammonia (1.5±0.5mg L^{-1}), nitrite (0.67±0.65mg L^{-1}), nitrate (1.96±1.44mg L^{-1}), pH (7.83±0.39) and temperature (27.5±2.46°C) were within the fish culture limit (BOYD, 1990). Fish mortality was verified throughout the trial period. The survival at the final of the experiment was: 95.84% fish fed a supplemented probiotic diet; 97.92% vaccinated fish; and 100% vaccinated fish fed a supplemented probiotic diet; and 93.75 untreated fish (control).

There were no interaction between the factors vaccination and probiotic feeding on the haematological parameters. The lymphocyte number in blood from the two vaccinated groups was lower when compared to unvaccinated fish (Table 1). Comparing with other studies, intraperitoneal vaccination in Nile tilapia (***O. niloticus***) led to an increase in lymphocyte number, after 7 and 21 days post-vaccination, and also after 21 days when the fish were orally vaccinated (SILVA et al., 2009). In this case, the decrease of lymphocytes could be due to their migration to the tissues as a response to the vaccination.

Table 1. Hematological parameters (mean and standard error) of hybrid surubim (*Pseudoplatystoma reticulatum*♀, Spix and Agassiz x *P. corruscans*♂, Eigenmann and Eigenmann) fed diet containing probiotic (Probiotic); vaccinated against ***A. hydrophila*** and fed diet containing probiotic (Probiotic + vaccine) and vaccinated fish (Vaccine) against ***A. hydrophila***; and untreated fish (Control), after 41 days of treatment.

Treatments	RBC ($x10^6$ mL^{-1})	Thrombocytes ($x10^3$ mL^{-1})	WBC ($x10^3$ mL^{-1})	Lynphocytes ($x10^3$ mL^{-1})	Monocytes ($x10^3$ mL^{-1})
Control	1.64 ± 0.07	36.9 ±1.21^a	79.5 ±5.95	63.9 ±5.23^b	1.91± 0.03
Probiotic	1.85 ± 0.02	44.1 ±4.68^b	83.0 ±8.08	81.8 ±5.08^b	1.32 ± 0.01
Vaccine	1.66 ± 0.03	23.8 ±4.31^a	83.4 ±4.50	60.1 ±5.33^a	1.30 ± 0.02
Probiotic + vaccine	1.67 ± 0.06	46.4 ±5.04^b	86.4 ±4.75	57.3 ±8.14^a	1.52 ± 0.03
P of probiotic factor	0.068446	0.001678	0.595889	0.229583	0.550917
P of vaccine fator	0.170268	0.203834	0.547952	0.030445	0.525978
P of interaction	0.077059	0.074783	0.972808	0.103493	0.213258
	Eosinophils ($x10^3$ mL^{-1})	Basophils ($x10^3$ mL^{-1})	Neutrophils ($x10^3$ mL^{-1})	LG - PAS ($x10^3$ mL^{-1})	Hematocrit (%)
Control	0.67 ± 0.23	21.8 ± 0.35	3.02 ± 0.81	0.00 ± 0.00	24.9 ± 1.40
Probiotic	0.70 ± 0.29	15.7 ± 0.30	3.07 ± 0.55	0.00 ± 0.00	26.0 ± 0.80
Vaccine	0.29 ± 0.13	22.9 ± 0.50	2.81 ± 0.61	0.07 ± 0.07	23.3 ± 0.57
Probiotic + vaccine	0.41 ± 0.10	23.9 ± 0.57	3.77 ± 0.70	0.00 ± 0.00	24.3 ± 0.82
P of probiotic factor	0.729999	0.570656	0.468069	0.329257	0.281385
P of vaccine fator	0.124594	0.310845	0.719312	0.329257	0.101812
P of interaction	0.824305	0.437895	0.515520	0.329257	0.914649

*Different letters indicate significant difference by Tukey test (P<0.05).

Table 2. Plasmatic immunological parameters: lysozyme, total protein, total immunoglobulin (mean ± standard error), percentage of phagocytic leukocytes, agglutination titer and antimicrobial activity in hybrid surubim (*Pseudoplatystoma reticulatum*♀, Spix and Agassiz x *P. corruscans*♂, Eigenmann and Eigenmann) fed diet containing probiotic (Probiotic); vaccinated against ***A. hydrophila*** and fed diet containing probiotic (Probiotic + vaccine) and vaccinated fish (Vaccine) against ***A. hydrophila***; and untreated fish (Control), after 41 days of treatment.

Treataments	Lysozyme (µg mL^{-1})	Total protein (mg mL^{-1})	Total imunoglobulin (mg mL^{-1})	Phagocytosis (%)
Control	10.32 ± 0.39^a	23.64 ± 0.65	0.96 ± 0.39	7.22 ± 0.72
Probiotic	10.71 ± 0.61^b	24.60 ± 0.59	1.61 ± 0.50	9.75 ± 1.16
Vaccine	10.79 ± 0.47^b	24.04 ±0.59	1.38 ± 0.34	10.92 ± 1.01
Probiotic + vaccine	12.93 ± 0.26^b	23.83 ±0.53	1.05 ± 0.30	9.75 ± 1.44
P of probiotic factor	0.018765	0.256662	0.674756	0.558108
P of vaccine factor	0.024562	0.827398	0.853697	0.117067
P of interaction	0.092330	0.655884	0.230215	0.117067

	Agglutination titer	Antimicrobial activity	
	A. hydrophila	**A. hydrophila**	*E. durans*
Control	7.33 ± 0.21^a	10.40 ± 0.39^a	9.80 ± 0.00
Probiotic	8.17 ± 0.30^a	10.80 ± 0.19^a	10.00 ± 0.24
Vaccine	9.50 ± 0.42^b	12.20 ± 0.67^b	10.20 ± 0.58
Probiotic + vaccine	9.66 ± 0.21^b	11.60 ± 0.73^b	9.40 ± 0.37
P of probiotic factor	0.114263	0.857213	0.425920
P of vaccine factor	0.000006	0.030488	0.789424
P of interaction	0.283987	0.374978	0.192460

*Different letters indicate significant difference by Tukey test (P<0.05).

The supplemented probiotic diet increased thrombocyte number in the circulating blood of the hybrid surubim (Table 1). This result is in agreement with those obtained by JATOBÁ et al. (2011) where an increase in thrombocyte number in ***O. niloticus*** fed with lactic acid bacteria ***L. plantarum*** was reported. In vaccinated fish, it is expected that the white blood cells counts would be altered. However, in this case, the supplementation of probiotic could not improve the cells count. Future works should be performed to evaluate the haematological parameters, in the same conditions, however after a bacterial challenge. It is important to notice that the probiotic and the vaccine had influence in different cells groups. As well as the haematological parameters, the use of probiotic and vaccine did not presented interaction in the fish immunology in the present work, however we can see some differences when analysed them separately. Fish fed the supplemented probiotic diet and vaccinated fish had a higher plasma lysozyme concentration (Table 2), corroborating the study of MERRIFIELD et al. (2010) in ***O. mykiss***. The present study shows that both, supplemented diet and vaccinated fish had a higher plasmatic lysozyme than untreated fish. It was expected that vaccinated fish would present an increase in the concentration of total immunoglobulins in plasma, because these defence proteins are activated after immunization, as observed by MIKKELSEN et al. (2011) in ***Gadus morhua***, L. vaccinated against ***Vibrio anguillarum***. However, the increase in immunoglobulins observed in those studies was after infection, thereby enhancing the immune response and consequently the concentration of total and specific immunoglobulins.

The plasma antimicrobial activity of the surubim showed significant difference between treatments against ***A. hydrophila*** (Table 2). According to SILVA et al. (2013) the serum antimicrobial activity of intraperitoneally vaccinated hybrid surubins was higher than that in fish vaccinated by bath or unvaccinated surubim. In fact, these results have shown that vaccination increases the antimicrobial capacity of fish serum, furthermore against the bacterium ***A. hydrophila***, which the vaccine was previously prepared.

Moreover, the agglutination titer was higher in both vaccinated fish groups (Table 2). In the study of SILVA et al. (2013), the serum agglutination titer in vaccinated hybrid surubim by intraperitoneal way, before challenge with ***A. hydrophila***, was also higher than unvaccinated fish. Those results indicate that vaccination induced plasmatic agglutination in fish and enhanced the protection against possible subsequent infections, compared with unvaccinated fish or fish fed a supplemented probiotic diet.

Besides the fact that no interaction was observed in both haematological and immunological parameters, it can be attested that the probiotic and the vaccination helped the fish protection against pathogens however in different parameters. This remark can highlight a positive result, such as the probiotic and the vaccination indeed worked together in different and specific parts of the immune system in the fish, so they complement each other because they increase separate parameters.

The ***W. cibaria*** supplementation in diet had no interaction with vaccination in the hybrid surubim against ***A. hydrophila***; however, probiotic and vaccination promoted a response of different immunological and hematological parameters in these fish.

ACKNOWLEDGEMENTS

The authors thank Coordenação de Aperfeiçoamento de Pessoal de Nível Superior (CAPES) for scholarship to G.V. Pereira and the company Mar & terra S/A de pescados financial for support during the analyses.

REFERENCES

1. AMAR, E.C. et al. Effects of dietary β-carotene on the immune response of rainbow trout Oncorhynchus mykiss. Fisheries Science, v. 66, n.6, p.1068-1075, 2000. ISSN 1444-2906. Available from: <Available from:http://dx.doi.org/10.1046/j.1444-2906.2000.00170.x >. Accessed: Nov. 21, 2014. doi: 10.1046/j.1444-2906.2000.00170.x.
2. BALCÁZAR, J.L. et al. Enhancement of the immune response and protection induced by probiotic lactic acid bacteria against furunculosis in rainbow trout (Oncorhynchus mykiss). FEMS Immunology & Medical Microbiology, v.51, n.1, p.185-193, 2007. ISSN 1574-695X. Available from: <Available from:http://dx.doi.org/10.1111/j.1574-695X.2007.00294.x >. Accessed: Oct. 20, 2014. doi: 10.1111/j.1574-695X.2007.00294.x.
3. BOYD, C. E. Water quality in ponds for aquaculture. Alabama: Birmingham Publishing Company, 1990. 482p.
4. CAMPOS, J.L. Pintado culture in Brazil. Global Aquaculture Advocate, v.42, p.42-43, 2004. Available from: <pdf.gaalliance.org/pdf/GAA-Campos-Apr04.pdf>. Accessed: Oct. 03, 2014.
5. CHENG, S. et al. Analysis of the vaccine potential of a laboratory Escherichia coli strain in a Japanese flounder model. Fish & Shellfish Immunology, v.28, n.2, p.275-280, 2010. ISSN 1050-4648. Available from: <Available from: http://www.sciencedirect.com/science/article/pii/S105046480900357X >. Accessed: Oct. 04, 2014. doi: 10.1016/j.fsi.2009.11.003.
6. DA COSTA, M.M. et al. Antimicrobial sensibility of bacterial isolates from jundia (Rhamdia quelen). Pesquisa Veterinaria Brasileira, v.28, n.10, p.477-480, 2008. ISSN 0100-736X. Available from: <Available from:http://www.scielo.br/scielo.php?script=sci_arttext&pid=S0100-736X2008001000006 >. Accessed: Aug. 27, 2014.
7. GOLDENFARB, P.B. et al. Reproducibility in the hematology laboratory: the microhematocrit determination. American Journal of Clinical Pathology, v.56, n.1, p.35-39, 1971. ISSN 0002-9173. Available from: <Available from: http://pubget.com/paper/5556212/reproducibility-in-the-hematology-laboratory-the-microhematocrit-determination >. Accessed: Oct. 20, 2014.
8. ISHIKAWA, N.M. et al. Total leukocyte counts methods in fish, Oreochromis niloticus Archives of Veterinary Science, v.13, n.1, p.54-63, 2008. ISSN 1517-784X.
9. JATOBA, A. et al. Diet supplemented with probiotic for Nile tilapia in polyculture system with marine shrimp. Fish Physiology and Biochemistry, v.37, n.4, p.725-732, 2011. ISSN 0920-1742. Available from: <Available from:http://link.springer.com/article/10.1007%2Fs10695-011-9472-5 >. Accessed: Nov. 21, 2014. doi: 10.1007/s10695-011-9472-5.
10. MAGNADOTTIR, B. Immunological control of fish diseases. Marine Biotechnology, v.12, n.4, p.361-379, 2010. ISSN 1436-2228. Available from: <Available from: http://dx.doi.org/10.1007/s10126-010-9279-x >. Accessed: Nov. 13, 2014. doi: 10.1007/s10126-010-9279-x.

11. MARTINS, M.L. et al. Physiological and haematological response of Oreochromis niloticus (Osteichthyes: Cichlidae) exposed to single and consecutive stress of capture. Acta Scientiarum. Animal Sciences, v.26, n.4, p.449-456, 2004. ISSN 1807-8672.

12. MERRIFIELD, D. L. et al. Probiotic applications for rainbow trout (Oncorhynchus mykiss Walbaum) II. Effects on growth performance, feed utilization, intestinal microbiota and related health criteria postantibiotic treatment. Aquaculture Nutrition, v.16, n.5, p.496-503, 2010. ISSN 1353-5773. Available from: <Available from:http://onlinelibrary.wiley.com/doi/10.1111/j.1365-2095.2009.00688.x/abstract >. Accessed: Nov. 21, 2014. doi: 10.1111/j.1365-2095.2009.00688.x.

13. MIKKELSEN, H. et al. Vibriosis vaccines based on various sero-subgroups of Vibrio anguillarum O2 induce specific protection in Atlantic cod (Gadus morhua L.) juveniles. Fish & Shellfish Immunology, v.30, n.1, p.330-339, 2011. ISSN 1050-4648. Available from: <Available from:http://www.sciencedirect.com/science/article/pii/S1050464810003499 >. Accessed: Aug. 21, 2014. doi: 10.1016/j.fsi.2010.11.007.

14. MOURIÑO, J.L.P. et al. Effect of dietary supplementation of inulin and W. cibaria on haemato-immunological parameters of hybrid surubim (Pseudoplatystoma sp). Aquaculture Nutrition, v.18, n.1, p.73-80, 2012. ISSN 1365-2095. Available from: <Available from: http://dx.doi.org/10.1111/j.1365-2095.2011.00879.x >. Accessed: Aug. 15, 2014. doi: 10.1111/j.1365-2095.2011.00879.x.

15. PEREIRA, G. D. V. et al. Vaccination strategies with oral booster for surubim hybrid (Pseudoplatystoma corruscans × P. reticulatum) against haemorrhagic septicaemia. Aquaculture Research, v. 46, n. 8, p. 1831-1841, 2015. ISSN 1365-2109. Disponível em: < http://dx.doi.org/10.1111/are.12339 >.

16. PIAIA, R. et al. Growth and survival of fingerlings of silver catfish exposed to different photoperiods. Aquaculture International, v.7, n.3, p.201-205, 1999. ISSN 0967-6120. Available from: <Available from:http://dx.doi.org/10.1023/A%3A1009299830102 >. Accessed: Aug. 14, 2014. doi: 10.1023/A:1009299830102.

17. ROSENFELD, G. Pancromic stain for haematology and clinical cytology. A new combination of the components May-Grünwald and Giemsa in just one formula for rapid staining. Memórias do Instituto Butantan, v.20, p.329-334, 1947.

18. SANKARAN, K.; GURNANI, S. On the variation in the catalytic activity of lysozyme in fishes. Indian Journal of Biochemistry & Biophysics, v.9, n.2, p.162-165, 1972. ISSN 0301-1208.

19. SASAN, H. et al. Construction of vaccine from Lactococcus lactis bacteria using Aeromonas hydrophila virulent Aerolysin gene. Iranian Journal of Fisheries Sciences, v.10, n.1, p.143-154, 2011. Available from: <Available from: http://www.jifro.ir/browse.php?a_code=A-10-1-69&slc_lang=en&sid=1 >. Accessed on: Nov. 10, 2014.

20. SILVA, B.C. et al. Hematological and immunological responses of Nile tilapia after polyvalent vaccine administration by different routes. Pesquisa Veterinaria Brasileira, v.29, n.11, p.874-880, 2009. ISSN 0100-736X. Available from: <Available from: http://www.scielo.br/scielo.php?pid=S0100-736X2009001100002&script=sci_arttext >. Accessed: Aug. 04, 2014.

21. SILVA, B.C. et al. Haemorrhagic septicaemia in the hybrid surubim (Pseudoplatystoma corruscans×Pseudoplatystoma fasciatum) caused by Aeromonas hydrophila. Aquaculture Research, v.43, n.6, p.908-916, 2012. ISSN 1365-2109. Available at: <Available at: http://dx.doi.org/10.1111/j.1365-2109.2011.02905.x >. Accessed: Aug. 04, 2014. doi: 10.1111/j.1365-2109.2011.02905.x.

22. SILVA, B.C. et al. Immunization of hybrid surubim (Pseudoplatystoma corruscans x P. fasciatum) against Motile Aeromonas hydrophila septicemia. Brazilian Archives of Biology and Technology, v.56, p.81-84, 2013. ISSN 1516-8913. Available at: <Available at: http://www.scielo.br/scielo.php?script=sci_arttext&pid=S1516-89132013000100010&nrm=iso >. Accessed: Aug. 05, 2014.

CHAPTER 10

FLAVOBACTERIUM PSYCHROPHILUM VACCINE DEVELOPMENT

Esther Gómez, Jessica Méndez, Desirée Cascales, and José A. Guijarro

INTRODUCTION

Flavobacterium psychrophilum is the etiological agent of bacterial cold water disease (BCWD). Juvenile rainbow trout *(Oncorhynchus mykiss)* are particularly affected, and, in this case, the disease is called rainbow trout fry syndrome.

The disease affecting rainbow trout and Coho salmon (*Oncorhynchus kisutch*) was known only in North America (Borg, 1948). The first cases in Europe were reported in France and Germany in mid-1980s (von Weis, 1987; Bernardet *et al.*, 1988; Bernardet and Kerouault, 1989). Shortly after this, more cases of infection in Coho salmon and ayu (*Plecoglossus altivelis*) were reported in Japan (Wakabayashi *et al.*, 1991). Subsequently, the disease appeared in all parts of the world (Bernardet and Bowman, 2006), becoming one of the main causes of economic losses in salmonid aquaculture.

Although external signs of infection are variable depending on both the fish species and their development stage, in larger fish they typically consist of skin erosion and muscle degeneration followed by progressive tissue necrosis (Cipriano and Holt, 2005; Barnes and Brown, 2011). In addition, septicaemia can develop, without any of these characteristic external signs mainly in fry (Cipriano and Holt, 2005; Barnes and Brown, 2011).

The disease appears most frequently when water temperatures range between 10°C and 14°C (Borg, 1960). As there is no commercial vaccine, treatment with antibiotics is the only means for its eradication. However, resistance to these compounds is increasing (Cipriano and Holt, 2005), and the number of them available for the treatment of BCWD is limited. These are the main reasons for interest in the development of an effective vaccine against BCWD.

Flavobacterium psychrophilum is considered a 'fastidious' bacterium because it is difficult to culture, isolate and manipulate, but, in spite of this, the need for an effective vaccine has motivated research into this organism. Thus, in the last few years, the development of genetic techniques (Alvarez *et al.*, 2004; Perez-Pascual *et al.*, 2011; Gomez *et al.*, 2012) and culture media (Alvarez and Guijarro, 2007; Perez-Pascual *et al.*, 2010) together with the availability of the complete genome sequence of the bacterium (Duchaud *et al.*, 2007) have facilitated its study and, in particular, that of its virulence determinants. In this respect, some factors associated with pathogenesis have been described. Among these are extracellular proteases (Bertolini *et al.*, 1994; Ostland *et al.*, 2000), an iron acquisition system (Alvarez *et al.*, 2008), an adhesin (Nematollahi *et al.*, 2003), haemolytic activities (Lorenzen and Olesen, 1997; Lorenzen *et al.*, 1997; Moller *et al.*, 2003) and a thiol: disulphide oxidoreductase (Alvarez *et al.*, 2006). Besides, putative virulence genes have been identified in the genome analysis of this bacterium (Duchaud *et al.*, 2007). Additionally, different studies in bacterial surface structures have identified major antigens such as lipopolysaccharide (LPS) O-antigens (MacLean *et al.*, 2001) and surface proteins (Crump *et al.*, 2001; 2005; Massias *et al.*, 2004; Dumetz *et al.*, 2007), which open new perspectives for vaccine development. However, mechanisms enabling the bacterium to evade the fish immune system have been described for *F. psychrophilum*; it was, for example, able to suppress the non-specific humoral defence mechanisms (Siwicki *et al.*, 2004). Indeed, viable bacteria have been found in spleen phagocytes, where they are protected against lysozyme activity, complement and exposure to the immune system (Decostere *et al.*, 2000). The bacterium must have additional mechanisms of protection against the complement system because high complement activity

Citation: Gómez, E., Méndez, J., Cascales, D., and Guijarro, J. A. (2014), Flavobacterium psychrophilum vaccine development. Microbial Biotechnology, 7: 414–423. DOI:10.1111/1751-7915.12099.

in rainbow trout serum was unable to reduce the number of *F. psychrophilum* cells (Wiklund and Dalsgaard, 2002). Therefore, specific immune responses must exist in order to avoid the dispersion of the infection.

To obtain an effective vaccine for *F. psychrophilum*, besides the understanding of virulence factors and the identification of putative antigens, it is essential to understand certain other aspects of the infection process, such as the route by which the bacterium enters the fish, serotype variability and the immune response of the fish after both natural and experimental infections. All these questions, especially the latter, are the main subjects that will be analysed in this review.

WHY IS THE DEVELOPMENT OF A VACCINE A DIFFICULT TASK?

There are still a number of fundamental points that have to be resolved before a vaccine for BCWD can be developed and commercialized. A key question for vaccine development is the identification of the point of entry of the bacterium. In the case of *F. psychrophilum*, these data have not yet been established, although most authors suggest that BCWD could be a water-borne infection, and the bacterium could penetrate fish via damaged skin or gills (Dalsgaard, 1993). Infected fish could also act as a reservoir in a carrier state, constantly shedding bacteria into the water and transmitting the bacteria by horizontal transfer (Madsen and Dalsgaard, 1999; Madetoja *et al.*, 2000; Madsen *et al.*, 2005). Additionally, some authors described the presence of the bacterium in healthy fish, as part of the microbiota of the fins, gills and opercula of adult salmonids (Nematollahi*et al.*, 2003). When the fish immune system is not in optimal conditions, the bacterium might cause infection. The presence of the bacterium in the perivitellin space of the eggs, milt and ovarian fluids has been reported, and vertical transmission of *F. psychrophilum* via intra-ovum infection has already been suggested by several authors (Brown *et al.*, 1997; Izumi and Wakabayashi, 1997; Rangdale *et al.*, 1997; Vatsos *et al.*, 2001; 2006; Cipriano, 2005). Therefore, prevention measures such as iodophor treatment of the eggs are not totally effective because the bacterium is protected within the perivitellin space (Kumagai *et al.*, 2000), which represents a significant source of infection from the earliest life stages. Nevertheless, it should be taken into account that some authors (Madsen *et al.*, 2005; Madsen and Dalsgaard, 2008) did not find the bacterium inside eggs, but still found that disinfection could be an effective method for the elimination of the bacteria.

Bearing all this in mind, vaccination against BCWD should be carried out as early as possible. However, a major problem remains: the bacterium may cause the young fish problems when they are as small as 0.2–0.5 g and consequently, their immune system is not fully developed, which compromises the usefulness of the vaccine. In salmonid fish, the onset of adaptive immunity occurs when they weigh from 0.5 to 2.5 g, according to the results found in vaccination by immersion against *Yersinia ruckeri* (Ellis, 1988) and *Vibrio anguillarum* (Johnson *et al.*, 1982; Tatner and Horne, 1983). At these weights, there is another question to consider for vaccination may only be possible by immersion or in-feed administration because the small size of the immature fish makes other routes of administration unviable for economic and practical reasons. However, administration of the vaccine by these routes does not produce a long-term immune response, and if the fish were vaccinated early, they probably would not have enough protection. The injection route induces the strongest and longest immune response; however, it should only be used with fish weighing over 15 g (Hastein *et al.*, 2005), and the vaccine needs to be combined with an adjuvant to induce a long-term immunological response. This adjuvant has a double function: it improves the presentation of the antigen to the immune cells and keeps it within the fish tissues, thus increasing the length of protection. However, adjuvants cause undesirable side effects such as reactions in the abdominal cavity, so they have to be previously tested to minimize these problems (Hastein *et al.*, 2005). It should be borne in mind that most of the fish vaccination studies cited in this review do not clearly state the role of the innate immune system in protection. In this sense, it would be useful for studies on fish vaccination to distinguish between the effect of adjuvants and other immunostimulants on the activation of the innate immune system and, on the other hand, the adaptive immune responses to the specific antigens.

In addition, reproducibility of experimental *F. psychrophilum* vaccination is not yet well established, and there is no standardized vaccination model (Decostere *et al.*, 2000). Immersion trials are the most difficult to reproduce, and experimental conditions have to be established in order to obtain reliable results. By contrast, although reproducible intraperitoneal infection models have been established (Madsen and Dalsgaard, 1999; Garcia *et al.*, 2000), the intraperitoneal route bypasses the natural barrier of the fish to external bacterial threats and is therefore not the best model to analyse vaccine effectiveness.

Another factor that should be studied in depth before designing an effective vaccine is the serotyping within the bacterial species. One of the first serotype classifications (Lorenzen and Olesen, 1997) identified only three serotypes when Danish isolates were compared with others from the rest of Europe (Table 1). These serotypes were 'Th', which was predominant among Danish isolates; 'Fd', a minority serotype, and 'Fp^T', found in isolates that did not come from clinical outbreaks. Izumi and Wakabayashi (1999) defined a similar number of serotypes (Table 1). However, the authors of both classifications pointed out that besides the three defined serotypes, there might be others that were host specific. In fact, Mata *et al.*, identified seven host-specific serotypes in 2002 (Table 1), and Izumi and colleagues (2003) detected a novel serotype isolated from amago (*Oncorhynchus masou rhodurus*). Therefore, differences in serotype should be determined, and cross protection experiments have to be carried out in order to design a vaccine with broad-spectrum effectiveness. The development of increasingly sophisticated molecular techniques made it possible to distinguish different fish species-associated ribotypes (Chakroun *et al.*, 1998). In addition, by multilocus sequence typing analysis on isolates from 10 different host species and four continents, Nicolas and colleagues (2008) showed that there were some clonal complexes with worldwide geographic distribution showing a high level of association with particular fish species. Thus, it seems that there is a correlation among host and particular serotypes, ribotypes and also clonal complexes. This means an additional difficulty for the creation of a vaccine for preventing infection in a wide range of *F. psychrophilum* strains.

Table 1. Comparison of the most representative serological classifications of *F. psychrophilum*

Host	Serotype classification		
	Mata and colleagues (2002)	Lorenzen and Olesen (1997)	Izumi and Wakabayashi (1999)
Salmon	1	Fp^T	O1
Trout	2a	Fd	
Trout	2b	Th-2	
Trout	3	Th-1	O3
Eel	4		
Carp	5		
Tench	6		
Ayu	7		O2

Finally, as a minor consideration, technical complications may arise when large amounts of bacteria or particular antigens have to be produced because of the handling difficulties and the fastidious nature of *F. psychrophilum*.

POTENTIAL VACCINE TARGETS AND VACCINATION ASSAYS PERFORMED WITH *F. PSYCHROPHILUM*

The investigation of virulence factors in *F. psychrophilum* has led to the definition of potential vaccine candidates, from proteins with immunogenic properties, to attenuated strains resulting from mutations in genes related to the infection process. This review includes the principal vaccination trials with subunit, inactivated, attenuated and even wild-type *F. psychrophilum* strains, as well as studies in which potential vaccine targets are suggested for further investigations.

Potential Subunit Vaccines

Studies on the surface characteristics of *F. psychrophilum* have led to the identification of several antigens that may be potential vaccine candidates.

In 2001, Crump and colleagues, in an extensive antigen characterization of several strains, identified three immunogenic cell surface molecules of *F. psychrophilum* that elicited a strong immunological response in fish. Rainbow trout convalescent antiserum, obtained from fish that had survived challenge with *F. psychrophilum* 259–93, recognized, apart from a 20 kDa protein, two molecular mass classes of LPS: a low-molecular-mass LPS antigen of 16 kDa present in the outer layer and a high-molecular-mass LPS (~100 kDa) containing O antigen, which was abundant in culture supernatants (Crump *et al.*, 2001). The authors concluded that these could be potential vaccine candidates. Two years later, Merle and colleagues (2003) also identified (by Western blotting) about 20 major cell envelope antigens which reacted with rabbit antibodies produced against membrane proteins. Among them, the glycoprotein P60 was one of the predominant antigens in the cell envelope of *F. psychrophilum*. This protein was selectively extracted with Triton X-100 and easily purified, and it was proposed as one of the clear candidates for use as a subunit vaccine. Dumetz and colleagues (2007) identified this P60 membrane glycoprotein, called OmpA, as one of the immunodominant antigens of the bacterium. Anti-OmpA antibodies were able to inhibit *F. psychrophilum* growth, showing that OmpA is readily accessible on the cell surface and could be important to the physiology of the bacterium.

Another surface-located *Flavobacterium*-specific protein antigen (FspA) strongly recognized by convalescent rainbow trout sera was characterized by Crump and colleagues (2005). This FspA protein was identified by two-dimensional gel electrophoresis and immunoblotting, and further extensively characterized by nanospray-tandem mass spectrometry and peptide sequencing. This antigen was also expressed in *Escherichia coli* as a fusion protein able to retain its immunoreactivity, representing a well-characterized subunit vaccine candidate against BCWD.

More recently, Sudheesh and colleagues (2007) compared proteins from cellular and extracellular products of a virulent (CSF-259-93) and non-virulent (ATCC 49418) strains of *F. psychrophilum* in order to identify antigens associated with either virulence or host immunity. By Western blot analysis combined with two-dimensional gel electrophoresis, several immunogenic proteins reacting with rainbow trout anti-CSF-259-93 serum and shared by both strains were identified. However, at least 15 immunogenic proteins seemed to be unique for the virulent strain, and four proteins were exclusive for the non-virulent one. By liquid chromatography-mass spectrometry analysis, some of them were identified, and two heat shock proteins, HSP60 and HSP70 were found to be of particular interest. These cellular proteins were expressed in both virulent and non-virulent strains of *F. psychrophilum* and were highly reactive against trout antiserum, indicating that they were indeed immunodominant antigens of the bacteria. However, Plant and colleagues (2009) showed, in an experimental challenge, that vaccination of rainbow trout by recombinant and DNA vaccines based on these proteins resulted in no protection against *F. psychrophilum*.

Apart from all the above-described components representing possible candidates for a *F. psychrophilum* subunit vaccine, Dumetz and colleagues (2008) defined the outer membrane subproteome of the bacterium in order to identify the dominant antigens targeted by the rainbow trout immune system during infection. An immunoproteomic approach using antibodies from BCWD-convalescent rainbow trout allowed the identification of 25 immunoreactive antigens. These included not only previously characterized antigens such as OmpA and FspA, but also newly described antigen proteins and lipoproteins of unknown function, providing a useful reservoir from which to design new vaccines. Recently, LaFrentz and colleagues (2011) performed an immunoproteomic analysis of *F. psychrophilum*. They used two-dimensional polyacrylamide gel electrophoresis and Western blotting with sera from fish immunized with high- and mid-molecular mass fraction proteins of the bacterium. Mass spectrometry was used to determine the protein identity. Fifteen immunogenic proteins were positively identified following Mascot searches in the *F. psychrophilum* genome. Results included outer membrane protein OmpA (P60), trigger factor, ClpB, elongation factor G, gliding motility protein GldN and a conserved hypothetical protein.

Most of these described *F. psychrophilum* antigens were not tested as potential vaccines in fish. Nevertheless, some *F. psychrophilum* preparations have been shown to increase the relative protection of fish against bacterial infection. Thus, Rahman and colleagues (2002) tested a vaccine preparation consisting of the outer membrane fraction of *F. psychrophilum*, obtained from a Sarkosyl cell lysate after ultracentrifugation, and including proteins with a range of molecular masses from 25 to 120 kDa. This preparation was injected intraperitoneally into juvenile rainbow trout and ayu, which produced high antibody titres and resulted in a high relative percent survival (RPS), significantly greater than those in fish immunized with a formalin-killed bacterin (FKB). The increased protection was probably due to the outer membrane proteins and LPS present in a highly enriched preparation. Therefore, this kind of preparation could be useful for further studies using immersion or in-feed administration of vaccine.

LaFrentz and colleagues (2004) evaluated the protective immune response that a specific range of molecular mass proteins from *F. psychrophilum*, when emulsified with Freund's complete adjuvant (FCA), elicited in rainbow trout. The authors demonstrated that protective antigens of *F. psychrophilum* were primarily within the molecular weight range of 60–200 kDa and could involve LPS O-polysaccharide chains and/or proteins. Immunization using the 70–100 kDa region resulted in nearly complete fish protection with a mean RPS of 94% at the lowest challenge dose (6.25×10^6 colony-forming units per fish). Western blot analysis using sera from fish immunized with the 70–100 kDa region showed that high molecular weight proteins and carbohydrates are recognized by serum antibodies. Initially, it was hypothesized that the carbohydrates corresponded to high molecular mass LPS containing O-polysaccharide (LaFrentz *et al.*, 2004). However, some years later, LaFrentz and colleagues (2007) found that the antigens were not LPS, but the components of the glycocalyx of *F. psychrophilum*. These antigens may be involved in eliciting the highly protective immune response. Therefore, they are important candidates for consideration during the development of vaccines against BCWD.

Another candidate for the development of a vaccine is the 18 kDa outer membrane-associated OmpH-like protein identified by Massias and colleagues (2004) and, subsequently, characterized by Dumetz and colleagues (2006). This protein might have different roles, including a chaperone periplasmic function. Vaccination trials by intraperitoneal injection using a fraction highly enriched with the OmpH-like protein with or without FCA induced significant protective immunity in fish, with 54.1% and 88.5% RPS respectively. This protection is probably the result of the relatively high titres of antibodies produced against the flavobacterial OmpH-like protein, which also has antimicrobial activities (Dumetz *et al.*, 2006).

Crump and colleagues (2007) implemented a new strategy to find a subunit vaccine for *F. psychrophilum*. A recombinant vaccine in *E. coli* was generated to avoid the difficulties of handling *F. psychrophilum*. A highly expressed 166-amino-acid fusion protein in *E. coli*, homologous to the bacterial ribosomal protein L10, reacted with convalescent rainbow trout serum. When this chimaera, adjuvanted with FCA, was injected intraperitoneally into rainbow trout, an 82% RPS was obtained, indicating a high level of protection against virulent *F. psychrophilum* challenge. However, it should be taken into account that control fish injected with FCA adjuvant reached a 47% RPS, while saline injected fish presented values below 15%, which clearly demonstrates that the adjuvant has a pronounced effect on the level of protection, probably though the innate immune system. This adjuvant effect should be always assessed in fish vaccination studies because it seems that it could be quantitatively important in protection. Although it is not expected that a ribosomal protein located inside the bacterium could be a main inducer of the adaptive immune system, there are other described cases in *Brucella abortus* (Oliveira *et al.*, 1996; Ribeiro *et al.*, 2002) and *Leishmania* species (Iborra *et al.*, 2005; Stober *et al.*, 2006).

Finally, Plant and colleagues (2011) cloned and purified three recombinant proteins to be used as vaccines in rainbow trout: elongation factor-Tu, SufB Fe-S assembly protein and ATP synthase β of *F. psychrophilum*. Unfortunately, the results indicated that they are not suitable vaccine candidates, at least when administered alone as purified proteins via intraperitoneal injection.

Potential Inactivated Cell Vaccines

Inactivated bacterial whole-cell vaccines are the most commonly used to induce immunity because of their safety and effectiveness when they are administered via injection. There are different methods for obtaining non-viable cells without causing cellular lysis, such as formalin or heat inactivation. From the studies, a significant level of fish immunization can be inferred when adjuvanted-inactivated *F. psychrophilum* cells were used as vaccines. Consequently, they should be considered as good candidates for an inactivated vaccine.

In juvenile rainbow trout sera and mucosae, LaFrentz and colleagues (2002) tested antibody responses after immunizing fish with an average weight of 10 g by the intraperitoneal route using *F. psychrophilum* FKB preparations, together with FCA. In these conditions, a strong serum antibody response was found after 6 weeks. However, 9 weeks was needed to detect significant mucosal antibody titres. In addition, following a 12-week immunization, subcutaneous injection challenge with live *F. psychrophilum* cells was performed. The results showed that 83% RPS protective immunity was conferred to fish immunized with FKB+FCA. Interestingly, a 51% RPS was obtained when saline solution+FCA was used. These results showed that the stimulation of non-specific immune factors increase the immune response. However, specific antibodies were necessary to provide nearly complete protection.

Nevertheless, immersion vaccination trials did not confer any protection.

The efficacy of *F. psychrophilum* FKB was also evaluated in juvenile ayu with an average weight 0.5 g, but in this case, the vaccine was orally administered (Kondo *et al.*, 2003). Juvenile ayu were fed with dry pellets mixed with the FKB, either every day for 2 weeks or on 5 days over 2 weeks. At 3 and 7 weeks after vaccination, experimental immersion challenges were performed. The results showed, in both challenge tests, a significant increase in survival rates in orally vaccinated fish compared with control fish. However, the daily-vaccinated fish group had lower survival rates than the other group with 5 days administration over 2 weeks. The authors pointed out that this could be due to an immunological tolerance probably caused by formalin, as Joosten and colleagues (1995) and Patrie-Hanson and Ainsworth (1999) reported for *V. anguillarum* in carp (*Cyprinus carpio*) and *Edwardsiella ictaluri* in channel catfish (*Ictalurus punctatus*) respectively.

Heat- or formalin-inactivated *F. psychrophilum* of Fd or Th serotypes, adjuvanted with mineral oil, were intraperitoneally injected into rainbow trout with an average weight of 50 g, in order to test the immune response generated (Madetoja *et al.*, 2006). After fish vaccination, antibody titres were significantly higher in vaccinated fish sera than in non-vaccinated ones. In addition, vaccinated fish, when challenged against *F. psychrophilum* under laboratory conditions, showed a significantly higher survival rate compared with non-vaccinated fish (Madetoja *et al.*, 2006). However, in the case of skin mucus, specific antibody production was not increased (Madetoja *et al.*, 2006).

Interestingly, orally administered *F. psychrophilum* FKB from logarithmic phase cultures were able to induce immunity in rainbow trout of 1.6 g average weight (Aoki *et al.*, 2007). However, when FKB for vaccination were obtained from stationary phase cultures, immunization was significantly lower. In logarithmic phase cultures, *F. psychrophilum* cells had many membrane vesicles on the surface that were released into the medium during the stationary phase. The presence of these vesicles in the vaccine preparation seemed to play an important role because in rainbow trout immunized with stationary phase cells combined with a membrane vesicle-rich supernatant, the protection resulted in an RPS of 94–100%, although immunization with membrane vesicle-rich supernatant alone resulted in no protection against *F. psychrophilum* infection (Aoki *et al.*, 2007). Although the nature of the vesicle components was not analysed by the authors, they suggested that a combination of these membrane vesicles, which probably consist of outer membrane lipids, outer membrane proteins and soluble periplasmic components, with FCB was the best choice to obtain high protection levels against the disease.

An intraperitoneal injection vaccine has been developed by Fredriksen and colleagues (2013a). Whole *F. psychrophilum* formalin-inactivated cells of an isolated strain cross binding the Th and Fd serotypes were used as a vaccine in a water-in-oil formulation. Intramuscular infection challenges in rainbow trout (average weight 36.6 g) using a highly virulent *F. psychrophilum* strain demonstrated a high level of protection against BCWD (Fredriksen *et al.*, 2013a).

Similarly, Fredriksen and colleagues (2013b), showed high protection levels (RPS higher than 77.5%) in rainbow trout (average weight 33.1 g), with both divalent and multivalent (named FLAVO IPN and FLAVO AVM6 respectively) intraperitoneally administered vaccines containing whole-cell antigens of *F. psychrophilum*.

Potential Attenuated Vaccines

Live attenuated strains of *F. psychrophilum* may offer the advantage of enhancing the stimulation of the immune system for long periods, whereas the subunit and whole cell-inactivated vaccines have a shorter-lasting effect. However, safety issues as well as environmental aspects must be considered before releasing this kind of vaccine. These risks include reversion to virulence, the spreading of antibiotic resistance genes, exchange of genetic information with other bacteria, changes in host tropism, etc. Therefore, this type of vaccine has the disadvantage of having to be submitted to strict controls and regulations, thus making its application difficult in certain countries. Therefore, up to now only a few examples of this kind of vaccine can be found.

Alvarez and colleagues (2008) obtained an attenuated *F. psychrophilum* mutant in which the gene coding for the *ExbD2* protein was disrupted. This protein is part of the TonB complex and is involved in iron uptake. Intramuscular injection of this mutant followed by challenge experiments with a virulent *F. psychrophilum* strain showed a significant protective immune response in rainbow trout fry. Indeed, 6 weeks after intramuscular immunization using the attenuated strain, an RPS value of 81.8% was obtained. It should be pointed out that some vaccinated fish developed skin lesions after challenge, but, in contrast to the normal progression of the disease, these lesions gradually healed in a few days, at least on visual examination, probably as a consequence of the immune response generated. This healing phenomenon was also observed by LaFrentz and colleagues (2008).

In vaccination studies with attenuated strains, LaFrentz and colleagues (2008) found that a selected rifampicin-resistant *F. psychrophilum* strain, called 259-93B.17, was highly attenuated. This mutant presented five differentially expressed proteins in relation to the wild-type strain, but the role of these proteins, as well as the mechanisms involved in its antibiotic resistance, were still unknown. Rainbow trout immunization by intraperitoneal injection with the 259-93B.17 strain resulted in significant protection against *F. psychrophilum* wild-type strain and produced high specific antibody titres at 8 and 15 weeks after immunization. However, a more relevant result is that fish that had undergone immersion-immunization with the 259-93B.17 strain developed protective immune responses at 10 weeks post-immunization. Consequently, this strain may be useful as an attenuated vaccine against BCWD. Further studies identified a mutation in the *rpoB* gene, which encodes the β-subunit of the RNA polymerase of the 259-93B.17 attenuated strain (Gliniewicz *et al.*, 2012). In this study, eight proteins specific to the wild-type strain and six specific to the attenuated strain were identified (Gliniewicz *et al.*, 2012). Among them, a putative antigen (FP1493) was distinguished by Western blotting. This antigen was later expressed as a recombinant protein and tested as a subunit vaccine. Unfortunately, it did not confer protection against the parental strain. So other differentially expressed and immunoreactive proteins should be evaluated for their effectiveness as subunit vaccines.

Potential Wild-Type Based Vaccines

The use of wild-type strains of the bacterium for developing potential vaccines has been evaluated. Thus, rainbow trout fries were immersed with two pathogenic Danish isolates at the same time and, after a 26-day post-intraperitoneal challenge with the more pathogenic isolate, RPS values were 88.2% and 25% of the fry were positive in specific antibody titres (Lorenzen *et al.*, 2010). However, despite their apparently being very effective in generating immunity, the use of wild-type-based vaccines leads to the risk of virulent strains being released into the environment.

A different vaccination approach was undertaken by Sugahara and Eguchi (2012). Instead of studying the effects of an inactivated vaccine, they investigated the effectiveness of warmed water as a treatment against *F. psychrophilum* in ayu fish. To this end, after 1, 6 or 24 h immersion in a live bacterial suspension, ayu were treated with warmed water (28°C) for three days. Fourteen days after immersion, high antibody titres against *F. psychrophilum* and cumulative mortalities of 36%, 30% and 18% were obtained for warmed water-treated fish after 1, 6 and 24 h immersion respectively. These values were significantly lower than those obtained in control fish (90%). From the results, they concluded that warmed water treatment could cure BCWD in ayu. The authors considered that warmed water probably acted as an inactivation treatment that led *F. psychophilum* cells into the viable but non-culturable state inside the fish organs, having a vaccinating effect (Sugahara *et al.*, 2010). However, this method should only be applied to naturally infected fish as a treatment, but not as a means of vaccination by immersion with live cells because of the risk of environmental contamination with these strains. Additionally, it could not be applied to cold-water fish because of the temperature needed for this procedure.

Passive Immunization

Anti-*F. psychrophilum* sera from either convalescent or immunized adult trout or goat were injected intraperitoneally into rainbow trout fry. Then they were challenged by subcutaneous injection with a virulent strain of *F. psychrophilum*. Higher RPS rates were obtained when fry were administered with sera from immunized adult rainbow trout instead of with sera from convalescent fish. However, goat anti-*F. psychrophilum* serum did not confer protection to fry. Taking into account these results, the authors suggested that antibodies are not the only factors playing a relevant role in conferring protection against *F. psychrophilum* (LaFrentz *et al.*, 2003).

CONCLUSIONS AND FUTURE APPROACHES

The readers of this review may observe that during the last 20 years, multiple attempts have been made to obtain a commercial vaccine for preventing BCWD. A significant number of antigens have been identified that confer immunization in particular conditions. In addition, cross protection between Fd and Th isolated serotypes has been demonstrated, suggesting that some of the major antigens could be shared by both serotypes. In the same way, inactivated vaccines were able to induce a good immunological response. However, most of these experiments were carried out using antigen administration routes that are impractical for field application. The antigenic components of *F. psychrophilum* are well defined, but a procedure for an efficient immersion or oral administration system must still be developed. Attenuated- and virulent-based vaccines are seen to be promising candidates for obtaining a strong immune response induction. However, the use of this kind of vaccine means some environmental and safety risks have to be managed, and most countries, particularly those in the European Union, have restrictive regulations on their use. Moreover, the use of bacteria as delivery vectors for recombinant heterologous antigens implies the construction of recombinant strains and their release into nature. This situates attenuated and live recombinant vaccines in the debate, and there is not yet a consensus as to whether they should be used.

Another important field in fish vaccination is the development of immunostimulants and adjuvants that are administrated before, with or after the vaccines and which could amplify the immune response by increasing the antibody titres in sera and mucus without having inappropriate effects on the fish.

Despite the important advances that have been made in the last few years in relation to different aspects of vaccination against *F. psychrophilum*, new strategies and initiatives are needed in order to develop a definitive commercial vaccine. Studies on oral vaccine administration based on the stimulation of the gut-associated lymphoid system of fish should be conducted. This involves the use and development of an encapsulation method to protect antigens against degradation in the foregut. Nevertheless, it should be established, according to previous experiences, that this procedure may result in a low immune response and an inadequate duration of protection. Concerning immersion vaccination trials, emphasis should be put on the factors influencing the uptake of a particular antigen by fish in order to develop mechanisms to improve this uptake. Consequently, the most appropriate immersion times, water temperatures, antigen concentrations and the tissues participating in the uptake should be established. The lack of normalized values for these parameters is a serious obstacle to further progress, and it is probable that vaccination against *F. psychrophilum* will not be achieved successfully without addressing these issues.

ACKNOWLEDGEMENTS

Esther Gómez and Desireé Cascales were the recipients of grants from the Spanish Ministerio de Ciencia e Innovación (MICINN).

REFERENCES

1. Alvarez, B., and Guijarro, J.A. (2007) Recovery of *Flavobacterium psychrophilum* viable cells using a charcoal-based solid medium. *Lett Appl Microbiol* 44: 569–572.
2. Alvarez, B., Secades, P., McBride, M.J., and Guijarro, J.A. (2004) Development of genetic techniques for the psychrotrophic fish pathogen *Flavobacterium psychrophilum*. *Appl Environ Microbiol* 70: 581–587.
3. Alvarez, B., Secades, P., Prieto, M., McBride, M.J., and Guijarro, J.A. (2006) A mutation in *Flavobacterium psychrophilum* tlpB inhibits gliding motility and induces biofilm formation. *Appl Environ Microbiol* 72:4044–

4053.

4. Alvarez, B., Alvarez, J., Menendez, A., and Guijarro, J.A. (2008) A mutant in one of two exbD loci of a TonB system in *Flavobacterium psychrophilum* shows attenuated virulence and confers protection against cold water disease. *Microbiology* 154: 1144–1151.

5. Aoki, M., Kondo, M., Nakatsuka, Y., Kawai, K., and Oshima, S. (2007) Stationary phase culture supernatant containing membrane vesicles induced immunity to rainbow trout *Oncorhynchus mykiss* fry syndrome. *Vaccine*25: 561–569.

6. Barnes, M.E., and Brown, M.L. (2011) A review of *Flavobacterium psychrophilum* biology, clinical signs, and bacterial cold water disease prevention and treatment. *Open Fish Sci J* 4: 40–48.

7. Bernardet, J., and Kerouault, B. (1989) Phenotypic and genomic studies of '*Cytophaga psychrophila*' isolated from diseased rainbow trout (*Oncorhynchus mykiss*) in France. *Appl Environ Microbiol* 55: 1796–1800.

8. Bernardet, J., Baudin-Laurencin, F., and Tixerant, G. (1988) First identification of *Cytophaga psychrophila* in France. *Bull Eur Assoc Fish Pathol* 8: 104–105.

9. Bernardet, J.-F., and Bowman, J.P. (2006) The genus Flavobacterium. In The Prokaryotes 7. Dworkin, M., Falkow, S., Rosemberg, E., Schleifer, K.H., and Stackebrant, E. (eds). New York, NY, USA: Springer, pp. 481–531.

10. Bertolini, J., Wakabayashi, H., Watral, V., Whipple, M., and Rohovec, J. (1994) Electrophoretic detection of proteases from selected strains of *Flexibacter psychrophilus* and assessment of their variability. *J Aquat Anim Health* 6: 224–233.

11. Borg, A.F. (1948) Studies on Myxobacteria Associated with Diseases in Salmonid Fishes. Seattle, WA, USA: University of Washington.

12. Borg, A.F. (1960) Studies on myxobacteria associated with diseases in salmonid fishes. Washington, DC, USA: American Association for the Advancement of Science, Wildlife Disease.

13. Brown, L., Cox, W., and Levine, R. (1997) Evidence that the causal agent of bacterial cold-water disease*Flavobacterium psychrophilum* is transmitted within salmonid eggs. *Dis Aquat Organ* 29: 213–218.

14. Chakroun, C., Grimont, F., Urdaci, M.C., and Bernardet, J.-F. (1998) Fingerprinting of *Flavobacterium psychrophilum* isolates by ribotyping and plasmid profiling. *Dis Aquat Organ* 33: 167–177.

15. Cipriano, R.C. (2005) Intraovum infection caused by *Flavobacterium psychrophilum* among eggs from captive Atlantic salmon broodfish. *J Aquat Anim Health* 17: 275–283.

16. Cipriano, R.C., and Holt, R.A. (2005) *Flavobacterium psychrophilum*, cause of bacterial cold-water disease and rainbow trout fry syndrome. US Department of the Interior, US Geological Survey, National Fish Health Research Laboratory.

17. Crump, E.M., Perry, M.B., Clouthier, S.C., and Kay, W.W. (2001) Antigenic characterization of the fish pathogen*Flavobacterium psychrophilum*. *Appl Environ Microbiol* 67: 750–759.

18. Crump, E.M., Burian, J., Allen, P.D., and Kay, W.W. (2005) Identification and expression of a host-recognized antigen, FspA, from *Flavobacterium psychrophilum*. *Microbiology* 151: 3127–3135.

19. Crump, E.M., Burian, J., Allen, P.D., Gale, S., and Kay, W.W. (2007) Identification of a ribosomal L10-like protein from *Flavobacterium psychrophilum* as a recombinant vaccine candidate for rainbow trout fry syndrome. *J Mol Microbiol Biotechnol* 13: 55–64.

20. Dalsgaard, I. (1993) Virulence mechanisms in *Cytophaga psychrophila* and other *Cytophaga*-like bacteria pathogenic for fish. *Annu Rev Fish Dis* 3: 127–144.

21. Decostere, A., Lammens, M., and Haesebrouck, F. (2000) Difficulties in experimental infection studies with*Flavobacterium psychrophilum* in rainbow trout (*Oncorhynchus mykiss*) using immersion, oral and anal challenges. *Res Vet Sci* 69: 165–169.

22. Duchaud, E., Boussaha, M., Loux, V., Bernardet, J.F., Michel, C., Kerouault, B., *etal.* (2007) Complete genome sequence of the fish pathogen *Flavobacterium psychrophilum*. *Nat Biotechnol* 25: 763–769.

23. Dumetz, F., Duchaud, E., LaPatra, S.E., Le Marrec, C., Claverol, S., Urdaci, M.C., and Le Henaff, M. (2006) A protective immune response is generated in rainbow trout by an OmpH-like surface antigen (P18) of*Flavobacterium psychrophilum*. *Appl Environ Microbiol* 72: 4845–4852.

24. Dumetz, F., Lapatra, S.E., Duchaud, E., Claverol, S., and Le Henaff, M. (2007) The *Flavobacterium psychrophilum*OmpA, an outer membrane glycoprotein, induces a humoral response in rainbow trout. *J Appl Microbiol* 103:1461–1470.

25. Dumetz, F., Duchaud, E., Claverol, S., Orieux, N., Papillon, S., Lapaillerie, D., and Le Henaff, M. (2008) Analysis of the *Flavobacterium psychrophilum* outer-membrane subproteome and identification of new antigenic targets for vaccine by immunomics. *Microbiology* 154: 1793–1801.

26. Ellis, A.E. (1988) Ontogeny of the immune system in teleost fish. In Fish Vaccination. Ellis, A.E. (ed.). London, UK: Academic Press, pp. 1–50.

27. Fredriksen, B.N., Furevik, A., Gauthier, D., Egenberg, M., Paulsen, E.D., and Brudeseth, B. (2013a) Intramuscular challenge of rainbow trout (*Oncorhynchus mykiss*) with two Norwegian field strains of *Flavobacterium psychrophilum*. *Fish Shellfish Immunol* 35: 595–598.

28. Fredriksen, B.N., Olsen, R.H., Furevik, A., Souhoka, R.A., Gauthier, D., and Brudeseth, B. (2013b) Efficacy of a divalent and a multivalent water-in-oil formulated vaccine against a highly virulent strain of *Flavobacterium psychrophilum* after intramuscular challenge of rainbow trout (*Oncorhynchus mykiss*). *Vaccine* 31: 1994–1998.

29. Garcia, C., Pozet, F., and Michel, C. (2000) Standardization of experimental infection with *Flavobacterium psychrophilum*, the agent of rainbow trout *Oncorhynchus mykiss* fry syndrome. *Dis Aquat Organ* 42: 191–197.

30. Gliniewicz, K., Plant, K.P., LaPatra, S.E., LaFrentz, B.R., Cain, K., Snekvik, K.R., and Call, D.R. (2012) Comparative proteomic analysis of virulent and rifampicin-attenuated *Flavobacterium psychrophilum. J Fish Dis* 35: 529–539.

31. Gomez, E., Perez-Pascual, D., Fernandez, L., Mendez, J., Reimundo, P., Navais, R., and Guijarro, J.A. (2012)Construction and validation of a GFP-based vector for promoter expression analysis in the fish pathogen*Flavobacterium psychrophilum. Gene* 497: 263–268.

32. Hastein, T., Gudding, R., and Evensen, O. (2005) Bacterial vaccines for fish – an update of the current situation worldwide. *Dev Biol (Basel)* 121: 55–74.

33. Iborra, S., Carrion, J., Anderson, C., Alonso, C., Sacks, D., and Soto, M. (2005) Vaccination with the *Leishmania infantum* acidic ribosomal P0 protein plus CpG oligodeoxynucleotides induces protection against cutaneous leishmaniasis in C57BL/6 mice but does not prevent progressive disease in BALB/c mice. *Infect Immun* 73:5842–5852.

34. Izumi, S., and Wakabayashi, H. (1997) Use of PCR to detect *Cytophaga psychrophila* from apparently healthy juvenile ayu [*Plecoglossus altivelis*] and coho salmon [*Oncorhynchus kisutch*] eggs. *Fish Pathol* 32: 169–173.

35. Izumi, S., and Wakabayashi, H. (1999) Further study on serotyping of *Flavobacterium psychrophilum. Fish Pathol*34: 89–90.

36. Izumi, S., Liu, H., Aranishi, F., and Wakabayashi, H. (2003) A novel serotype of *Flavobacterium psychrophilum*detected using antiserum against an isolate from amago, *Oncorhynchus masou rhodurus* Jordan & Gilbert, in Japan. *J Fish Dis* 26: 677–680.

37. Johnson, K., Flynn, J., and Amend, D. (1982) Onset of immunity in salmonid fry vaccinated by direct immersion in *Vibrio anguillarum* and *Yersinia ruckeri* bacterins. *J Fish Dis* 5: 197–205.

38. Joosten, P.H., Aviles-Trigueros, M., Sorgeloos, P., and Rombout, J. (1995) Oral vaccination of juvenile carp (*Cyprinus carpio*) and gilthead seabream (*Sparus aurata*) with bioencapsulated *Vibrio anguillarum* bacterin.*Fish Shellfish Immunol* 5: 289–299.

39. Kondo, M., Kawai, K., Okabe, M., Nakano, N., and Oshima, S. (2003) Efficacy of oral vaccine against bacterial coldwater disease in ayu *Plecoglossus altivelis. Dis Aquat Organ* 55: 261–264.

40. Kumagai, A., Yamaoka, S., Takahashi, K., Fukuda, H., and Wakabayashi, H. (2000) Waterborne transmission of*Flavobacterium psychrophilum* in Coho salmon eggs. *Fish Pathol* 35: 25–28.

41. LaFrentz, B.R., LaPatra, S.E., Jones, G.R., Congleton, J.L., Sun, B., and Cain, K.D. (2002) Characterization of serum and mucosal antibody responses and relative per cent survival in rainbow trout, *Oncorhynchus mykiss*(Walbaum), following immunization and challenge with *Flavobacterium psychrophilum. J Fish Dis* 25: 703–713.

42. LaFrentz, B.R., LaPatra, S.E., Jones, G.R., and Cain, K.D. (2003) Passive immunization of rainbow trout,*Oncorhynchus mykiss* (Walbaum), against *Flavobacterium psychrophilum*, the causative agent of bacterial coldwater disease and rainbow trout fry syndrome. *J Fish Dis* 26: 371–384.

43. LaFrentz, B.R., LaPatra, S.E., Jones, G.R., and Cain, K.D. (2004) Protective immunity in rainbow trout*Oncorhynchus mykiss* following immunization with distinct molecular mass fractions isolated from*Flavobacterium psychrophilum. Dis Aquat Organ* 59: 17–26.

44. LaFrentz, B.R., Lindstrom, N.M., LaPatra, S.E., Call, D.R., and Cain, K.D. (2007) Electrophoretic and Western blot analyses of the lipopolysaccharide and glycocalyx of *Flavobacterium psychrophilum. Fish Shellfish Immunol* 23:770–780.

45. LaFrentz, B.R., LaPatra, S.E., Call, D.R., and Cain, K.D. (2008) Isolation of rifampicin resistant *Flavobacterium psychrophilum* strains and their potential as live attenuated vaccine candidates. *Vaccine* 26: 5582–5589.

46. LaFrentz, B.R., LaPatra, S.E., Call, D.R., Wiens, G.D., and Cain, K.D. (2011) Identification of immunogenic proteins within distinct molecular mass fractions of *Flavobacterium psychrophilum. J Fish Dis* 34: 823–830.

47. Lorenzen, E., and Olesen, N.J. (1997) Characterization of isolates of *Flavobacterium psychrophilum* associated with coldwater disease or rainbow trout fry syndrome I: serological studies. *Dis Aquat Organ* 31: 209–220.

48. Lorenzen, E., Dalsgaard, I., and Bernardet, J.-F. (1997) Characterization of isolates of *Flavobacterium psychrophilum* associated with coldwater disease or rainbow trout fry syndrome I: phenotypic and genomic studies. *Dis Aquat Organ* 31: 197–208.

49. Lorenzen, E., Brudeseth, B.E., Wiklund, T., and Lorenzen, N. (2010) Immersion exposure of rainbow trout (*Oncorhynchus mykiss*) fry to wildtype *Flavobacterium psychrophilum* induces no mortality, but protects against later intraperitoneal challenge. *Fish Shellfish Immunol* 28: 440–444.

50. MacLean, L.L., Vinogradov, E., Crump, E.M., Perry, M.B., and Kay, W.W. (2001) The structure of the lipopolysaccharide O-antigen produced by *Flavobacterium psychrophilum* (259-93). *Eur J Biochem* 268:2710–2716.

51. Madetoja, J., Nyman, P., and Wiklund, T. (2000) *Flavobacterium psychrophilum*, invasion into and shedding by rainbow trout *Oncorhynchus mykiss. Dis Aquat Organ* 43: 27–38.

52. Madetoja, J., Lonnstrom, L.G., Bjorkblom, C., Ulukoy, G., Bylund, G., Syvertsen, C., *etal.* (2006) Efficacy of injection vaccines against *Flavobacterium psychrophilum* in rainbow trout, *Oncorhynchus mykiss* (Walbaum). *J Fish Dis* 29: 9–20.

53. Madsen, L., and Dalsgaard, I. (1999) Reproducible methods for experimental infection with *Flavobacterium psychrophilum* in rainbow trout *Oncorhynchus mykiss. Dis Aquat Organ* 36: 169.

54. Madsen, L., and Dalsgaard, I. (2008) Water recirculation and good management: potential methods to avoid disease outbreaks with *Flavobacterium psychrophilum.*

J Fish Dis 31: 799–810.

55. Madsen, L., Moller, J.D., and Dalsgaard, I. (2005) *Flavobacterium psychrophilum* in rainbow trout, *Oncorhynchus mykiss* (Walbaum), hatcheries: studies on broodstock, eggs, fry and environment. *J Fish Dis* 28: 39–47.
56. Massias, B., Dumetz, F., Urdaci, M.C., and Le Henaff, M. (2004) Identification of P18, a surface protein produced by the fish pathogen *Flavobacterium psychrophilum*. *J Appl Microbiol* 97: 574–580.
57. Mata, M., Skarmeta, A., and Santos, Y. (2002) A proposed serotyping system for *Flavobacterium psychrophilum*. *Lett Appl Microbiol* 35: 166–170.
58. Merle, C., Faure, D., Urdaci, M.C., and Le Henaff, M. (2003) Purification and characterization of a membrane glycoprotein from the fish pathogen *Flavobacterium psychrophilum*. *J Appl Microbiol* 94: 1120–1127.
59. Moller, J.D., Larsen, J.L., Madsen, L., and Dalsgaard, I. (2003) Involvement of a sialic acid-binding lectin with hemagglutination and hydrophobicity of *Flavobacterium psychrophilum*. *Appl Environ Microbiol* 69: 5275–5280.
60. Nematollahi, A., Decostere, A., Pasmans, F., and Haesebrouck, F. (2003) *Flavobacterium psychrophilum infections in salmonid fish*. *J Fish Dis* 26: 563–574.
61. Nicolas, P., Mondot, S., Achaz, G., Bouchenot, C., Bernardet, J.-F., and Duchaud, E. (2008) Population structure of the fish-pathogenic bacterium *Flavobacterium psychrophilum*. *Appl Environ Microbiol* 74: 3702–3709.
62. Oliveira, S.C., Harms, J.S., Banai, M., and Splitter, G.A. (1996) Recombinant *Brucella abortus* proteins that induce proliferation and gamma-interferon secretion by CD4+ T cells from *Brucella*-vaccinated mice and delayed-type hypersensitivity in sensitized guinea pigs. *Cell Immunol* 172: 262–268.
63. Ostland, V., Byrne, P., Hoover, G., and Ferguson, H. (2000) Necrotic myositis of rainbow trout, *Oncorhynchus mykiss* (Walbaum): proteolytic characteristics of a crude extracellular preparation from *Flavobacterium psychrophilum*. *J Fish Dis* 23: 329–336.
64. Patrie-Hanson, L., and Ainsworth, A.J. (1999) Humoral immune responses of channel catfish (*Ictalurus punctatus*) fry and fingerlings exposed to *Edwardsiella ictaluri*. *Fish Shellfish Immunol* 9: 579–589.
65. Perez-Pascual, D., Menendez, A., Fernandez, L., Mendez, J., Reimundo, P., Navais, R., and Guijarro, J.A. (2010)Spreading versus biomass production by colonies of the fish pathogen *Flavobacterium psychrophilum*: role of the nutrient concentration. *Int Microbiol* 12: 207–214.
66. Perez-Pascual, D., Gomez, E., Alvarez, B., Mendez, J., Reimundo, P., Navais, R., *et al.* (2011) Comparative analysis and mutation effects of fpp2-fpp1 tandem genes encoding proteolytic extracellular enzymes of*Flavobacterium psychrophilum*. *Microbiology* 157: 1196–1204.
67. Plant, K.P., LaPatra, S.E., and Cain, K.D. (2009) Vaccination of rainbow trout, *Oncorhynchus mykiss* (Walbaum), with recombinant and DNA vaccines produced to *Flavobacterium psychrophilum* heat shock proteins 60 and 70. *J Fish Dis* 32: 521–534.
68. Plant, K.P., Lapatra, S.E., Call, D.R., and Cain, K.D. (2011) Immunization of rainbow trout, *Oncorhynchus mykiss*(Walbaum), with *Flavobacterium psychrophilum* proteins elongation factor-Tu, SufB Fe-S assembly protein and ATP synthasebeta. *J Fish Dis* 34: 247–250.
69. Rahman, H., Kuroda, A., Dijkstra, J.M., Kiryu, I., Nakanishi, T., and Ototake, M. (2002) The outer membrane fraction of *Flavobacterium psychrophilum* induces protective immunity in rainbow trout and ayu. *Fish Shellfish Immunol* 12: 169–179.
70. Rangdale, R., Richards, R., and Alderman, D. (1997) Colonisation of eyed rainbow trout ova with *Flavobacterium psychrophilum* leads to rainbow trout fry syndrome in fry. *Bull Eur Assoc Fish Pathol* 17: 108–111.
71. Ribeiro, L.A., Azevedo, V., Le Loir, Y., Oliveira, S.C., Dieye, Y., Piard, J.C., *et al.* (2002) Production and targeting of the *Brucella abortus* antigen L7/L12 in *Lactococcus lactis*: a first step towards food-grade live vaccines against brucellosis. *Appl Environ Microbiol* 68: 910–916.
72. Siwicki, A.K., Pozet, F., Morand, M., Kazuñ, K., Glabski, E., and Trapkowska, S. (2004) Atypical bacterial gill disease in rainbow trout (*Oncorhynchus mykiss*): influence on non specific humoral defence mechanisms. *Archiv Polish Fish* 12: 5–12.
73. Stober, C.B., Lange, U.G., Roberts, M.T., Gilmartin, B., Francis, R., Almeida, R., *et al.* (2006) From genome to vaccines for leishmaniasis: screening 100 novel vaccine candidates against murine *Leishmania* major infection. *Vaccine* 24: 2602–2616.
74. Sudheesh, P.S., LaFrentz, B.R., Call, D.R., Siems, W.F., LaPatra, S.E., Wiens, G.D., and Cain, K.D. (2007) Identification of potential vaccine target antigens by immunoproteomic analysis of a virulent and a non-virulent strain of the fish pathogen *Flavobacterium psychrophilum*. *Dis Aquat Organ* 74: 37–47.
75. Sugahara, K., and Eguchi, M. (2012) The use of warmed water treatment to induce protective immunity against the bacterial cold-water disease pathogen *Flavobacterium psychrophilum* in ayu (*Plecoglossus altivelis*). *Fish Shellfish Immunol* 32: 489–493.
76. Sugahara, K., Fujiwara-Nagata, E., Fukuda, A., and Eguchi, M. (2010) Viable but non-culturable state of bacterial cold-water disease pathogen *Flavobacterium psychrophilum* at various temperatures. *Fish Pathol* 45: 158–163.
77. Tatner, M.F., and Horne, M.T. (1983) Susceptibility and immunity to *Vibrio anguillarum* in post-hatching rainbow trout fry, *Salmo gairdneri* Richardson 1836. *Dev Comp Immunol* 7: 465–472.
78. Vatsos, I.N., Thompson, K.D., and Adams, A. (2001) Adhesion of the fish pathogen *Flavobacterium psychrophilum* to unfertilized eggs of rainbow trout (*Oncorhynchus mykiss*) and n-hexadecane. *Lett Appl Microbiol* 33: 178–182.
79. Vatsos, I.N., Thompson, K.D., and Adams, A.

(2006) Colonization of rainbow trout, *Oncorhynchus mykiss*(Walbaum), eggs by *Flavobacterium psychrophilum*, the causative agent of rainbow trout fry syndrome. *J Fish Dis* 29: 441–444.

80. Wakabayashi, H., Horiuchi, M., Bunya, T., and Hoshiai, G. (1991) Outbreaks of cold-water disease in coho salmon in Japan. *Fish Pathol* 26: 211–212.

81. von Weis, J. (1987) Über das Vorkommen einer Kaltwasserkrankheit bei Regenbogenforellen, Salmo gairdneri.*Tieraerztl Umsch* 42: 575–577.

82. Wiklund, T., and Dalsgaard, I. (2002) Survival of *Flavobacterium psychrophilum* in rainbow trout (*Oncorhynchus mykiss*) serum in vitro. *Fish Shellfish Immunol* 12: 141–153.

CHAPTER 11

FISH CYTOKINES AND IMMUNE RESPONSE

Sebastián Reyes-Cerpa, Kevin Maisey, Felipe Reyes-López, Daniela Toro-Ascuy, Ana María Sandino, and Mónica Imarai

INTRODUCTION

The immune system can be defined as a complex system that protects the organism against organisms or substances that might cause infection or disease. One of the most fascinating characteristics of the immune system is its capability to recognize and respond to pathogens with significant specificity. Innate and adaptive immune responses are able to recognize foreign structures and trigger different molecular and cellular mechanisms for antigen elimination. The immune response is critical to all individuals; therefore numerous changes have taken place during evolution to generate variability and specialization, although the immune system has conserved some important features over millions of years of evolution that are common for all species. The emergence of new taxonomic categories coincided with the diversification of the immune response. Most notably, the emergence of vertebrates coincided with the development of a novel type of immune response. Apparently, vertebrates inherited innate immunity from their invertebrate ancestors [1].

In higher vertebrates, the immune system consists of primary and secondary lymphoid organs with distinct compartments and morphology located in anatomically distinct sites. The thymus and bone marrow constitute the primary lymphoid organs, while the spleen, lymph nodes, and mucosal associated lymphoid tissue (MALT) comprise the secondary lymphoid organs [2].

Fish are a heterogeneous group divided into three classes: Agnatha (jawless fish such as the hagfish and lampreys), Chrondrichthyes (cartilaginous fish such as sharks, rays and skates) and Osteichthyes (bony fish) [3]. As in all vertebrates, fish have cellular and humoral immune responses and organs, the main function of which is immune defence. Most generative and secondary lymphoid organs in mammals are also found in fish, except for lymphatic nodules and bone marrow [3].

The head kidney or pronephros has hematopoietic functions [3, 4], and unlike in higher vertebrates, it is the immune organ involved in phagocytosis [5], antigen processing, production of IgM [6, 7] and immune memory through melanomacrophagic centres [8, 9]. The thymus, another lymphoid organ situated near the opercular cavity in teleosts, produces T lymphocytes involved in allograft rejection, stimulation of phagocytosis and antibody production by B cells [10, 11]. The spleen is a large, blood-filtering organ that undergoes increasing structural complexity in order to augment its efficiency in trapping and processing antigens [12-15]. Melanomacrophage centres are present for clearance of ingested material and can be surrounded by immunoglobulin-positive cells, especially after immunization [8]. Proliferation of granular cells has also been observed in association with ellipsoids and melanomacrophage centres after immunization [16].

Innate and Adaptive Immune Response

The development of an immune system is essential for the survival of living organisms. In vertebrates, immunity can be divided into two components, the innate immune response and the adaptive immune response. The innate immune response is the initial line of defence against infection, which includes physical barriers and cellular response. The adaptive immune response is capable of specific antigen recognition and is responsible for the secondary immune response.

The innate immune system recognizes conserved molecular structures common to pathogenic microorganisms such as polysaccharides, lipopolysaccharides (LPS), peptidoglycans, bacterial DNA, and double-strand viral RNA, among others, through their interaction with specific receptors

Citation: Sebastián Reyes-Cerpa, Kevin Maisey, Felipe Reyes-López, Daniela Toro-Ascuy, Ana María Sandino and Mónica Imarai (2012). Fish Cytokines and Immune Response, New Advances and Contributions to Fish Biology, Prof. Hakan Turker (Ed.), InTech, DOI: 10.5772/53504.

like toll receptors (TLRs). These mechanisms of recognition may lead directly to successful removal of pathogens, for instance by phagocytosis, or may trigger additional protective responses through induction of adaptive immune responses [17]. Cells of the innate immune system have a diverse array of functions. Some cells are phagocytic, allowing them to engulf and degrade pathogenic particles. Other cells produce and secrete cytokines and chemokines that can stimulate and help guide the migration of cells and further direct the immune response [18].

The adaptive system recognizes foreign structures by means of two cellular receptors, the B cell receptor (BCR) and the T cell receptor (TCR). Adaptive immunity is highly regulated by several mechanisms. It increases with antigen exposure and produces immunological memory, which is the basis of vaccine development and the preventive function of vaccines [19, 20]. The adaptive response generally starts days after infection and is capable of recognizing specific protein motifs of peptides, which leads to a response that increases in both speed and magnitude with each successive exposure [21]. The main effector cells of the adaptive immune response are the lymphocytes, specifically B cells and T cells. When B cells are activated, they are capable of differentiating into plasma cells that can secrete antibodies. Upon activation T cells differentiate into either helper T cells or cytotoxic T cells. Helper T cells are capable of activating other cells of the adaptive immune response such as B cells and macrophages, while cytotoxic T cells upon activiation are able to kill cells that have been infected [22].

Fish Immune Response

Immune responses in fish have not been as well characterized as they have in higher vertebrates. Consequently, there is not enough information about the components of the fish immune system and its function and regulation. Key immune mammalian homologous genes have been identified in several fish species, suggesting that the fish immune system shares many features with the mammalian system. For example, the identification of α and β T cell receptor genes (TCR) [23], key T cell markers such as CD3, CD4, CD8, CD28, CD40L, and a great number of cytokines and chemokines [24-26] suggest that T helper (Th)1, Th2 and Th17 and the regulatory subset Treg are present in fish. Some cell subsets have been better studied mainly because their activity can be easily differentiated and measured, as in the case of cytotoxic cells [27] and macrophages [28, 29]. Finally, B cells have been much more studied due to the availability of monoclonal antibodies that have been isolated and identified by a number of techniques [30, 31]. Phenotypic characterization of leukocytes has been hampered mainly by the lack of membrane cell markers [32, 33]. Researchers anticipate developing antibodies for cell lineage markers of fish immunocompetent cells that can be used to isolate and characterize immune cells to obtain insights into their regulation and role in immune response [34-36].

Antibodies in teleosts play a key role in the immune response. In general, IgM is the main immunoglobulin in teleosts that can elicit effective specific humoral responses against various antigens. For IgM, one gene alone can generate as many as six structural isoforms. Therefore, diversity is the result of structural organization rather than genetic variability [37]. Recently, several reports have provided evidence for the existence of IgD/IgZ/IgT in fish [38-41]. Interestingly, B cells from rainbow trout and salmon have high phagocytic capacity, suggesting a transition in B lymphocyte during evolution in which a key cell type of the innate immunity and phagocytosis evolved into a highly specialized component of the adaptive immune response in higher vertebrates [42, 43].

Fish Cytokines

Cytokines are secreted proteins with growth, differentiation, and activation functions that regulate the nature of immune responses. Cytokines are involved in several steps of the immune response, from induction of the innate response to the generation of cytotoxic T cells and the production of antibodies. In higher vertebrates, the combination of cytokines that are secreted in response to an immune stimulation induces the expression of immune-related genes through multiple signalling pathways, which contributes to the initiation of the immune response. Cytokines can modulate immune responses through an autocrine or paracrine manner upon binding to their corresponding receptors [44].

Cytokines have overlapping and sometimes contradictory pleiotropic functions that make their classification difficult. Cytokines are produced by macrophages, lymphocytes, granulocytes, DCs, mast cells, and epithelial cells, and can be divided into interferons (IFNs), interleukins (ILs), tumor necrosis factors (TNFs), colony stimulating factors, and chemokines [45]. They are secreted by activated immune-related cells upon induction by various pathogens, such as parasitic, bacterial, or viral components [46]. Macrophages can secret IL-1, IL-6, IL-12, TNFα, and chemokines such as IL-8 and MCP-1, all of which are indispensable for macrophage, neutrophil, and lymphocyte recruitment to the infected tissues and their activation as pathogen eliminators [47]. Meanwhile, cytokines released by phagocytes in tissues can also induce acute phase proteins, including mannose-binding lectin (MBL) and C-reactive protein (CRP), and promote migration of DCs [48].

Fish appear to possess a repertoire of cytokines similar to those of mammals. To date several cytokine homologues and suppressors have been cloned in fish species [24, 25, 49]. Some cytokines described in fish are TNFα, IL-1β, IL-6 or IFN.

Current knowledge of fish cytokines is based on mammal models of the cytokines network and their complex interactions. In this review we included the pro-inflammatory cytokines associated with innate and adaptive immunity, regulatory cytokines and anti-inflammatory cytokines.

Pro-Inflammatory Fish Cytokines

Tumour Necrosis Factor α *(TNF*α*)*

TNFα (tumour necrosis factor alpha) is a pro-inflammatory cytokine that plays an important role in diverse host responses, including cell proliferation, differentiation, necrosis, apoptosis, and the induction of other cytokines. TNFα can induce either NF-kB mediated survival or apoptosis, depending on the cellular context [50]. TNFα mediates powerful anti-microbial responses, including inducing apoptosis, killing infected cells, inhibiting intracellular pathogen replication, and up-regulating diverse host response genes. Many viruses have evolved strategies to neutralize TNFα by direct binding and inhibition of the ligand or its receptor or modulation of various downstream signalling events [51].

TNFα has been identified, cloned, and characterized in several bony fish, including Japanese flounder [52], rainbow trout [53, 54], gilthead seabream [55], carp [56] catfish [57], tilapia [58], turbot [59] and goldfish [60]. These studies have revealed the existence of some obvious differences from their mammalian counterpart, such as the presence of multiple isoforms of TNFα in some teleost species [54,56] the high constitutive expression of this gene in different tissues of healthy fish and its relatively poor up-regulation by immune challenge *in vitro* and *in vivo* [53, 55, 57]. However, the most unexpected and interesting difference between fish and mammal TNFα concerns the weak *in vitro*effects of TNFα on phagocyte activation in goldfish [60], rainbow trout [57], turbot [59] and gilthead seabram [61]. This weak *in vitro* activity of fish TNFα sharply contrasts with the powerful actions exerted by the i.p. injection of recombinant TNFα in gilthead seabream, which includes the recruitment of phagocytes to the injection site, with a concomitant strong increase in their respiratory burst [61]. Apparently endothelial cells are the main target cells of fish TNFα, suggesting that TNFα is mainly involved in the recruitment of leukocytes to the inflammatory foci rather than in their activation [62]. Despite the above, differential expression has been observed in studies with rainbow trout leucocytes, which have shown increased response to different pro-inflammatory stimuli, as human recombinant TNFα [63], LPS [53, 64], zimosan and muramyl dipeptide as a peptidoglycan constituent of both gram-positive and gram-negative bacteria [64]. Moreover, it is known that Infectious Pancreatic Necrosis Virus (IPNV)-mediated up-regulation of TNFα regulates both the Bad/Bid-mediated apoptotic pathway and the RIP1 (receptor-interacting protein-1)/ROS-mediated secondary necrosis pathway [65].

Interleukin 1 Family

In mammals, the 11 members of the Interleukin-1 family include IL-1α (IL-1F1), IL-1β (IL-1F2), IL-1 receptor antagonist (IL-1ra/IL-1F3), IL-18 (IL-1F4), IL-1F5-10 and IL-33 (IL-1F11). These molecules tend to be either pro-inflammatory or act as antagonists that inhibit the activities of particular family members [66]. Despite these semantic issues, to date only two clear homologues of these molecules have been discovered in fish, IL-1β and IL-18 [24].

Interleukin 1

IL-1β is one of the earliest expressed pro-inflammatory cytokines and enables organisms to respond promptly to infection by inducing a cascade of reactions leading to inflammation. Many of the effector roles of IL-1β are mediated through the up- or down-regulation of expression of other cytokines and chemokines [67]. Mammalian IL-1β is produced by a wide variety of cells, but mainly by blood monocytes and tissue macrophages. IL-1β was the first interleukin to be characterized in fish and has since been identified in a number of fish species, such as rainbow trout [68], carp [69], sea bass [70], gilt head seabram [71], haddock [72], tilapia [73]. A second IL-1β gene (IL-1beta2] has been identified in trout [74]

In mammals pro-IL-1β remains cytosolic and requires cellular proteases to release the mature peptide. It is known that the peptide is cleaved by the IL-1β converting enzyme (ICE) [75]. However, the aspartic acid residue for which this enzyme has specificity is not present in all fish genes sequenced to date. Nevertheless, using a combination of multiple alignments and analysis of the N-terminal sequences of known mature peptides, it is possible to predict fish gene cutting sites. In trout, this gives a mature peptide of 166 and 165 aminoacids for IL-1β1 and IL-1β2 [76].

Like its mammalian counterpart, teleost IL-1β has been found to be regulated in response to various stimuli, such as LPS or poly I:C [68, 70-74, 77-81]. The biological activity of recombinant IL-1β (rIL-1β) has been studied in several fish species, indicating that fish IL-1β is involved in the regulation of immune relevant genes, lymphocyte activation, migration of leucocytes, phagocytosis and bactericidal activities [77, 81-84].

Interleukin 18

In mammals, IL-18 is mainly produced by activated macrophages. It is an important cytokine with multiple functions in innate and acquired immunity [85-87]. One of its primary biological properties is to induce interferon gamma (IFNγ) synthesis in Th1 and NK cells in synergy with IL-12 [88, 89]. It promotes T and NK cell maturation, activates neutrophils and enhances Fas ligand-mediated cytotoxicity [90-92]. Like IL-1β, it is synthesized as an inactive precursor of approximately 24 kDa and is stored intracellularly. Activation and secretion of IL-18 is mainly effected through specific cleavage of the precursor after D35 by caspase 1, also termed the IL-1β-converting enzyme (ICE), which is believed to be one of the key processes regulating IL-18 bioactivity [93, 94]. Some other enzymes, including caspase 3 and neutrophil proteinase 3, also cleave the IL-18 precursor to generate active or inactive mature molecules [95, 96].

IL-18 was discovered in fish by analysis of sequenced fish genomes (fugu) and EST databases (medaka) [97, 98]. An alternative splicing form of the IL-18 mRNA was discovered in trout that may have an important role in regulating IL-18 expression and processing in this species. This form shows a lower constitutive expression relative to the full length transcript, but unlike the full length transcript, it increases in response to LPS and polyI:C stimulation in the RTG-2 fibroblast cell line [98]. The expression level of the full length transcript can increase in response to LPS plus IL- 1b in head kidney leucocyte cultures, and by IFNγ in RTS-11 cells [99].

Other Pro-Inflammatory Cytokines

Interleukin 6

A number of other interleukins are considered pro-inflammatory, some of which are released during the cytokine cascade that follows bacterial infection. Of these IL-6 is one of the best known, and is itself a member of the IL-6 family of cytokines that includes IL-11 and IL-31, as well as cytokines such as mammalian CNTF, LIF, OSM, CT-1 and CT-2 [24]. Whilst the homology of known fish molecules with many of these IL-6 family members is not conclusive [100], true homologues appear to be present in at least in the cases of IL-6 and IL-11

[24]. IL-6 is produced by a diverse group of cells including T lymphocytes, macrophages, fibroblasts, neurons, endothelial and glial cells. The pleiotropic effects of IL-6 are mediated by a 2-subunit receptor [101] and include the regulation of diverse immune and neuro-endocrine processes. IL-6 has been implicated in the control of immunoglobulin production, lymphocyte and monocyte differentiation, chemokine secretion and migration of leukocytes to inflammation sites [102-104].

IL-6 was first discovered in fugu by analysis of the genome sequence [105] and subsequently in other species as part of EST analysis of immune gene-enriched cDNA libraries [106-108]. However, little is known about the function and signalling pathways of IL-6 in fish. Interestingly, trout IL-6 expression in macrophages is reported to be induced by LPS, poly I:C and IL-1β in the macrophage cell line RTS-11, as well as in head kidney macrophages [109]. Moreover, IL-6 induces the expression of itself, so it can act in an autocrine and paracrine fashion to increase its expression, with the potential to both amplify and exacerbate the inflammatory response. However, IL-6 can significantly down-regulate the expression of trout TNFα1, TNFα2, and IL-1β, suggesting a potential role of trout IL-6 in limiting host damage during inflammation [109].

Interleukin 11

In mammals, IL-11 is produced by many cell types throughout the body. Basal and inducible IL-11 mRNA expression can be detected in fibroblasts, epithelial cells, chondrocytes, synoviocytes, keratinocytes, endothelial cells, osteoblasts and certain tumour cells and cell lines [110]. Viral [111] and bacterial infection [112] and cytokine stimulation (IL-1, TNFα and TGF-β1] induce IL-11 expression. IL-11 acts on multiple cell types, including hemotopoietic cells, hepatocytes, adipocytes, intestinal epithelial cells, tumour cells, macrophages, and both osteoblasts and osteoclasts. In the hematopoietic compartment IL-11 supports multilineage and committed progenitors, contributing to myeloid, erythroid, megakaryocyte and lymphoid lineages [113]. IL-11 is also an anti-inflammatory cytokine that inhibits the production of pro-inflammatory cytokines from lipopolysaccharde (LPS)-stimulated macrophages [114]. In combination with its trophic effects on the gastrointestinal epithelium, IL-11 plays an important role in the protection and restoration of gastrointestinal mucosa [115, 116].

The teleostean IL-11 orthologue has been found to consist of duplicate IL-11 genes, named IL-11a and IL-11b [117], with expression patterns indicating that both divergent forms of teleostean IL-11 play roles in antibacterial and antiviral defence mechanisms of fish [117-119]. In trout, IL-11 molecule is grouped with IL-11a and is constitutively expressed in intestine and gills and is highly up-regulated at other immune sites (spleen, head kidney, liver) following bacterial infection. *In vitro*, the macrophage-like RTS-11 cell line has shown enhanced IL-11 expression in response to LPS, bacteria, poly I:C and rIL-1β [118]. In carp, IL-11a is modulated by LPS, ConA and peptidoglycan in head kidney macrophages [117, 120] and cortisol has been found to inhibit IL-11 expression on its own and in combination with LPS [117]. In contrast to carp IL-11a, which shows low levels of constitutive expression in blood leucocytes, IL-11b in Japanese flounder shows higher expression at this site, and strong up-regulation was found in response to rhabdovirus infection in kidney cells [119]. This suggests that these paralogues have some complementarity of function related to their differential expression, although study of both forms in a single experiment is still required [24].

Chemokines

Chemokines are a superfamily of approximately 40 different small secreted cytokines that direct the migration of immune cells to infection sites. Their activity is coordinated by binding to G-protein-linked receptors with seven transmembrane domains. Four distinct subgroups make up the chemokine superfamily. These are designated as CXC (or a), CC (or b), C (or g) and CX_3C (or d), which are defined by the arrangement of the first two cysteine residues within their peptide structure. The CC subfamily can be further subdivided according to the total number of cysteine residues, as some members of this group contain four cysteines whilst the remainder possesses six (and are known as the C6-b group). Similarly, the CXC subfamily contains two subgroups based on whether or not the first two cysteines are preceded by a Glu-Leu-Arg (ELR) motif associated with specificity to neutrophils [76, 121].

Interleukin 8

An important chemokine related to the pro-inflammatory process is CXCL-8, also called interleukin 8, this chemokine is a member of the CXC chemokine subfamily and attracts neutrophils, T lymphocytes and basophils *in vitro,* but not macrophages or monocytes [122]. Many cell-types, including macrophages, produce IL-8 in response to a variety of stimuli (LPS, cytokines and viruses). The neutrophil-attracting ability of IL-8 can be attributed to the presence of the ELR motif adjacent to the CXC motifs at its N-terminus, presumably by affecting its binding to specific receptors [123, 124]. In contrast, CXC chemokines lack an ELR motif and specifically attract lymphocytes but not neutrophils. The biological effects of IL-8 on neutrophils include increased cytosolic calcium levels, respiratory burst, a change in neutrophil shape and chemotaxis[125].

The fish IL-8 has been found in flounder [126], trout [125, 127], catfish [128], and lamprey [129]. *In vitro* stimulation of a trout macrophage cell line (RTS-11) [125] or *in vivo* intraperitoneal challenge [78] with either LPS or poly I:C did result in clear up-regulation of IL-8 expression. Moreover, induction of IL-8 expression in primary cultures of rainbow trout leukocytes stimulated for 24 hours with LPS and TNFα confirms that this fish chemokine is associated with inflammatory response, as has been suggested in mammals [127]. Interestingly, the ELR motif associated with the neutrophil-attracting ability is absent from the lamprey molecule and it is similar in flounder, where CXCL8 also lacks the ELR motif and appears to be regulated by a bacterial mechanism, since its transcript has only been detected in the major immune organs (spleen and head kidney) of an LPS stimulated flounder. The case of the trout is different, although there is also no ELR preceding the CXC motif, it has a very similar motif (DLR) in this position [130]. The human CXCL8 molecule, where the ELR motif has been mutated to DLR, retains neutrophil-attracting ability, albeit at lower potency [123]. Consequently, it is possible that the trout molecule has similar chemotactic activity to that of mammalian CXCL8 [130].

The Interleukin 2 Family

The IL-2 subfamily of cytokines signals via the common gamma chain (gC or CD132), a member of the type I cytokine receptor family expressed in most leucocytes.These cytokines in mammals include IL-2, IL-4, IL-7, IL-9, IL-15 and IL-21. IL-2, IL-4, IL-9 and IL-21 are all cytokines released from Th cells, which affect their responses [24], whilst IL-7 and IL-15 are particularly important for the maintenance of T cell memory [131]. To date molecules with homology to all of these have been found in fish, except IL-9 [24].

Interleukin 2

Interleukin-2 (IL-2 is an important immunomodulatory cytokine that primarily promotes proliferation, activation and differentiation of T cells [132]. IL-2, initially known as T-cell growth factor (TCGF), is synthesized and secreted mainly by Th1 cells that have been activated by stimulation by certain mitogens or by interaction of the T-cell receptor with the antigen/MHC complex on the surface of antigen-presenting cells [133-135]. Although CD4 T cells are the major source of IL-2 production in response to TCR stimulation, transient induction of IL-2 mRNA and production of the protein has been detected in murine dendritic cells activated by gram-negative bacteria [136]. IL-2 can also be produced by B cells in certain situations [137, 138]. The produced IL-2 promotes the expansion and survival of activated T cells and is also required for the activation of natural killer (NK) cells [139] and for immunoglobulin (Ig) synthesis by B cells [140].

The IL-2 gene has been detected only recently in fish by analysis of the fugu genome sequence, which also identified IL-21 as a neighbouring gene, as in mammals, providing the first direct evidence for the existence of a true IL-2 homologue in bony fish [141]. The gene has a 4 exon/3 intron organisation, as in mammals, and showed no constitutive expression in a range of tissues examined. However, injection of Fugu with poly I:C induced expression of IL-2 in the gut and gills [141]. Moreover, IL-2 could be induced in head kidney cell cultures stimulated with PHA, and in T-cell enriched cultures isolated from PBL when stimulated with B7-H3 or B7- H4 Ig fusions proteins in the presence of PHA [24, 142]. IL-2 has since been cloned in rainbow trout [143, 144]. The trout IL-2 was significantly up-regulated in head kidney leucocytes by the T cell mitogen PHA and in classical mixed leucocyte reactions and *in vivo*following infection with bacteria (*Y. ruckeri*) or the parasite *Tetracapsuloides bryosalmonae*. More importantly, the recombinant trout IL-2 produced in *Escherichia coli* was shown to induce expression of two transcription factors (STAT5 and Blimp-1) known to be involved in IL-2 signalling in mammals [143], as well as interferon-g (IFNγ) and IL-2 itself, and a CXC chemokine known to be induced by IFNγ, termed a IFNγ-inducible protein (γIP) [145].

Interleukin 4

Interleukin-4 IL-4 is a pleiotropic cytokine produced by T cells, mast cells, and basophils and is known to regulate an array of functions in B cells, T cells, macrophages, hematopoietic and non-hematopoietic cells [146, 147]. IL-4 serves as a key cytokine in driving Th2 differentiation and mediating humoral immunity, allergic responses and certain autoimmune diseases [148]. The IL-4 gene is conserved evolutionally in the animal kingdom and has been isolated from various animals including humans [149], mice [150, 151] and bovines [152], in which the IL-4 locus has been mapped in a region adjacent to those of IL-5 and IL-13 on the same chromosome [153, 154].

Teleost fish have two genes of the IL-4/13 family, IL-4/13A and IL-4/13B, which are situated on separate chromosomes in regions that duplicated during the fish-specific whole genome duplication (FS-WGD) around 350 million years ago [155, 156]. A few IL-4-like genes have been found in fish to date. The first was discovered by searching the *Tetraodon nigroviridis* genome [157]. In this work, IL-4 was constitutively expressed in head kidney, spleen, liver, brain, gill, muscle and heart. The ubiquitous expression of IL-4 is consistent with a postulated role in immune cytokines regulation. Stimulating the fish with a mixed stimulant containing ConA, PHA and PMA significantly up-regulated the expression of IL-4, which suggests that IL-4 is involved in the immune inflammatory responses triggered by mitogens [157], as in mammals, where it has been observed that this mitogen increases IL-4 expression [158]. However, the homology (amino acid identity) of this molecule was very low [12–15%), making it difficult to be sure it is an IL-4 homologue, although clearly related to Th2-type cytokines [24]. In fugu, T cell enriched PBL was found to express more IL-4/13A and IL-4/13B after stimulation with recombinant B7 molecules [142]. In zebrafish a recombinant IL4/13B was shown to increase the number of IgT-positive and CD209-positive cells in blood [159, 160], and in zebrafish spleen the expression of IL-4/13B and transcription factor related to Th2 immune response as GATA-3, and STAT6 was simultaneously enhanced after PHA stimulation [161]. The IL-4/13A gene was identified in trout and salmon [162], where the tissue distribution of salmonid IL-4/13A and GATA-3 expression were compared to the expression of IL-4, IL-13, and GATA-3 in mice. High levels of these transcripts were found in both salmonid and murine thymus, while constitutive IL-4/13A richness of skin and respiratory tissue was found in salmonids but not in mice. Experiments with isolated cells from gill and pronephros (head kidney) indicated that trout IL-4/13A is mainly expressed by surface IgM-negative cells, readily inducible by PHA but not by poly I:C, and regulated differently from the Th1 cytokine IFNγ gene. In mammals, IL-5 is also considered a Th-2 type cytokine and along with IL-3 and GM-CSF it signals through receptors with a common γ-chain (γC). None of these cytokines have been discovered in fish to date [24].

Interleukin 7

The cytokine IL-7 plays several important roles during lymphocyte development, survival, and homeostatic proliferation [163]. It is produced by many different stromal cell types, including epithelial cells of the thymus and the intestine [164-166]. There is only one report on IL-7 in fish, for the fugu molecule that was discovered using a gene synteny approach by searching with the mammalian IL-7 gene neighbours C8orf70 and PKIA. Fugu IL-7 shows constitutive expression in head kidney, spleen, liver, intestine, gill and muscle, with expression shown to increase in head kidney cultures stimulated with LPS, poly I:C or PHA [24, 167].

Interleukin 15

The central action of IL-15 cytokine is on T-cells, dendritic cells and NK cells. IL- 15 is an important regulator of the innate immune response to infection and autoimmune disease conditions. This gene shares activities with IL-2 and utilizes IL-2R β and γ units [45].

Two genes with homology to IL-15 have been discovered in fish. One shows similar gene organisation and synteny to mammalian and chicken IL-15, and has been termed IL-15. The second gene, which has a 4-exon structure and is in a different genome location, has been termed IL-15-like [168-170]. They show differential expression patterns in terms of the tissues where constitutive expression is apparent, and in terms of inducibility in PBL, with IL-15L being refractory to induction [168]. Two alternative splice variants of IL-15L (IL-15La and IL-15Lb) have also been described [170]. Trout IL-15, which has subsequently been cloned and sequenced, was strongly induced by rIFNγ in two trout cell lines (RTS-11 and RTG-2). rIL-15 could up-regulate IFNγ expression in splenic leucocytes, suggesting a positive feedback loop exists in fish between these two cytokines. Interestingly, unstimulated head kidney leucocytes were not responsive to rIL-15, at least in terms of the IFNγ expression level [171].

Interleukin 21

Interleukin 21 (IL-21) is a newly recognized member of IL-2 cytokine family that utilizes the common γ-chain receptor subunit for signal transduction [172-174]. In humans and other mammals, IL-21 is produced by both Th1 and Th2 cells [172, 175, 176]. IL-21 has pleiotropic effects on both innate and adaptive immune responses and can act on CD4+ and CD8+ T cells, B cells, NK cells, dendritic cells (DC), myeloid cells, and other tissue cells. IL-21 enhances the proliferation of anti–CD3-stimulated T cells and acts in concert with other γc cytokines to enhance the growth of CD4+ T cells [177]. IL-21–producing CD4+ T cells exhibit a stable phenotype of IL-21 production in the presence of IL-6 but retain the potential to produce IL-4 under Th2-polarizing conditions and IL-17A under Th17-polarizing conditions [178]. IL-21 stimulates CD8+ T cell proliferation and synergizes with IL-15 in promoting CD8+ T cell expansion in vitro and their antitumor effects in vivo [177, 179]. B cells that encounter IL-21 in the context of Ag-specific (BCR) stimulation and T cell co/stimulation undergo class-switch recombination and differentiate into Ab-producing plasma cells. In contrast, B cells encountering IL-21 during nonspecific TLR stimulation or without proper T cell help undergo apoptosis [180].

Since its discovery in fugu as a gene neighbour of IL-2 [141], IL-21 has been reported in tetraodon [181, 182] and rainbow trout [182]. Fugu IL-21 shows low constitutive expression. However, stimulation of isolated kidney leucocytes with PHA induced IL-21 expression. IL-21 was also up-regulated at mucosal sites as gill and gut when fish were injected with LPS or poly I:C [141]. Similarly, in tetraodon IL/21 expression is low but detectable in the gut, gonad and gills of healthy fish, and is induced in the kidney, spleen and skin following LPS injection [181]. In trout IL-21 expression is highest in gills and intestine, and is induced *in vivo* by bacterial (*Y. ruckeri*) and viral (VHSV) infection [182]. Relative to IL-2, induction of IL-21 expression in head kidney cells appears more rapidly but has shorter duration after stimulation. The trout rIL-21 has also been produced and shown to increase the expression of IL-10, IL-22 and IFNγ, and to a lesser extent IL-21, and to maintain the expression levels of key lymphocyte markers in primary cultures [182]. Thus, IL-21 may act as a survival factor for fish T and B cells [24].

The Interleukin 10 Family

Interleukin-IL-10 is an anti-inflammatory cytokine and a member of the class II cytokine family that also includes IL-19, IL-20, IL-22, IL-24, IL-26 and the interferons [183]. Although the predicted helical structure of these homodimeric molecules is conserved, certain receptor-binding residues are variable and define the interaction with specific heterodimers of different type-2 cytokine receptors. This leads to diverse biological effects through the activation of signal transducer and activator of transcription (STAT) factors [184].

Interleukin 10

Interleukin-10 (IL-10) was discovered initially as an inhibitory factor for the production of Th1 cytokines. Subsequently, pleiotropic inhibitory and stimulatory effects on various types of blood cells were described for IL-10, including its role as a survival and differentiation factor for B cells. IL-10, which is produced by activated monocytes, T cells and other cell types like keratinocytes, appears to be a crucial factor for at least some forms of peripheral tolerance and a major suppressor of the immune response and inflammation. The inhibitory function of IL-10 is mediated by the induction of regulatory T cells [185].

IL-10 was discovered in fish by searching the fugu genome. The translation showed 42–45% similarity to mammalian molecules with very low constitutive expression in tissues [186]. IL-10 has since been cloned in several other fish species including carp [187] zebrafish [188], rainbow trout [189], sea bass [190, 191] and cod [79]. Such studies have shown that IL-10 expression can be increased by LPS stimulation, by bacterial infection, by bath administration of immunostimulants [192] and by IPNV infection which may be associated with mechanisms of immune evasion [78].

Interleukin 20 (IL-20like)

In mammals, IL-20 was discovered as a new member of the IL-10 family of cytokines. IL-20 shares the highest amino-acid sequence identity with IL-10, IL-24 and IL-19. It is secreted by immune cells and activated epithelial cells like keratinocytes. A high expression of the corresponding IL-20 receptor chains has been detected on epithelial cells. In terms of function, IL-20 might therefore mediate crosstalk between epithelial cells and tissue-infiltrating immune cells under inflammatory conditions [193].

In fish, the gene of IL-20 has been described in putterfish [183], zebrafish [194] and trout [195]. In the latter work, the IL-20 gene, called IL-20-like (IL-20L) has been described as having a high level of expression in immune related tissues and in the brain, suggesting an important role of the fish IL-

20L molecule in both the immune and nervous systems. Although the exact cell types expressing IL-20L have yet to be defined, macrophages express IL-20L. Moreover, IL-20L expression in the macrophage cell line RTS-11 is modulated by pro-inflammatory cytokines, signalling pathway activators, microbial mimics and the immuno-suppressor dexamethasone. These data suggest that trout IL-20L plays an important role in the cytokine network. The increased expression of IL-20L was only detected at late stages (4–24 h) of LPS stimulation in RTS-11 cells and in spleen 24–72 h after infection with *Yersinia ruckeri*, which suggests that the increased expression of IL-20L by LPS and infection is via the rapid increase of pro-inflammatory cytokines (e.g., IL-1β) and other factors known to occur [195].

Interleukin 22/26

In mammals, interleukin-22 is secreted by Th17 cells [196], as well as by a subset of NK cells, designated as NK22 [197]; and even by some Th1 cells [198]. Studies have suggested there is a distinct Th22 cell lineage [199, 200]. Many of the same cytokines that induce differentiation and proliferation of IL-17-producing cells also lead to the secretion of IL-22 by Th17 cells, NK22 cells, and putative Th22 cells, including IL-6, IL-23, IL- 1β, TGF-β, and TNFα [201]. IL-17 and IL-22 are therefore frequently produced together in response to infections [202]. Interleukin-22 interacts with a heterodimeric receptor, IL-10R2/IL-22R1 [203], which is expressed on a variety of non-lymphoid cells, especially epithelial cells. Ligation of this receptor leads to both protective and detrimental effects. In synergy with IL-17, IL-22 induces pro-inflammatory cytokines in human bronchial epithelial cells against *Klebsiella pneumoniae* infection [204] and in colonic myofibroblasts [205]. Independently or in synergy with IL-17, IL-22 acts in defence against intestinal infection of mice with *Citrobacter rodentium* [206]. Moreover, IL-22 has been implicated in intestinal homeostasis keeping commensal bacteria contained in anatomical niches, which is key to our symbiotic relationship and normal intestinal physiology. However, the mechanisms that restrict colonization to specific niches are unclear. David Artis and colleagues have described a crucial role for IL-22-producing innate lymphoid cells (ILCs) in preventing lymphoid-resident commensal bacteria from escaping their niche and causing inflammation [207].

IL-26 can be produced by primary T cells, NK cells and T cell clones following stimulation with specific antigen or mitogenic lectins. IL-26 was initially shown by several groups to be co-expressed with IL-22 [208]. IL-26 is co-expressed with IFNγ and IL-22 by human Th1 clones, but not by Th2 clones. It was subsequently found that IL-26 is co-expressed with IL-17 and IL-22 by Th17 cells, an important subset of CD4+ T-helper cells that are distinct from Th1 and Th2 cells [209-211]. More recently, a novel subset of CD56+ NKp44+ NK cells was identified that co-expresses IL-22 and IL-26, especially following treatment with IL-23 [212]. Furthermore, a different subset of immature NK cells was described that do not express CD56 or NKp44 but do express CD117 and CD161 and constitutively express IL-22 and IL-26 [213].

The mechanisms that regulate transcription of the human IL-26 gene are so far largely undefined. It is possible and perhaps likely that expression of the IL-26 gene is induced in an IL-23-dependent manner because IL-23 is known to induce differentiation of Th17 cells, and IL-23 amplifies expression of IL-17 and IL-22 by Th17 cells [214].

In fish, the IFNγ locus was discovered using a gene synteny approach, and was first reported for fugu [215]. It contained a homologue of IL-22/26, that later studies of the zebrafish genome revealed to be two genes, one with clear homology to IL-22 and one with somewhat less clear homology to IL-26 [216]. The IL-22 gene was expressed constitutively in intestine and gills in all the treated and non-treated tissues. The gene was also expressed in kidney and spleen in LPS and PolyI:C-treated tissues, respectively, while IL-26 was expressed only in intestine treated with PolyI:C without expression [216]. IL-22 expression has been correlated with disease resistance in haddock vaccinated against *V. anguillarum,* with a strong constitutive expression in gills in vaccinated fish but not in control fish 24 hours post bath challenge, resulting in complete protection in fish vaccinated [217]. Moreover, IL-22, a cytokine released by Th-17 cells in mammals, is also interesting, and such responses are thought to be crucial for protection against extracellular microbes and at mucosal sites [218]. This coupled with the recent discovery of novel gill-associated immune tissue in fish [219] may provide a clue to a potential mechanism of resistance elicited by the *V. anguillarum* vaccination [24].

The Interleukin 17 Family

Interleukin-17 and a related family of genes are known to have pro-inflammatory actions and are associated with diseases [220]. After the discovery of the human IL- 17 gene [221], five cellular paralogs of IL-17 were identified, namely IL-17B, C, D, E and F [222-227]. These paralogs, identified by ESTs, genomics and proteomic databases, share identities of 20–50% with IL-17A gene. Human IL-17 A and F are present in tandem in opposite transcriptional orientation on the same chromosome 6p12, while IL-17B (Chr 5q24), IL-17C (Chr 16q24), IL-17D (Chr 13q11) and IL-17E (Chr 14q11) are dispersed. The structural similarities lead to the classification of IL-17 A, B, C, D, E, and F genes to a larger IL-17 sub-family [45]. Several IL-17 family members have been discovered in teleost fish, but homology to mammalian genes has not always been easy to assign. Two IL-17A or F homologue genes (IL-17A/F) have been found on the same chromosome. However, it has been difficult to determine which gene codes IL-17A and F. This gene in zebrafish was named IL-17A/F1 and 2. Furthermore, another IL-17A or F homologue gene (IL-17A/F3) has been found in zebrafish localized on a chromosome different from that of IL-17A/F1 and 2 [228]. In addition to those in zebrafish, IL-17A or F homologue genes have been found in rainbow trout [229], Atlantic salmon [230], pufferfish (IL-17A/F1, 2 and 3) [231], and medaka (IL-17A/F1, 2 and 3) [232].

The tissue distribution of the fugu IL-17 gene family also differs. In particular, IL-17 family genes are highly expressed in the head kidney and gills. Moreover, expression of IL-17 family genes is significantly up-regulated in the lipopolysaccharide-stimulated head kidney, suggesting that Fugu IL-17 family members are involved in inflammatory responses [231]. In Atlantic salmon IL-17D expression is widely distributed in tissues, with the highest levels of expression in testis, ovary and skin. Infection with *A. salmonicida* by injection increases IL-

17D expression levels in the head kidney (but not the spleen) in a time-dependent manner. Skin and kidney showed an increased IL-17D expression level in fish given a cohabitation challenge with *A. salmonicida* [230]. The two trout IL-17C genes show some degree of differential expression within tissues, with IL-17C1 being more dominant in the gills and skin, whilst IL-17C2 is more dominant in the spleen, head kidney and brain. Expression of both genes increases significantly with bacterial infection, although the increased expression of IL-17C2 is greater in terms of fold change. Similarly, both genes could be up-regulated in the trout RTS-11 cell line by LPS, poly I:C, calcium ionophore and rIL-1β, with IL-17C2 showing higher fold increases in all cases [229].

Interleukin 12

IL-12 is a heterodimeric cytokine composed of p35 and p40 subunits. It can mediate a number of different activities, including stimulation of IFNγ secretion from resting lymphocytes, NK cell stimulation and cytolytic T cell maturation. Perhaps most crucially, IL-12 also affects the progression of uncommitted T cells to either the Th1 lineage, which in general is characterized by secretion of lymphokines associated with cell-mediated rather than humoral immunity [233].

The p35 and p40 subunits were discovered in fish by analysis of the fugu genome [234]. The p35 locus is quite well conserved, with Schip1 being the immediate neighbour in all cases. This association has allowed p35 to be cloned by gene walking from Schip1 from fish species for which no genome sequence is available [24, 235]. The p40 subunit in fugu is constitutively expressed in all the tissues examined, except muscle, and no increases in expression were seen 3 h after injection with poly I:C or LPS. This constitutive and broad expression distribution of the p40 subunit suggests that it may be expressed in most cell types. The expression of the p35 subunit is more limited in its tissue expression and is induced after injection with poly I:C in the head kidney and the spleen, but not after injection with LPS. These results show that there are differences from the mammalian data in fugu IL-12 subunit expression. Further investigation will be required to show whether this is unique to fugu, if IL-12 is involved more in antiviral defence in fish and if the two subunits are regulated differently from their regulation in the mammalian system [234].

Transforming Growth Factor β (TGF-β)

TGF-β is a pleiotropic cytokine that regulates cell development, proliferation, differentiation, migration, and survival in various leukocyte lineages including lymphocytes, dendritic cells, NK cells, macrophages and granulocytes [236, 237]. In the mammalian immune system, TGF-β1 is a well-known suppressive cytokine and its dominant role is to maintain immune tolerance and suppress autoimmunity [238, 239]. The potent immunosuppressive effects of TGF-β1 are mediated predominantly through its multiple effects on T cells: TGF-β1 suppress Th1 and Th2 cell proliferation, while it promotes T regulatory cell generation by inducing Foxp3 expression. On the other hand, TGF-β also promotes immune responses by inducing the generation of Th17 cells [236, 240, 241]. Therefore, the regulatory roles of TGF-β as a positive or negative control device in immunity are widely acknowledged in mammals [238, 240, 241].

In teleost, despite the lack of extensive investigation on the functional role of TGF-β, some recent studies have revealed that TGF-β1 also exterts powerful immune depressing effects on activated leukocytes, as it does in mammals. For instance, TGF-β1 significantly blocks TNFα-induced activation of macrophage in goldfish and common carp, but induces the proliferation of the goldfish fibroblast cell line CCL71 [242, 243]. In grass carp, TGF-β1 down-regulates LPS/PHA-stimulated the proliferation of peripheral blood lymphocyte by contrast with the stimulatory effect of TGF-b1 alone in the same cells [244]. In red sea bream, similar phenomenon was observed during leukocyte migration under TGF-β1 treatment, with or without LPS challenges [245]. These findings not only define TGF-β1 as an immune regulator in teleost, but also indicate that TGF-β1 may have retained similar functions in immunity during the evolution of vertebrates [246].

Interferons

Interferons genes are involved in mediating cellular resistance against viral pathogens and modulating innate and adaptive immune systems. Broadly, IFNs are classified into two main groups called type I and type II [45]. Type I IFN includes the classical IFNα/β, which is induced by viruses in most cells, whereas type II IFN is only composed of a single gene called IFNγ and is produced by NK cells (NK cells) and T lymphocytes in response to interleukin-12 (IL-12), IL-18, mitogens or antigens [247]. Structurally both IFN types belong to the class II a-helical cytokine family, but have different 3-dimensional structures and bind to different receptors [248].

Two IFNs (IFNα1 and IFNα2) have been cloned from Atlantic salmon and characterized with respect to sequence, gene structure, promoter, antiviral activity and induction of ISGs [249-252]. Salmon IFNα1 induces both Mx and ISG15 proteins in TO cells and thus has properties similar to mammalian IFNα/β and IFNλ [251, 252]. Furthermore, salmon IFNα1 induces potent antiviral activity against the IPNV *in vitro* [251], but this protection has not been observed *in vivo,* despite a high level of expression of IFNα detected in spleen and head kidney of Atlantic salmon challenged intraperitoneally with IPNV [78].

At least three type I IFNs have been discovered in rainbow trout. The IFN1 (rtIFN1) and rtIFN2 show high sequence similar to Atlantic salmon IFNα1 and IFNα2, which contains two cysteines. On the other hand, rtIFN3 contains four cysteines, which further confirms the relationship between mammalian IFNα and fish IFNs. Recombinant rtIFN1 and rtIFN2 have both been shown to up-regulate expression of Mx and inhibit VHSV replication in RTG-2 cells. In contrast, recombinant rtIFN3 has been found to be a poor inducer of Mx and antiviral activity. Interestingly, the three rtIFNs show differential expression in cells and tissues [253]. This suggests that the three trout IFNs have different functions in the immune system of fish, which is an interesting subject for further research [254].

IFNγ has been identified in several fish species, including rainbow trout and Atlantic salmon [215, 216,248, 255-257]. In contrast to the type I IFNs, fish and mammalian IFNγ are similar in exon/intron structure and display gene synteny. However, some fish species also possess a second IFNγ subtype named

IFN gamma rel, which is quite different from the classical IFNγ [258]. Rainbow trout and carp IFNγ have several functional properties in common with mammalian IFNγ, including the ability to enhance respiratory burst activity, nitric oxide production, and phagocytosis of bacteria in macrophages [257-259]. Far less is known about the antiviral properties of fish IFNγ. However, it has been reported that it induces antiviral activity against both IPNV and the Salmon Alpha Virus (SAV) in salmon cell lines [260]

Tools for Fish Cytokine Analysis

The major strategy of functional genomics is to identify the types of responses to specific pathogens based on cytokines expression as a predictor of profile immune response, which began by using suppressive subtractive hybridization as major tools at the beginning of the immunogenomics and upgrade to platforms of wide screening that allow identify thousands of EST's that are differentially regulated in their expression and that allow identifying potential candidates as biomarkers in the progression of the immune response at differential environmental conditions, not only against pathogens, but also in captivity stress conditions that affect the fisheries production.

Suppressive Subtractive Hybridization (SSH)

One of the most important biological processes in higher eukaryotes against external stimuli is the response mediated by differential gene expression. To understand the molecular regulation of these processes, the relevant subsets of differentially expressed genes of interest must be identified, cloned, and studied in detail using specific molecular techniques. In this matter, subtractive cDNA hybridization has been a powerful approach to identify and isolate cDNAs of differentially expressed genes [261-263]. Numerous cDNA subtraction methods have been reported. In general, they involve hybridization of cDNA from one population (tester) to excess of mRNA (cDNA) from another population (driver) and then separation of the unhybridized fraction (target) from the common hybridized sequences. One of these tools is a PCR-based technique called representational difference analysis, which does not require physical separation of single-strand (ss) and double-strand (ds) cDNAs. Representational difference analysis has been applied to enrich genomic fragments that differ in size or representation [264] and to clone differentially expressed cDNAs [265]. However, representational difference analysis has the problem of the wide differences in abundance of individual mRNA species so that multiple rounds of subtraction are needed [265]. Other strategies, such as mRNA differential display [266] and RNA fingerprinting by arbitrary primed PCR [267], are potentially faster methods for identifying differentially expressed genes, but both of these methods have high levels of false positives [268] that bias high-copy-number mRNA [269], which can inappropriate in experiments where only a few genes are expected to vary [268]. One of the techniques most often used to establish differential expression pattern between two conditions is suppression subtractive hybridization (SSH), which selectively amplifies target cDNA fragments (differentially expressed) and simultaneously suppresses non-target DNA amplification. The method is based on the fact that long inverted terminal repeats attached to DNA fragments can selectively suppress amplification of undesirable sequences in PCR procedures [270]. This method overcomes the problem of differences in mRNA abundance by incorporating a hybridization step that normalizes (equalizes) sequence abundance during the course of subtraction by standard hybridization kinetics [271]. Two types of SSH are possible: forward SSH, when the reaction involves the hybridization of cDNA from one population indicated as the evaluated phenotype (tester) to excess of mRNA (cDNA) from a control phenotype (driver); and reverse SSH, when the conditions described above are inverted. Together, the two processes are called reciprocal SSH.

Different works have been done with SSH to evaluate fish immune response at the gene expression level against challenges with bacteria-derived pathogen-associated molecular patterns (PAMP) like LPS [272, 273] and whole bacteria like *Aeromonas salmonicida* [274, 275], *Listonella anguillarum* [276],*Edwardsiella tarda* [277], and *Vibrio parahaemolyticus* [278] (Table 1).

A critical step in any immune response is the recognition of invading organisms. This is mediated by many proteins, including pattern recognition receptors (PRR), which recognize and bind to molecules present on the surface of microorganisms. LPS is an essential cell wall component of gram-negative bacteria and is recognized by PRR, triggering a series of responses that lead to the activation of the host defence system. These PRRs include a number of toll-like receptors, as well as other cell-surface and cytosolic receptors that, upon stimulation, modulate immunity [279, 280]. In LPS-stimulated yellow grouper spleen a subtracted cDNA library was constructed using SSH. The contigs and singlets obtained were analyzed and a low number of immune-related genes were found [272]. In Asian seabass the up-regulation of differentially expressed genes like pro-inflammatory cytokines and related receptors, such as TNF receptor super family member 14 (TNFRSF14), IL-31 receptor A (IL31RA), chemokine receptor-like 1 (CMKLR1), chemokine (C-X-C motif) receptor 3 (CXCR3), chemokine (C-C motif) receptor 7 (CCR7) and chemokine (C-C motif) ligand 25 (CCL25), was identified at 24h post-challenge by bacterial LPS in spleen Complement components were also identified [273]. These genes are a solid basis for a better understanding of immunity in Asian seabass and for developing effective strategies for immune protection against infections in that species.

Infection of Atlantic salmon by *A. salmonicida* was observed to stimulate an acute-phase response (APR) as part of the innate immune defence system to infection, whose gene expression pattern was remarkably observed in liver at 7 days post-infection [275] indicating that the liver appears to be the main source of APPs in fish, as in mammals. Not surprisingly, the liver gene expression pattern observed in other fish species against *L. anguillarum* [276], *E. tarda* [277], and *V. parahaemolyticus*[278]. The APR is characterized by alterations in the levels of plasma proteins referred to as acute-phase proteins (APPs), as well as the secretion of some other innate defence molecules important for innate immunity, such as complementary systems [281-283]. In Atlantic cod stimulated with atypical *A. salmonicida* (formalin-killed) interleukin-1β (IL-1- β), interleukin-8 (IL-8), CC chemokine type 3, interferon regulatory factor 1 (IRF1), ferritin heavy subunit, cathelicidin, and hepcidin were identified in the forward spleen SSH library.

Atlantic cod IRF1 was constitutively expressed at low levels, and expression was significantly elevated in spleen and head kidney at 24 h following A. salmonicida stimulation, with the highest levels of induction observed in the spleen [274]. The target IRFI genes, as well as their importance in innate immune responses in fish, have not yet been determined, although the expression of IRF1 in teleost macrophages can be induced by both IFNγ and IL-1β, with IFNγ being a much more potent inducer of IRF1 than IL-1β [99].

Table 1. Transcriptomics studies on fish after treatments with bacteria or virus *in vivo* analyzed with SSH. LPS: Lipopolysaccharide; ISKNV: Infectious spleen and kidney necrosis virus; poly I:C: polyriboinosinic polyribocytidylic acid; SBNNV: Sea bass nervous necrosis virus; ACNNV: Atlantic cod nervous necrosis virus; SGIV: Singapure grouper iridovirus; PBL: Peripheral blood leukocytes.

Microorganism	Fish	Pathogen	Tissue/Cell type	Infection route	Reference
Bacteria	Atlantic salmon	*Aeromonas salmonicida*A449	Liver, head kidney and spleen	Intraperitoneal	Tsoi et al., 2004
	Yellow grouper	*LPS (E. coli*)	Spleen	Intraperitoneal	Wang et al., 2007
	Atlantic cod	*Aeromonas salmonicida (formalin-killed*)	Head kidney, Spleen	Intraperitoneal	Feng et al., 2009
	Asian seabass	*LPS (E. coli*)	Spleen	Intraperitoneal	Xia and Yue 2010
	Ayu	*Listonella anguillarum*	Liver	Intraperitoneal	Li et al., 2011
	Japanese flounder	*Edwardsiella tarda* (NE8003)	PBL	Intraperitoneal	Matsuyama et al., 2011
	Marine medaka	*Vibrio parahaemolyticus*	Liver	Intraperitoneal	Bo et al., 2012
Virus	Mandarian fish	ISKNV	Spleen	Intramuscular	He et al., 2006
	Sea bream	Nodavirus (strain 475-9/99)	Brain	Intramuscular	Dios et al., 2007
	Atlantic cod	poly I:C	Spleen	Intraperitoneal	Rise et al., 2008
	Sea bass	SBNNV	Head kidney	Intramuscular	Poisa-Beiro et al., 2009
	Atlantic cod	ACNNV	Brain	Intraperitoneal	Rise et al., 2010
	Orange-spotted grouper	SGIV	Spleen	Intraperitoneal	Xu et al., 2010

SSH has been in several investigations to evaluate fish gene expression patterns against challenges with PAMPs, such as polyriboinosinic polyribocytidylic acid (poly I:C) [284], Infectious Spleen and Kidney Necrosis Virus (ISKNV) [285], Nodavirus [286-288], and Singapore grouper iridovirus (SGIV) [289].

Spleen gene expression in mandarin fish at 4 days post-infection with ISKNV of Mx protein, interferon-inducible protein Gig-2, and viperin (interferon-inducible and antiviral protein) was up-regulated, suggesting IFN pathway stimulation after ISKNV infection [285]. Also, two inflammatory cytokine genes, CC chemokine and IL-8, were found in the forward SSH library, whereas the CD59/Neurotoxin/Ly-6-like protein gene was down-regulated. In mammals, CD59 is a complement regulatory protein, which can inhibit complement activation and membrane attack complex (MAC) formation on autologous cells [290], suggesting that down-regulation in the ISKNV-infected host cells may make these cells more sensitive to complement attack, mounting an anti-virus mechanism of the host [285]. In orange-spotted grouper after 5 days of infection with Singapore grouper iridovirus (SGIV) novel genes were annotated as immune-related, such as C-type lectin, epinecidin, and complement components C3 and C9. Interestingly, the most abundant clone was C-type lectin, and the microarray results at 1, 5 and 9 days post-infection indicated that its expression was up-regulated in liver, spleen and kidney [289]. Lectins are multivalent carbohydrate-binding proteins that function as important pattern-recognition receptors (PRR) and have been isolated and characterized in fish [291-294]. C-type lectin represents a very large family, most members of which are able to bind PAMP and microorganisms themselves through sugar moieties and play important roles in non-self recognition and clearance of invading microorganisms. The up-regulation of C-type lectin in different organs with immunological functions confirmed as SSH as microarrays suggest an important role in the development of control strategies against SGIV infection. The SSH method was used to generate a subtracted cDNA library enriched in gene transcripts differentially expressed after 1 day post-infection in the brains of sea bream infected with nodavirus. Most of the expressed sequence tags (ESTs) differentially expressed in infected tissues fell into gene categories related to cell structure, transcription, cell signalling or different metabolic routes. Other interesting putative homologies corresponded to genes expressed in stress responses, such as heat shock proteins (Hsp-70) and to immune-related genes such as the Fms-interacting protein, TNFα-induced protein, interferon-induced with helicase C domain protein (mda-5), which in mammals play an important role in the synthesis and secretion of IFN type I [295]. Another nodavirus, sea bass nervous necrosis virus (SBNNV) was studied to identify genes potentially involved in antiviral immune defence in sea bass head kidney using the SSH technique [287]. The results of up-regulated EST from sea bass head kidney SSH showed significant similarities with immune genes, such as β-2 microglobulin, heat shock protein 90 (Hsp-90), IgM, MHC class I and class II, and β-galactoside-binding lectin, identified as a member of the galectin family and closely related to the galectin-1 group (Sbgalectin-1). When the recombinant protein (rSbgalectin-1) was produced and functional assays were conducted, a decrease in IL-1β, TNFα, and Mx expression was observed in the brain of sea bass simultaneously injected with nodavirus and rSbgalectin-1 compared to those infected with the nodavirus alone, suggesting a potential anti-inflammatory protective role of Sbgalectin-1 during viral infection. A similar nodavirus, the Atlantic cod nervous necrosis virus (ACNNV), was studied to evaluate the transcript expression responses in the Atlantic cod (*Gadus morhua*) brain to asymptomatic high nodavirus carrier state [288]. In the forward brain SSH library was identified with significant similarity to genes with immune-relevant functional annotations the interferon stimulated gene 15 (ISG15), IL-8 variant 5, DEXH (Asp-Glu-X-His) box polypeptide 58 (DHX58; LGP2), radical Sadenosyl methionine domain-containing 2 (RSAD2; viperin), β-2-microglobulin (B2M), chemokine CXC-like protein, signal transducer and activator of transcription 1 (STAT1), and CC chemokine type 2. Interestingly, ISG15, DHX58, RSAD2, and sacsin (SACS) transcripts are all strongly upregulated by both high nodavirus carriage and intraperitoneal poly I:C stimulation, suggesting a similar host response is significantly induced in the brain by both nodavirus and poly I:C. This expression pattern is corroborated when the response of Atlantic cod spleen is evaluated against poly I:C stimulation, showing the up-regulation of ISG15, RSAD2, LGP2 and other transcripts such as MHC class I, and IRF1, 7, and 10, indicating that Atlantic cod recognize dsRNA and mount a interferon pathway response [284].

Microarrays

Microarray analysis measures the expression of large numbers of genes in parallel. This methodology, which combines hypotheses-driven and hypotheses-free research strategies, is used to infer molecular mechanisms, classify samples, and diagnose and search for novel biomarkers. With the use of standard platforms, laboratory protocols and procedures for processing of primary data, the results of microarrays analyses are well suited for database management and meta-analysis across multiple experiments, whilst data mining is based on powerful statistical procedures with support from functional and structural annotations of genes [296].

The Atlantic salmon is of particular importance to the global aquaculture industry. Salmonid cDNA microarrays were constructed shortly after large-scale sequencing of salmon and trout cDNA libraries by several research institutes. One of the projects related to salmon sequencing is GRASP (Genomics Research on Atlantic Salmon Project), an initiative funded by Genome Canada that is intended to improve understanding of physiological and evolutionary processes influencing the survival and phenotype of salmonids and other fish in natural and aquaculture environments. The first salmonid GRASP microarray platform (GRASP-1), containing 7356 salmonid elements representing 3557 different cDNAs (3.7K), was obtained from 80,388 ESTs, principally from cDNA libraries [298] of different salmon species such as Atlantic salmon, rainbow trout, Chinook salmon, sockeye salmon, and lake whitefish cDNA libraries. The second version of the GRASP microarray platform (GRASP-2) was developed and contained cDNAs representing 16,006 genes (16K). The genes identified in the array have been stringently selected from Atlantic salmon and rainbow trout EST databases representing a wide variety of different classes of genes [297]. Finally, a new expanded salmonid cDNA microarray (GRASP-3) of 32,000 features (32K) was created where 69% of the total EST collection used was from Atlantic salmon [298]. The Aleksei Krasnov's group designed the rainbow trout microarray (SFA1.0) by identifying a relatively small number of genes (1300 genes; 1.3K) using clones from normalized and subtracted cDNA libraries, as well as genes selected by the functional categories of Gene Ontology for inclusion in a microarray aimed at characterizing transcriptome responses to environmental stressors [299] to maximize the presence of transcripts related directly to immune response in rainbow trout, because of which this platform is also called Immunochip (SFA1.0 immunochip). The updated SFA platform (1.8K; SFA2.0 immunochip) was specially designed for studies of responses to pathogens and stressors and has substantially improved coverage of immune genes [300]. Another cDNA platform in commercial fish species has been designed in Japanese flounder [301] and European flounder [302], turbot [303], and sole [304].However despite impressive achievements, cDNA platforms suffer from limitations and disadvantages. At present most research groups working with salmonids and other aquaculture species do not have full access to clones required for fabrication of cDNA microarrays. Maintenance and PCR amplification of large clone sets is expensive and time consuming, while the risk of errors is high [296]. Probably the most important drawback of cDNA microarrays is their limited ability to discriminate paralogs since long probes cross-hybridize with highly similar transcripts from members of multi-gene families [305]. In salmonids this problem is aggravated by the large number of expressed gene duplicates. These complications can be resolved with oligonucleotide microarrays (ONM) that also provide greater accuracy and reproducibility of analyses. Until recently, the use of ONM platforms was hampered by the cost, but they are now rapidly replacing cDNA platforms. Construction of ONM platforms begins with establishment of mRNA sequence sets for comprehensive coverage of transcriptomes with low redundancy. The next stage is identifying genes by searching protein databases and annotating them according to functions, pathways and structural features. For successful development and use of ONM, it is necessary to define the gene composition and optimum number of spot replicates and to choose criteria for quality assessment [296]. Because of the commercial importance of salmonid species, there is special interest gene expression pattern against different pathogens. Initially salmonid (rainbow trout) ONM contained 1672 elements, representing more than 1400 genes [306]. Currently, one of the most often used ONM platforms to evaluate the response against different conditions and pathogens is the custom salmon ONM (SIQ-3), based on the Agilent Technology system (21K in 4x44K format). Because limited availability of peripheral blood leukocyte (PBL) markers is a well-recognised problem of fish immunology, this platform compares the transcriptomes of PBL and other tissues to search for genes with preferential expression in leukocytes [296], making it a very significant tool to evaluate the response to pathogens in Atlantic salmon. Another ONM platform based on 500K ESTs Atlantic salmon and 250K ESTs rainbow trout [298] is the cGRASP 44K salmonid oligo array (Agilent eArray), although no studies employing this platform have been published yet. Another ONM has been designed in fish model organisms like zebrafish and in commercial fish species such as channel catfish and turbot [307].

Functional genomic studies based on evaluating immune responses, also called immunogenomics, have been conducted *in vivo* to evaluate the response to different pathogens at the systemic level in different organs, especially the liver and head kidney. The functional genomic approach has been used with*Salmo salar* and *Oncorhynchus mykiss*, where PAMPs, whole bacteria and viruses are the most studied pathogens. Here we present different works in fish challenged by bacteria or viruses where differential gene expression profiles were evaluated using microarray platforms with special emphasis on *in vivo*fish immune response (Table 2).

Studies with Bacterial Pathogens, Pamps and Cytokine Network Interactions

One of the most commonly studied bacterial pathogens is *Aeromonas salmonicida*, a gram-negative bacteria and the causative agent of furunculosis. In fact prior to the development of species-specific cDNA microarrays a preliminary study used a human microarray (GENEFILTERS GF211) to explore the liver response in Atlantic salmon infected using a cohabitation model [275]. Only 4 mRNAs were consistently up-regulated ($p < 0.01$) from the 241 positively identified spots with a clearly detectable hybridization signal, none of them related to cytokine expression. This was probably due to the lack of sequence homology, a problem commonly associated with cross-species cDNA hybridization. Thus the creation of species-specific platforms was a key step in fish immunology. Using a custom Atlantic salmon cDNA microarray (NRC-IMB) consisting of over 4,000 different cDNA amplicons, the first results for challenge with Aeromonas salmonicida were reported in 2005 [275]. The study described a cohabitation challenge and identified 16 up-regulated mRNAs in all three tissues studied (spleen, liver and head kidney), whereas 2 and 19 mRNAs were identified as down-regulated in the head kidney and liver, respectively. The authors found that genes related to the acute phase response were up-regulated in spleen and head kidney of infected salmon, indicating that the infected fish underwent a typical acute phase response to infection. The effects of an *Aeromonas salmonicida* infection were recently reported in turbot, *Scolphtalmus maximus*, [307]. Using a custom designed oligonucleotide-microarray (8x15K), the authors identified a set of 48 differentially regulated mRNAs in the spleen of challenged fish at 3 dpi, mostly related to the acute-phase and the stress/defence immune response. A study using channel and blue catfish explored the effects of a gram-negative bacterial infection on the acute phase response (APR) [308]. The authors showed up-regulation of mRNA transcripts involved in iron homeostasis, transport proteins, complement components and inflammatory and humoral immune response, indicating that conserved APR occurs as part of the innate immune response in both catfish species. Interestingly, a more acute response was observed composed of several immune pathways in the blue but not the channel catfish. More studies are required to elucidate expression patterns resulting from gram-negative bacterial infection of phylogenetically similar and different fish are required to describe common and divergent responses. This could lead to the development of marker systems, consensus on the APR in fish and treatments tailored to certain species, all of which have significant applied interest.

The activity of LPS from gram-negative bacteria, a common membrane-associated PAMP used in immunological studies, has been explored in several fish species. These studies include effects on the spleen in channel catfish [285], rainbow trout head kidney [309], and liver in the Senegalese sole [310]. Using a 19K oligonucleotide microarray (ONM) it was observed that some pro-inflammatory mRNAs in the catfish spleen were up-regulated very quickly, principally between 2 and 4 hours post-injection with LPS, whereas immunoglobulin- (2h post-injection) and antigenic presentation-related mRNA transcripts were repressed 24h post-injection [311]. A similar inhibition was reported in head kidney of rainbow trout, where the suppression of major cellular processes, including immune function and an initial stress reaction, was followed by a proliferative hematopoietic-type/biogenesis response 3 dpi [309]. However, in the Senegalese sole a clear up-regulation of transcripts related to the immune response was reported 24 hpi in the liver [310]. These results collectively highlight the diversity of responses observed at the tissue level and reflect the nature of the immune system that is diffusely located throughout many organ compartments.

Table 2. Transcriptomics studies on fish after treatments with bacteria or virus in vivo analyzed with microarrays. FDKC: Formaldehyde –killed cells; LPS: Lipopolysaccharide; FKC: Formalin-killed cells; ISAV: Infectious salmon anemia virus; PD: Pancreas disease; CMS: Cardiomyopathy syndrome; HSMI: Heart and skeletal muscle inflammation; IHNV: Infectious hematopoietic necrosis virus; VHSV: Viral hemorrhagic septicemia virus; HIRRV: Hirame rhabdovirus; poly I:C: polyriboinosinic polyribocytidylic acid; PBL: Peripheral blood leukocytes; ONM: oligonucleotide microarray.

Microorganism	Fish	Pathogen	Tissue/Cell type	Resource	Platform	Reference
Bacteria	Atlantic salmon	*Aeromonas salmonicida*(A449)	Liver	cDNA	GENE- FILTERS GF211	Tsoi et al., 2003
		Aeromonas salmonicida(A449)	Head kidney/liver/ spleen	cDNA	NRC-IMB	Ewart et al., 2005
		Aeromonas salmonicida	Liver/spleen	cDNA	SFA-2	Skugor et al., 2009
		Aeromonas salmonicida(Brivax II)	Liver	cDNA	TRAITS/SPG	Martin et al., 2010
		Piscirickettsia salmonis	Head kidney	cDNA	GRASP-1	Rise et al., 2004
	Chinook salmon	*Vibrio anguillarum*	Head kidney	cDNA	GRASP-1	Ching et al., 2010
	Rainbow Trout	*Vibrio anguillarum*(FDKC)	Liver	ONM	OSUrbt	Gerwick et al., 2007
		LPS (*E.coli*026:B6)	Head kidney	cDNA	SFA-1	MacKenzie et al., 2008
		LPS (*E. coli*026:B6)	HK macrophages	cDNA	SFA-1	Mackenzie et al., 2008
		LPS, PGN (*E. coli*0111:B4)	HK macrophages	cDNA	SFA-2	Boltaña et al., 2011
		PGN (*E. coli*0111:B4, K12)	HK macrophages	cDNA	SFA-2	Boltaña et al., 2011
		LPS, PGN (*E. coli*0111:B4)/poly I:C	Erythrocytes	cDNA	SFA-2	Morera et al., 2011
	Brook trout	LPS (*E. coli*026:B6)	HK macrophages	cDNA	SFA-1	Mackenzie et al., 2006
	Channel catfish	LPS (*E.coli*0127:B8)	Spleen	ONM	UMSMED-1	Li et al., 2006
		Edwardsiella ictaluri (MS-S97-773)	Gills/head kidney/liver/ skin/spleen	ONM	UMSMED-2	Peatman et al., 2007
	Blue catfish	*Edwardsiella ictaluri* (MS-S97-773)	Liver	ONM	UMSMED-2	Peatman et al., 2008
	Japanese flounder	*Edwardsiella tarda* (NE9505)	Head kidney	cDNA	Japanese flounder custom-3	Yasuike et al., 2010
		Streptococcus iniae (02; FKC)	Head kidney cells	cDNA	Japanese flounder custom-3	Dumrongphol et al., 2008
		Mycobacterium bovis (TUMSAT-Msp001) FKC	Kidney cells	cDNA	Japanese flounder custom-3	Kato el al., 2010
	Zebrafish	*Mycobacterium marinum* (M, E11)	Whole fish	ONM	MWG - Sigma Genosys - Affimetrix	Meijer et al., 2005
		Mycobacterium marinum (E11; Mma20)	Whole fish	ONM	ZF Agilent	Van der Sar et al., 2009
		Streptococcus suis (HA9801)	Whole fish	ONM	Affimetrix Zebrafish GeneChip	Wu et al., 2010
	Turbot	*Aeromonas salmonicida*	Spleen	ONM	Turbot custom Agilent	Millán et al., 2010
	Solea	LPS (*E. coli*011:B4)	Liver	cDNA	GENIPOL-1	Osuna-Jimenez et al., 2009

Virus	Atlantic salmon	ISAV (Glesvaer 2/90)	Gills/heart/liver/spleen	cDNA	SFA-2	Jorgensen et al., 2008
		ISAV (Glesvaer2/90)	Heart/PBL	ONM	SIQ-3	Krasnov et al., 2011
	Chinook salmon	ISAV NA-HRP4 (970)	Head kidney	cDNA	GRASP-3	Leblanc et al., 2010
	Rainbow Trout	IHNV (32/87)/attenuated IHNV	Head kidney	cDNA	SFA-1	MacKenzie et al., 2008
		IHNV (strain 220-90)	Head kidney	cDNA	GRASP-2	Purcell et al., 2011
	Japanese flounder	VHSV (KRRV9822)	Head kidney cells	cDNA	Japanese flounder custom-1	Byon et al., 2005
		VHSV (KRRV9822)	Head kidney cells	cDNA	Japanese flounder custom-2	Byon et al., 2006
		HIRRV (8601H)	Kidney cells	cDNA	Japanese flounder custom-1	Yasuike et al., 2007
		HIRRV (8601H)	Kidney cells	cDNA	Japanese flounder custom-3	Yasuike et al., 2010
	Zebrafish	VHSV (07.71)	Fin/head kidney/liver/spleen	ONM	ZF Agilent	Encinas et a., 2010
	Turbot	Nodavirus (AH95)/poly (I:C)	Kidney	cDNA	Turbot custom	Park et al., 2009

For gram-positive infections in fish at the level of transcriptome analyses, infection of zebrafish with *Streptococcus suis* is the only model reported [272]. *Streptococcus suis* is a pathogen associated with zoonosis reported in several countries [312, 313]. The Affymetrix Zebrafish GeneChip was used to identify 125 up-regulated transcripts where the most significant pathways were antigen processing and presentation, leukocyte trans-endothelial migration and the proteosome. The authors suggested that the target list obtained could serve as infection markers for gram-positive infection in fish.

Undoubtedly, the identification of prognostic biomarkers for disease resistance is a major aim for aquaculture. Functional genomics has the potential to identify such potential tools. Disease resistance is normally measured by challenge with the pathogen of interest and assessing the cumulative mortalities. Surviving fish or non-challenged siblings from the same family are then considered 'resistant'. Because this process is costly there is a need for non-lethal methodologies of measuring resistance, ideally based on molecular determinants of resistance. An initial example of this approach used the GRASP 3.7K cDNA array to identify *in vitro* macrophage and *in vivo* head kidney biomarkers in response to *Piscirickettsia salmonis* infection, yielding a number of 11 regulated genes common to both challenges. The researchers proposed 19 highly regulated transcripts as potential biomarkers to evaluate the efficacy of vaccines against *Piscirickettsia salmonis* [314]. C-type lectin 2-1, a gene whose product is involved in endocytosis and the C/EBP-driven inflammatory response [315] was identified and has been identified in almost all reports in which bacterial preparations have been used to challenge live fish [275, 309, 314, 316, 317]. Another study aimed at identifying biomarkers at the transcriptional level described differences between triploid and diploid Chinook salmon under live Vibrio anguillarum challenge using the GRASP 3.7K cDNA microarray [318]. Twelve annotated mRNAs were identified as showing significant differences between diploid and triploid fish. The authors however were unable to provide a description of the underlying mechanisms to explain the observed reduced immune function of triploid salmon.

Individual variation is a major hurdle for the development of prognostic markers as both genetic and epigenetic factors must be taken into account. The utility of, for example, C-type lectin in salmon, and other potential biomarkers in other species for bacterial disease resistance, requires further development. The future publication of several fish genomes coupled to array platforms with a much increased transcript representation could provide an exciting route to further develop this strategy by combining both functional and structural genomics for species of commercial interest with a sequenced genome. Several studies have attempted to correlate gene expression profiles with the activity of bacterins (killed bacteria preparations) used to vaccinate fish in culture. Most studies have concentrated on the rainbow trout and Japanese flounder [262, 277, 319, 320]. In trout, intraperitoneal administration of killed *V. anguillarum* resulted in identifying 36 differentially expressed transcripts [320]. Most of the identified targets are involved in inflammatory response and respond to a broad range of stimuli. This suggests that these targets have little use as markers for vaccination, contrary to previous descriptions in other studies. Both the second and fourth versions of the Japanese flounder cDNA microarray have been used to address vaccination [262, 277, 319]. The results of experimental infection with Gram-negative *E. tarda* indicated that a formalin-killed preparation reduced mortality in vaccinated fish from 90% to 20% [277]. However, a correlation between the transcriptome and the efficacy of vaccination could not be identified.

The effects of a commercial vaccine for Atlantic salmon (a six-component oil-adjuvant vaccine from PHARMAQ) were evaluated to correlate vaccine protection to high and low resistance to furunculosis. The authors did not find any association between either group and suggested that "outcomes of vaccination depend largely on the ability of host to prevent the negative impacts of immune response and to repair damages" [305]. Although this study did not identify correlations between vaccination and gene expression profiles, the potential of a functional genomics approach to evaluate the efficacy and underlying mechanisms of vaccination is highlighted. In terms of the immune response and the resulting complexity in expression patterns resulting from multiple cell types and different tissue responses, the investigator has the potential to obtain a clearer 'image' of the biological response from global expression data. A key objective is therefore to increase the available genomic resources much facilitated by next generation sequencing technologies to form a more robust representation of the immune system among different fish species. Furthermore, the increasing use of ONM platforms will also improve comparison across species as data sets become more easily comparable.

It remains difficult to compare microarray experiments across distinct platforms. In this respect, Meijer et al. 2005, evaluated host transcriptome profiling to *Mycobacterium marinum* infection of adult zebrafish employing three oligonucleotides platforms (MWG, Sigma Genosys, and Affimetrix). At a significance level of $P < 1.00E\text{-}5$, there were differences among the platforms in the total number of more than 2-fold up-regulated genes, whereas the 2-fold down-regulated genes were in a similar range. Evaluation of the distribution of infection-induced genes over different categories reveled some divergence in the set from MWG, probably due to the abundance of genes of the same UniGene cluster. As well, from the total overlap of 4,138 UniGene clusters among the three microarrays, only 66 and 93 genes were up- and down-regulated, respectively [321]. With this antecedent, the same group generated a new platform (Agilent 44K) that includes their 22K probes, a 16K set probes similar to the Sigma-Compugen oligonucleotide library, and 6K set of probes for selected genes of interest indentified by previous data mining of zebrafish transcript and genome databases [322], and they evaluated the transcriptome response to acute and chronic infection by *Mycobacterium marinum*. This important effort in combing different platforms makes it clear that not all relevant genes, including immune-related ones, are represented in all platforms. Consequntly, new efforts are necessary to broaden our understanding of the immune response in fish challenged with a pathogen of interest.

Wide Screening in Fish Challenged with Viral Pathogens

Two studies have reported host responses to IHNV with the SFA and GRASP platforms [309, 323]. The potential mechanisms responsible for host-specific virulence were assessed in rainbow trout infected with high (M) and low virulence (U) strains of IHNV. A marked down-regulation in biological processes, including the immune response, lymphocyte activation, response to stress, transcription and translation, together with a greater viral load (M), suggest that the higher virulence is due to the ability to suppress the immune response via the transcriptional and translational machinery of cell [323]. Furthermore, in rainbow trout was compared the expression profiles of IHNV and attenuated IHNV were compared in rainbow trout over a short time frame of one and three days post-challenge. At 3 dpi, a significant change in the transcriptional program of head kidney revealed an immunological shift orientated toward the activation of adaptive immunity. This shift was IHNV-dependent as determined by differences between the attenuated and virulent IHNV specific expression profiles. The rapid systemic spreading of IHNV inhibited TNFα, MHC class I, and several macrophage and cell cycle/differentiation markers and favored a MHC class II, immunoglobulin and MMP/TBX4-enhanced immune response [309].

Parallel studies were conducted with a cDNA microarray enriched with 213 immune-related genes to study the immune response and the efficacy of DNA vaccines containing the viral G proteins of VHSV and HIRRV administered intramuscularly in the Japanese flounder [301, 324]. As expected, all DNA vaccines containing the viral G glycoprotein conferred specific protection to the challenged fish one month after vaccination. It is suggested that the protection occurs via the IFN type I system due to the number of IFN-related genes up-regulated in both studies, ISG-15, interferon-stimulated gene 56kDa (ISG56) and the Mx protein. In both studies, VHSV and HIRRV in the Japanese flounder, the majority of differentially up-regulated genes were identified between 3 and 7 days post-vaccination (dpv), including the less effective DNA vaccine containing N protein of HIRRV. Interestingly, Mx, an antiviral protein commonly used as a marker for antiviral activity in animal species, was consistently up-regulated across vaccinations [301]. In a similar observation, IRF-3, Mx, Vig-1 and Vig-8 were up-regulated in trout at the site of DNA vaccination against IHNV at 7 dpv [323]. In turbot challenged with nodavirus, both Mx and IFN-inducible proteins were identified 24 hpi [303]. These and the previously described results suggest that both the host-expressed viral glycoprotein and the virulent rhadovirus induce a systemic anti-viral state indicative of non-specific IFN type1 innate immune response and that this canonical response is conserved among all fish. However, the mechanisms to develop a specific cytotoxic T or B lymphocyte-mediated humoral response in fish vaccinated with plasmid DNA-IHNV G that confers protective immunity have not been identified [323].

In direct relation to the above, a significant increase in transcript markers for adaptive immunity was reported in Atlantic salmon during ISA virus (ISAV) infection [325]. Importantly, a progressive increase was observed in IgZ mRNA parallel to a decrease in IgM expression that peaked > 30 days post-infection. This coordinated increase in a group of genes related to B lymphocyte differentiation and maturation and activation of T lymphocyte-mediated immunity, including CD4, TGF-β, CD8α and IFNγ, provides strong evidence for the coordinated regulation of the two arms of the immune system in response to viral infection. An important technological contribution derived from the above study was the development of an ONM for Atlantic salmon (SIQ-3). The first assessment of the performance of these arrays was carried out in Atlantic salmon for the study of virus-responsive genes from samples infected with ISA, salmonid alpha virus/

PD-virus, cardiomyopathy syndrome (CMS) agent, heart and skeletal muscle inflammation (HSMI) and PBL from fish infected with ISAV. Some 95 up-regulated transcripts were identified. Most of the regulated transcripts are related directly to the immune response or associated with antiviral response [296]. As previously mentioned, the creation of species-specific platforms has been a key challenge for the study of the immune response against pathogens. Despite impressive achievements, cDNA microarrays suffer from limitations and disadvantages, the most important drawback being the limited ability to discriminate between paralogs as long cDNA probes cross-hybridize with highly similar transcripts from members of multi-gene families [305]. Furthermore this information needs to be supplemented to establish if the increased level of detected transcripts is consistent with specific protein synthesis. Recently, a study employed a combined proteomic and transcriptomic approach to evaluate the immune response against VHS [326]. In the fins of infected fish a series of mRNA transcripts principally related to complement components, immunoglobulin-related proteins, and macrophages were up-regulated (> 2-fold), whereas in parallel using two dimensional differential gel electrophoresis (2D-DIGE), enzymes of the glycolytic pathway and some proteins related to cytoskeletal remodelling and apoptosis (such as annexin A1a) increased with infection. However, very few proteins related to anti-viral response were identified.

CONCLUDING REMARKS

A complex network exists to regulate the innate and adaptive immune responses of fish from the various cytokines that have been reported. The study of the functional activity of these cytokines is in progress and it will be interesting to know whether mammalian Th1, Th2, Th17 and Treg responses are present in fish, regulating specific cell-mediated immunity. The recombinant production of these cytokines and antibodies against them will be the next challenge in understanding the balance of such immune responses and aid in the effective design of therapeutic strategies to manipulate the fish immune system. towards humoral or cellular immunity in response to specific antigen stimulation, vaccine strategies, functional diets to increase the quality of fishery production and predict the health of cultured fish.

The study of functional genomics in fish has provided substantial data on species of commercial interest. The major aim has been to functionally identify the intensity of responses to specific pathogens and their associated molecular components and to identify transcripts in a whole organism or specific tissue that contribute to such responses. However, the complex biology of the immune response, in which different spatial-temporal expression occurs in multiple cell types at distinct body locations, makes complete mapping of a response difficult and expensive. Moreover, considering that arrays are only as good as the transcripts represented upon them. Thus, the representation of transcripts relevant to the immune response is intimately linked to gene discovery efforts through large-scale sequencing projects, where strategies like SSH contribute not only to understanding transcriptomic response against specific pathogens but also to gene discovery. In this area, access to high-throughput NGS technology has increased in recent years and promises to make an important contribution to understanding immune response in fish. The major task now is the meta-analysis of transcriptomic data to delineate responses common among fish species to specific pathogen groups and highly specific responses. This approach will reveal host specific expression profiles and facilitate the identification of prognostic markers for diseases.

ACKNOWLEDGMENTS

The authors are grateful for the support of VRID-USACH, INNOVA-CORFO 07CN13PBT-90, INNOVA-CORFO 09MCSS6691, INNOVA-CORFO 09MCSS6698 and CONICYT Fellowships to K. Maisey and D. Toro-Ascuy.

REFERENCES

1. Flajnik MF, Du Pasquier L. Evolution of the Immune System. 6 th ed. Paul WE, editor: Lippincott Williams & Wilkins; 2008.
2. Akirav EM, Liao S, Ruddle NH. Lymphoid Tissues and Organs. 6 th ed. Paul WE, editor: Lippincott Williams & Wilkins; 2008.
3. Zapata AG, Chibá A, Varas A. Cells and tissues of the immune system of fish. In: Iwama G, Nakanishi T, editors. The fish immune system Organism, pathogen and environment: Academic Press; 1996. p. 1-62.
4. Meseguer J, Lopez-Ruiz A, Garcia-Ayala A. Reticulo-endothelial stroma of the head-kidney from the seawater teleost gilthead seabream (Sparus aurata L.): an ultrastructural and cytochemical study. Anat Rec. 1995 Mar;241(3):303-9.
5. Danneving BH, Lauve A, Press MC, Landsverk T. Receptor-mediated endocytosis and phagocytosis by rainbow trout head kidney sinusoidal cells. Fish Shellfish Immunol. 1994;4:3-18.
6. Brattgjerd S, Evensen O. A sequential light microscopic and ultrastructural study on the uptake and handling of Vibrio salmonicida in phagocytes of the head kidney in experimentally infected Atlantic salmon (Salmo salar L.). Vet Pathol. 1996 Jan;33(1):55-65.
7. Kaattari SL, Irwin MJ. Salmonid spleen and anterior kidney harbor populations of lymphocytes with different B cell repertoires. Dev Comp Immunol. 1985 Summer;9(3):433-44.
8. Herraez MP, Zapata AG. Structure and function of the melano-macrophage centres of the goldfish Carassius auratus. Vet Immunol Immunopathol. 1986 Jun;12(1-4):117-26.
9. Tsujii T, Seno S. Melano-macrophage centers in the aglomerular kidney of the sea horse (teleosts): morphologic studies on its formation and possible function. Anat Rec. 1990 Apr;226(4):460-70.
10. Bowden TJ, Cook P, Rombout JH. Development and function of the thymus in teleosts. Fish Shellfish Immunol. 2005 Nov;19(5):413-27.
11. Zapata AG, Amemiya CT. Phylogeny of lower vertebrates and their immunological structures. Curr Top Microbiol Immunol. 1999;248:67-107.
12. Espenes A, Press CM, Dannevig BH, Landsverk T. Immune-complex trapping in the splenic ellipsoids of rainbow trout (Oncorhynchus mykiss). Cell Tissue Res.

1995 Oct;282(1):41-8.

13. Vallejo AN, Miller NW, Harvey NE, Cuchens MA, Warr GW, Clem LW. Cellular pathway(s) of antigen processing and presentation in fish APC: endosomal involvement and cell-free antigen presentation. Dev Immunol. 1992;3(1):51-65.
14. Vallejo AN, Miller NW, Clem LW. Cellular pathway(s) of antigen processing in fish APC: effect of varying in vitro temperatures on antigen catabolism. Dev Comp Immunol. 1992 Sep-Oct;16(5):367-81.
15. Ellis AE. Antigen trapping in the spleen and kidney of the plaice, Pleuronectes platessa. J Fish Dis. 1980;3:413-26.
16. Agius C, Roberts RJ. Melano-macrophage centres and their role in fish pathology. J Fish Dis. 2003 Sep;26(9):499-509.
17. Medzhitov R. The Innate Immune System. 6 th ed. Paul WE, editor: Lippincott Williams & Wilkins; 2008.
18. Janeway CA, Jr. How the immune system protects the host from infection. Microbes Infect. 2001 Nov;3(13):1167-71.
19. McHeyzer-Williams LJ, McHeyzer-Williams MG. Antigen-specific memory B cell development. Annu Rev Immunol. 2005;23:487-513.
20. Sallusto F, Lanzavecchia A. Heterogeneity of CD4+ memory T cells: functional modules for tailored immunity. Eur J Immunol. 2009 Aug;39(8):2076-82.
21. Dixon B, Stet RJ. The relationship between major histocompatibility receptors and innate immunity in teleost fish. Dev Comp Immunol. 2001 Oct-Dec;25(8-9):683-99.
22. Abbas AK, Lichtman AH, Pillai S. Cells and Tissues of the Adaptive Immune System. 6 th ed: Saunders; 2007.
23. Castro R, Bernard D, Lefranc MP, Six A, Benmansour A, Boudinot P. T cell diversity and TcR repertoires in teleost fish. Fish Shellfish Immunol. 2011 Nov;31(5):644-54.
24. Secombes CJ, Wang T, Bird S. The interleukins of fish. Dev Comp Immunol. 2011 Dec;35(12):1336-45.
25. Wang T, Gorgoglione B, Maehr T, Holland JW, Vecino JL, Wadsworth S, et al. Fish Suppressors of Cytokine Signaling (SOCS): Gene Discovery, Modulation of Expression and Function. J Signal Transduct. 2011;2011:905813.
26. Alejo A, Tafalla C. Chemokines in teleost fish species. Dev Comp Immunol. 2011 Dec;35(12):1215-22.
27. Nakanishi T, Toda H, Shibasaki Y, Somamoto T. Cytotoxic T cells in teleost fish. Dev Comp Immunol. 2011 Dec;35(12):1317-23.
28. Boltana S, Donate C, Goetz FW, MacKenzie S, Balasch JC. Characterization and expression of NADPH oxidase in LPS-, poly(I:C)- and zymosan-stimulated trout (Oncorhynchus mykiss W.) macrophages. Fish Shellfish Immunol. 2009 Apr;26(4):651-61.
29. Hanington PC, Hitchen SJ, Beamish LA, Belosevic M. Macrophage colony stimulating factor (CSF-1) is a central growth factor of goldfish macrophages. Fish Shellfish Immunol. 2009 Jan;26(1):1-9.
30. Salinas I, Zhang YA, Sunyer JO. Mucosal immunoglobulins and B cells of teleost fish. Dev Comp Immunol. 2011 Dec;35(12):1346-65.
31. Zhang YA, Salinas I, Oriol Sunyer J. Recent findings on the structure and function of teleost IgT. Fish Shellfish Immunol. 2011 Nov;31(5):627-34.
32. Randelli E, Buonocore F, Scapigliati G. Cell markers and determinants in fish immunology. Fish Shellfish Immunol. 2008 Oct;25(4):326-40.
33. Rombout JH, Huttenhuis HB, Picchietti S, Scapigliati G. Phylogeny and ontogeny of fish leucocytes. Fish Shellfish Immunol. 2005 Nov;19(5):441-55.
34. Araki K, Akatsu K, Suetake H, Kikuchi K, Suzuki Y. Characterization of CD8+ leukocytes in fugu (Takifugu rubripes) with antiserum against fugu CD8alpha. Dev Comp Immunol. 2008;32(7):850-8.
35. Toda H, Saito Y, Koike T, Takizawa F, Araki K, Yabu T, et al. Conservation of characteristics and functions of CD4 positive lymphocytes in a teleost fish. Dev Comp Immunol. 2011 Jun;35(6):650-60.
36. Takizawa F, Dijkstra JM, Kotterba P, Korytar T, Kock H, Kollner B, et al. The expression of CD8alpha discriminates distinct T cell subsets in teleost fish. Dev Comp Immunol. 2011 Jul;35(7):752-63.
37. Kaattari S, Evans D, Klemer J. Varied redox forms of teleost IgM: an alternative to isotypic diversity? Immunol Rev. 1998 Dec;166:133-42.
38. Wilson M, Bengten E, Miller NW, Clem LW, Du Pasquier L, Warr GW. A novel chimeric Ig heavy chain from a teleost fish shares similarities to IgD. Proc Natl Acad Sci U S A. 1997 Apr 29;94(9):4593-7.
39. Stenvik J, Jorgensen TO. Immunoglobulin D (IgD) of Atlantic cod has a unique structure. Immunogenetics. 2000 May;51(6):452-61.
40. Danilova N, Bussmann J, Jekosch K, Steiner LA. The immunoglobulin heavy-chain locus in zebrafish: identification and expression of a previously unknown isotype, immunoglobulin Z. Nat Immunol. 2005 Mar;6(3):295-302.
41. Hansen JD, Landis ED, Phillips RB. Discovery of a unique Ig heavy-chain isotype (IgT) in rainbow trout: Implications for a distinctive B cell developmental pathway in teleost fish. Proc Natl Acad Sci U S A. 2005 May 10;102(19):6919-24.
42. Li J, Barreda DR, Zhang YA, Boshra H, Gelman AE, Lapatra S, et al. B lymphocytes from early vertebrates have potent phagocytic and microbicidal abilities. Nat Immunol. 2006 Oct;7(10):1116-24.
43. Overland HS, Pettersen EF, Ronneseth A, Wergeland HI. Phagocytosis by B-cells and neutrophils in Atlantic salmon (Salmo salar L.) and Atlantic cod (Gadus morhua L.). Fish Shellfish Immunol. 2010 Jan;28(1):193-204.
44. Wang T, Huang W, Costa MM, Secombes CJ. The gamma-chain cytokine/receptor system in fish: more ligands and receptors. Fish Shellfish Immunol. 2011 Nov;31(5):673-87.
45. Savan R, Sakai M. Genomics of fish cytokines. Comp Biochem Physiol Part D Genomics Proteomics. 2006

Mar;1(1):89-101.

46. Salazar-Mather TP, Hokeness KL. Cytokine and chemokine networks: pathways to antiviral defense. Curr Top Microbiol Immunol. 2006;303:29-46.

47. Svanborg C, Godaly G, Hedlund M. Cytokine responses during mucosal infections: role in disease pathogenesis and host defence. Curr Opin Microbiol. 1999 Feb;2(1):99-105.

48. DeVries ME, Ran L, Kelvin DJ. On the edge: the physiological and pathophysiological role of chemokines during inflammatory and immunological responses. Semin Immunol. 1999 Apr;11(2):95-104.

49. Whyte SK. The innate immune response of finfish--a review of current knowledge. Fish Shellfish Immunol. 2007 Dec;23(6):1127-51.

50. Rahman MM, McFadden G. Modulation of tumor necrosis factor by microbial pathogens. PLoS Pathog. 2006 Feb;2(2):e4.

51. Benedict CA, Banks TA, Ware CF. Death and survival: viral regulation of TNF signaling pathways. Curr Opin Immunol. 2003 Feb;15(1):59-65.

52. Hirono I, Nam BH, Kurobe T, Aoki T. Molecular cloning, characterization, and expression of TNF cDNA and gene from Japanese flounder Paralychthys olivaceus. J Immunol. 2000 Oct 15;165(8):4423-7.

53. Laing KJ, Wang T, Zou J, Holland J, Hong S, Bols N, et al. Cloning and expression analysis of rainbow trout Oncorhynchus mykiss tumour necrosis factor-alpha. Eur J Biochem. 2001 Mar;268(5):1315-22.

54. Zou J, Wang T, Hirono I, Aoki T, Inagawa H, Honda T, et al. Differential expression of two tumor necrosis factor genes in rainbow trout, Oncorhynchus mykiss. Dev Comp Immunol. 2002 Mar;26(2):161-72.

55. Garcia-Castillo J, Pelegrin P, Mulero V, Meseguer J. Molecular cloning and expression analysis of tumor necrosis factor alpha from a marine fish reveal its constitutive expression and ubiquitous nature. Immunogenetics. 2002 Jun;54(3):200-7.

56. Saeij JP, Stet RJ, de Vries BJ, van Muiswinkel WB, Wiegertjes GF. Molecular and functional characterization of carp TNF: a link between TNF polymorphism and trypanotolerance? Dev Comp Immunol. 2003 Jan;27(1):29-41.

57. Zou J, Secombes CJ, Long S, Miller N, Clem LW, Chinchar VG. Molecular identification and expression analysis of tumor necrosis factor in channel catfish (Ictalurus punctatus). Dev Comp Immunol. 2003 Dec;27(10):845-58.

58. Praveen K, Evans DL, Jaso-Friedmann L. Constitutive expression of tumor necrosis factor-alpha in cytotoxic cells of teleosts and its role in regulation of cell-mediated cytotoxicity. Mol Immunol. 2006 Feb;43(3):279-91.

59. Ordas MC, Costa MM, Roca FJ, Lopez-Castejon G, Mulero V, Meseguer J, et al. Turbot TNFalpha gene: molecular characterization and biological activity of the recombinant protein. Mol Immunol. 2007 Jan;44(4):389-400.

60. Grayfer L, Walsh JG, Belosevic M. Characterization and functional analysis of goldfish (Carassius auratus L.) tumor necrosis factor-alpha. Dev Comp Immunol. 2008;32(5):532-43.

61. Garcia-Castillo J, Chaves-Pozo E, Olivares P, Pelegrin P, Meseguer J, Mulero V. The tumor necrosis factor alpha of the bony fish seabream exhibits the in vivo proinflammatory and proliferative activities of its mammalian counterparts, yet it functions in a species-specific manner. Cell Mol Life Sci. 2004 Jun;61(11):1331-40.

62. Roca FJ, Mulero I, Lopez-Munoz A, Sepulcre MP, Renshaw SA, Meseguer J, et al. Evolution of the inflammatory response in vertebrates: fish TNF-alpha is a powerful activator of endothelial cells but hardly activates phagocytes. J Immunol. 2008 Oct 1;181(7):5071-81.

63. Hardie LJ, Chappell LH, Secombes CJ. Human tumor necrosis factor alpha influences rainbow trout Oncorhynchus mykiss leucocyte responses. Vet Immunol Immunopathol. 1994 Jan;40(1):73-84.

64. Iliev DB, Liarte CQ, MacKenzie S, Goetz FW. Activation of rainbow trout (Oncorhynchus mykiss) mononuclear phagocytes by different pathogen associated molecular pattern (PAMP) bearing agents. Mol Immunol. 2005 Jun;42(10):1215-23.

65. Wang WL, Hong JR, Lin GH, Liu W, Gong HY, Lu MW, et al. Stage-specific expression of TNFalpha regulates bad/bid-mediated apoptosis and RIP1/ROS-mediated secondary necrosis in Birnavirus-infected fish cells. PLoS One. 2011;6(2):e16740.

66. Dinarello C, Arend W, Sims J, Smith D, Blumberg H, O'Neill L, et al. IL-1 family nomenclature. Nat Immunol. 2010 Nov;11(11):973.

67. Dinarello CA. Interleukin-1. Cytokine Growth Factor Rev. 1997 Dec;8(4):253-65.

68. Zou J, Grabowski PS, Cunningham C, Secombes CJ. Molecular cloning of interleukin 1beta from rainbow trout Oncorhynchus mykiss reveals no evidence of an ice cut site. Cytokine. 1999 Aug;11(8):552-60.

69. Fujiki K, Shin DH, Nakao M, Yano T. Molecular cloning and expression analysis of carp (Cyprinus carpio) interleukin-1 beta, high affinity immunoglobulin E Fc receptor gamma subunit and serum amyloid A. Fish Shellfish Immunol. 2000 Apr;10(3):229-42.

70. Scapigliati G, Buonocore F, Bird S, Zou J, Pelegrin P, Falasca C, et al. Phylogeny of cytokines: molecular cloning and expression analysis of sea bass Dicentrarchus labrax interleukin-1beta. Fish Shellfish Immunol. 2001 Nov;11(8):711-26.

71. Pelegrin P, Garcia-Castillo J, Mulero V, Meseguer J. Interleukin-1beta isolated from a marine fish reveals up-regulated expression in macrophages following activation with lipopolysaccharide and lymphokines. Cytokine. 2001 Oct 21;16(2):67-72.

72. Corripio-Miyar Y, Bird S, Tsamopoulos K, Secombes CJ. Cloning and expression analysis of two pro-inflammatory cytokines, IL-1 beta and IL-8, in haddock (Melanogrammus aeglefinus). Mol Immunol. 2007 Feb;44(6):1361-73.

73. Lee DS, Hong SH, Lee HJ, Jun LJ, Chung JK, Kim KH, et al. Molecular cDNA cloning and analysis of the organization and expression of the IL-1beta gene in the Nile tilapia, Oreochromis niloticus. Comp Biochem Physiol A Mol Integr Physiol. 2006 Mar;143(3):307-14.

74. Pleguezuelos O, Zou J, Cunningham C, Secombes CJ. Cloning, sequencing, and analysis of expression of a second IL-1beta gene in rainbow trout (Oncorhynchus mykiss). Immunogenetics. 2000 Oct;51(12):1002-11.

75. Cerretti DP, Kozlosky CJ, Mosley B, Nelson N, Van Ness K, Greenstreet TA, et al. Molecular cloning of the interleukin-1 beta converting enzyme. Science. 1992 Apr 3;256(5053):97-100.

76. Secombes CJ, Wang T, Hong S, Peddie S, Crampe M, Laing KJ, et al. Cytokines and innate immunity of fish. Dev Comp Immunol. 2001 Oct-Dec;25(8-9):713-23.

77. Lu DQ, Bei JX, Feng LN, Zhang Y, Liu XC, Wang L, et al. Interleukin-1beta gene in orange-spotted grouper, Epinephelus coioides: molecular cloning, expression, biological activities and signal transduction. Mol Immunol. 2008 Feb;45(4):857-67.

78. Reyes-Cerpa S, Reyes-Lopez FE, Toro-Ascuy D, Ibanez J, Maisey K, Sandino AM, et al. IPNV modulation of pro and anti-inflammatory cytokine expression in Atlantic salmon might help the establishment of infection and persistence. Fish Shellfish Immunol. 2012 Feb;32(2):291-300.

79. Seppola M, Larsen AN, Steiro K, Robertsen B, Jensen I. Characterisation and expression analysis of the interleukin genes, IL-1beta, IL-8 and IL-10, in Atlantic cod (Gadus morhua L.). Mol Immunol. 2008 Feb;45(4):887-97.

80. Zou J, Holland J, Pleguezuelos O, Cunningham C, Secombes CJ. Factors influencing the expression of interleukin-1 beta in cultured rainbow trout (Oncorhynchus mykiss) leucocytes. Dev Comp Immunol. 2000 Sep-Oct;24(6-7):575-82.

81. Buonocore F, Forlenza M, Randelli E, Benedetti S, Bossu P, Meloni S, et al. Biological activity of sea bass (Dicentrarchus labrax L.) recombinant interleukin-1beta. Mar Biotechnol (NY). 2005 Nov-Dec;7(6):609-17.

82. Hong S, Zou J, Crampe M, Peddie S, Scapigliati G, Bols N, et al. The production and bioactivity of rainbow trout (Oncorhynchus mykiss) recombinant IL-1 beta. Vet Immunol Immunopathol. 2001 Aug 30;81(1-2):1-14.

83. Peddie S, Zou J, Cunningham C, Secombes CJ. Rainbow trout (Oncorhynchus mykiss) recombinant IL-1beta and derived peptides induce migration of head-kidney leucocytes in vitro. Fish Shellfish Immunol. 2001 Nov;11(8):697-709.

84. Peddie S, Zou J, Secombes CJ. Immunostimulation in the rainbow trout (Oncorhynchus mykiss) following intraperitoneal administration of Ergosan. Vet Immunol Immunopathol. 2002 May;86(1-2):101-13.

85. Dinarello CA, Fantuzzi G. Interleukin-18 and host defense against infection. J Infect Dis. 2003 Jun 15;187 Suppl 2:S370-84.

86. Gracie JA, Robertson SE, McInnes IB. Interleukin-18. J Leukoc Biol. 2003 Feb;73(2):213-24.

87. Sugawara I. Interleukin-18 (IL-18) and infectious diseases, with special emphasis on diseases induced by intracellular pathogens. Microbes Infect. 2000 Aug;2(10):1257-63.

88. Micallef MJ, Ohtsuki T, Kohno K, Tanabe F, Ushio S, Namba M, et al. Interferon-gamma-inducing factor enhances T helper 1 cytokine production by stimulated human T cells: synergism with interleukin-12 for interferon-gamma production. Eur J Immunol. 1996 Jul;26(7):1647-51.

89. Okamura H, Nagata K, Komatsu T, Tanimoto T, Nukata Y, Tanabe F, et al. A novel costimulatory factor for gamma interferon induction found in the livers of mice causes endotoxic shock. Infect Immun. 1995 Oct;63(10):3966-72.

90. Dao T, Mehal WZ, Crispe IN. IL-18 augments perforin-dependent cytotoxicity of liver NK-T cells. J Immunol. 1998 Sep 1;161(5):2217-22.

91. Leung BP, Culshaw S, Gracie JA, Hunter D, Canetti CA, Campbell C, et al. A role for IL-18 in neutrophil activation. J Immunol. 2001 Sep 1;167(5):2879-86.

92. Xu D, Trajkovic V, Hunter D, Leung BP, Schulz K, Gracie JA, et al. IL-18 induces the differentiation of Th1 or Th2 cells depending upon cytokine milieu and genetic background. Eur J Immunol. 2000 Nov;30(11):3147-56.

93. Ghayur T, Banerjee S, Hugunin M, Butler D, Herzog L, Carter A, et al. Caspase-1 processes IFN-gamma-inducing factor and regulates LPS-induced IFN-gamma production. Nature. 1997 Apr 10;386(6625):619-23.

94. Gu Y, Kuida K, Tsutsui H, Ku G, Hsiao K, Fleming MA, et al. Activation of interferon-gamma inducing factor mediated by interleukin-1beta converting enzyme. Science. 1997 Jan 10;275(5297):206-9.

95. Akita K, Ohtsuki T, Nukada Y, Tanimoto T, Namba M, Okura T, et al. Involvement of caspase-1 and caspase-3 in the production and processing of mature human interleukin 18 in monocytic THP.1 cells. J Biol Chem. 1997 Oct 17;272(42):26595-603.

96. Sugawara S, Uehara A, Nochi T, Yamaguchi T, Ueda H, Sugiyama A, et al. Neutrophil proteinase 3-mediated induction of bioactive IL-18 secretion by human oral epithelial cells. J Immunol. 2001 Dec 1;167(11):6568-75.

97. Huising MO, Stet RJ, Savelkoul HF, Verburg-van Kemenade BM. The molecular evolution of the interleukin-1 family of cytokines; IL-18 in teleost fish. Dev Comp Immunol. 2004 May 3;28(5):395-413.

98. Zou J, Bird S, Truckle J, Bols N, Horne M, Secombes C. Identification and expression analysis of an IL-18 homologue and its alternatively spliced form in rainbow trout (Oncorhynchus mykiss). Eur J Biochem. 2004 May;271(10):1913-23.

99. Martin SA, Zou J, Houlihan DF, Secombes CJ. Directional responses following recombinant cytokine stimulation of rainbow trout (Oncorhynchus mykiss) RTS-11 macrophage cells as revealed by transcriptome profiling. BMC Genomics. 2007;8:150.

100. Wang T, Secombes CJ. Identification and expression analysis of two fish-specific IL-6 cytokine family members, the ciliary neurotrophic factor (CNTF)-like and M17 genes, in rainbow trout Oncorhynchus mykiss. Mol Immunol. 2009 Jul;46(11-12):2290-8.

101. Taga T, Kishimoto T. Gp130 and the interleukin-6 family of cytokines. Annu Rev Immunol. 1997;15:797-819.

102. Hirano T. Interleukin 6 and its receptor: ten years later. Int Rev Immunol. 1998;16(3-4):249-84.

103. Inoue K, Takano H, Shimada A, Morita T, Yanagisawa R, Sakurai M, et al. Cytoprotection by interleukin-6 against liver injury induced by lipopolysaccharide. Int J Mol Med. 2005 Feb;15(2):221-4.

104. Kaplanski G, Marin V, Montero-Julian F, Mantovani A, Farnarier C. IL-6: a regulator of the transition from neutrophil to monocyte recruitment during inflammation. Trends Immunol. 2003 Jan;24(1):25-9.

105. Bird S, Zou J, Savan R, Kono T, Sakai M, Woo J, et al. Characterisation and expression analysis of an interleukin 6 homologue in the Japanese pufferfish, Fugu rubripes. Dev Comp Immunol. 2005;29(9):775-89.

106. Iliev DB, Castellana B, Mackenzie S, Planas JV, Goetz FW. Cloning and expression analysis of an IL-6 homolog in rainbow trout (Oncorhynchus mykiss). Mol Immunol. 2007 Mar;44(7):1803-7.

107. Nam BH, Byon JY, Kim YO, Park EM, Cho YC, Cheong J. Molecular cloning and characterisation of the flounder (Paralichthys olivaceus) interleukin-6 gene. Fish Shellfish Immunol. 2007 Jul;23(1):231-6.

108. Castellana B, Iliev DB, Sepulcre MP, MacKenzie S, Goetz FW, Mulero V, et al. Molecular characterization of interleukin-6 in the gilthead seabream (Sparus aurata). Mol Immunol. 2008 Jul;45(12):3363-70.

109. Costa MM, Maehr T, Diaz-Rosales P, Secombes CJ, Wang T. Bioactivity studies of rainbow trout (Oncorhynchus mykiss) interleukin-6: effects on macrophage growth and antimicrobial peptide gene expression. Mol Immunol. 2011 Sep;48(15-16):1903-16.

110. Czupryn MJ, McCoy JM, Scoble HA. Structure-function relationships in human interleukin-11. Identification of regions involved in activity by chemical modification and site-directed mutagenesis. J Biol Chem. 1995 Jan 13;270(2):978-85.

111. Bartz H, Buning-Pfaue F, Turkel O, Schauer U. Respiratory syncytial virus induces prostaglandin E2, IL-10 and IL-11 generation in antigen presenting cells. Clin Exp Immunol. 2002 Sep;129(3):438-45.

112. Kernacki KA, Goebel DJ, Poosch MS, Hazlett LD. Early cytokine and chemokine gene expression during Pseudomonas aeruginosa corneal infection in mice. Infect Immun. 1998 Jan;66(1):376-9.

113. Nandurkar HH, Robb L, Begley CG. The role of IL-II in hematopoiesis as revealed by a targeted mutation of its receptor. Stem Cells. 1998;16 Suppl 2:53-65.

114. Trepicchio WL, Bozza M, Pedneault G, Dorner AJ. Recombinant human IL-11 attenuates the inflammatory response through down-regulation of proinflammatory cytokine release and nitric oxide production. J Immunol. 1996 Oct 15;157(8):3627-34.

115. Orazi A, Du X, Yang Z, Kashai M, Williams DA. Interleukin-11 prevents apoptosis and accelerates recovery of small intestinal mucosa in mice treated with combined chemotherapy and radiation. Lab Invest. 1996 Jul;75(1):33-42.

116. Liu Q, Du XX, Schindel DT, Yang ZX, Rescorla FJ, Williams DA, et al. Trophic effects of interleukin-11 in rats with experimental short bowel syndrome. J Pediatr Surg. 1996 Aug;31(8):1047-50; discussion 50-1.

117. Huising MO, Kruiswijk CP, van Schijndel JE, Savelkoul HF, Flik G, Verburg-van Kemenade BM. Multiple and highly divergent IL-11 genes in teleost fish. Immunogenetics. 2005 Jul;57(6):432-43.

118. Wang T, Holland JW, Bols N, Secombes CJ. Cloning and expression of the first nonmammalian interleukin-11 gene in rainbow trout Oncorhynchus mykiss. FEBS J. 2005 Mar;272(5):1136-47.

119. Santos MD, Yasuike M, Kondo H, Hirono I, Aoki T. Teleostean IL11b exhibits complementing function to IL11a and expansive involvement in antibacterial and antiviral responses. Mol Immunol. 2008 Jul;45(12):3494-501.

120. Ribeiro CM, Hermsen T, Taverne-Thiele AJ, Savelkoul HF, Wiegertjes GF. Evolution of recognition of ligands from Gram-positive bacteria: similarities and differences in the TLR2-mediated response between mammalian vertebrates and teleost fish. J Immunol. 2010 Mar 1;184(5):2355-68.

121. Murphy PM, Baggiolini M, Charo IF, Hebert CA, Horuk R, Matsushima K, et al. International union of pharmacology. XXII. Nomenclature for chemokine receptors. Pharmacol Rev. 2000 Mar;52(1):145-76.

122. Mukaida N, Harada A, Matsushima K. Interleukin-8 (IL-8) and monocyte chemotactic and activating factor (MCAF/MCP-1), chemokines essentially involved in inflammatory and immune reactions. Cytokine Growth Factor Rev. 1998 Mar;9(1):9-23.

123. Hebert CA, Vitangcol RV, Baker JB. Scanning mutagenesis of interleukin-8 identifies a cluster of residues required for receptor binding. J Biol Chem. 1991 Oct 5;266(28):18989-94.

124. Clark-Lewis I, Schumacher C, Baggiolini M, Moser B. Structure-activity relationships of interleukin-8 determined using chemically synthesized analogs. Critical role of NH2-terminal residues and evidence for uncoupling of neutrophil chemotaxis, exocytosis, and receptor binding activities. J Biol Chem. 1991 Dec 5;266(34):23128-34.

125. Laing KJ, Zou JJ, Wang T, Bols N, Hirono I, Aoki T, et al. Identification and analysis of an interleukin 8-like molecule in rainbow trout Oncorhynchus mykiss. Dev Comp Immunol. 2002 Jun;26(5):433-44.

126. Lee EY, Park HH, Kim YT, Choi TJ. Cloning and sequence analysis of the interleukin-8 gene from flounder (Paralichthys olivaceous). Gene. 2001 Aug 22;274(1-2):237-43.

127. Sangrador-Vegas A, Lennington JB, Smith TJ. Molecular

cloning of an IL-8-like CXC chemokine and tissue factor in rainbow trout (Oncorhynchus mykiss) by use of suppression subtractive hybridization. Cytokine. 2002 Jan 21;17(2):66-70.

128. Chen L, He C, Baoprasertkul P, Xu P, Li P, Serapion J, et al. Analysis of a catfish gene resembling interleukin-8: cDNA cloning, gene structure, and expression after infection with Edwardsiella ictaluri. Dev Comp Immunol. 2005;29(2):135-42.
129. Najakshin AM, Mechetina LV, Alabyev BY, Taranin AV. Identification of an IL-8 homolog in lamprey (Lampetra fluviatilis): early evolutionary divergence of chemokines. Eur J Immunol. 1999 Feb;29(2):375-82.
130. Laing KJ, Secombes CJ. Chemokines. Dev Comp Immunol. 2004 May 3;28(5):443-60.
131. Osborne LC, Abraham N. Regulation of memory T cells by gammac cytokines. Cytokine. 2010 May;50(2):105-13.
132. Sogo T, Kawahara M, Ueda H, Otsu M, Onodera M, Nakauchi H, et al. T cell growth control using hapten-specific antibody/interleukin-2 receptor chimera. Cytokine. 2009 Apr;46(1):127-36.
133. Morgan DA, Ruscetti FW, Gallo R. Selective in vitro growth of T lymphocytes from normal human bone marrows. Science. 1976 Sep 10;193(4257):1007-8.
134. Smith KA. T-cell growth factor. Immunol Rev. 1980;51:337-57.
135. Smith KA. Interleukin-2: inception, impact, and implications. Science. 1988 May 27;240(4856):1169-76.
136. Granucci F, Vizzardelli C, Pavelka N, Feau S, Persico M, Virzi E, et al. Inducible IL-2 production by dendritic cells revealed by global gene expression analysis. Nat Immunol. 2001 Sep;2(9):882-8.
137. Walker E, Leemhuis T, Roeder W. Murine B lymphoma cell lines release functionally active interleukin 2 after stimulation with Staphylococcus aureus. J Immunol. 1988 Feb 1;140(3):859-65.
138. Gaffen SL, Wang S, Koshland ME. Expression of the immunoglobulin J chain in a murine B lymphoma is driven by autocrine production of interleukin 2. Cytokine. 1996 Jul;8(7):513-24.
139. Yu TK, Caudell EG, Smid C, Grimm EA. IL-2 activation of NK cells: involvement of MKK1/2/ERK but not p38 kinase pathway. J Immunol. 2000 Jun 15;164(12):6244-51.
140. Gold MR, DeFranco AL. Biochemistry of B lymphocyte activation. Adv Immunol. 1994;55:221-95.
141. Bird S, Zou J, Kono T, Sakai M, Dijkstra JM, Secombes C. Characterisation and expression analysis of interleukin 2 (IL-2) and IL-21 homologues in the Japanese pufferfish, Fugu rubripes, following their discovery by synteny. Immunogenetics. 2005 Mar;56(12):909-23.
142. Sugamata R, Suetake H, Kikuchi K, Suzuki Y. Teleost B7 expressed on monocytes regulates T cell responses. J Immunol. 2009 Jun 1;182(11):6799-806.
143. Diaz-Rosales P, Bird S, Wang TH, Fujiki K, Davidson WS, Zou J, et al. Rainbow trout interleukin-2: cloning, expression and bioactivity analysis. Fish Shellfish Immunol. 2009 Sep;27(3):414-22.
144. Zhang YA, Hikima J, Li J, LaPatra SE, Luo YP, Sunyer JO. Conservation of structural and functional features in a primordial CD80/86 molecule from rainbow trout (Oncorhynchus mykiss), a primitive teleost fish. J Immunol. 2009 Jul 1;183(1):83-96.
145. Laing KJ, Bols N, Secombes CJ. A CXC chemokine sequence isolated from the rainbow trout Oncorhynchus mykiss resembles the closely related interferon-gamma-inducible chemokines CXCL9, CXCL10 and CXCL11. Eur Cytokine Netw. 2002 Oct-Dec;13(4):462-73.
146. Banchereau J, Bidaud C, Fluckiger AC, Galibert L, Garrone P, Malisan F, et al. Effects of interleukin 4 on human B-cell growth and differentiation. Res Immunol. 1993 Oct;144(8):601-5.
147. Mosmann TR, Cherwinski H, Bond MW, Giedlin MA, Coffman RL. Two types of murine helper T cell clone. I. Definition according to profiles of lymphokine activities and secreted proteins. J Immunol. 1986 Apr 1;136(7):2348-57.
148. Finkelman FD, Urban JF, Jr. The other side of the coin: the protective role of the TH2 cytokines. J Allergy Clin Immunol. 2001 May;107(5):772-80.
149. Yokota T, Otsuka T, Mosmann T, Banchereau J, DeFrance T, Blanchard D, et al. Isolation and characterization of a human interleukin cDNA clone, homologous to mouse B-cell stimulatory factor 1, that expresses B-cell- and T-cell-stimulating activities. Proc Natl Acad Sci U S A. 1986 Aug;83(16):5894-8.
150. Noma Y, Sideras P, Naito T, Bergstedt-Lindquist S, Azuma C, Severinson E, et al. Cloning of cDNA encoding the murine IgG1 induction factor by a novel strategy using SP6 promoter. Nature. 1986 Feb 20-26;319(6055):640-6.
151. Lee F, Yokota T, Otsuka T, Meyerson P, Villaret D, Coffman R, et al. Isolation and characterization of a mouse interleukin cDNA clone that expresses B-cell stimulatory factor 1 activities and T-cell- and mast-cell-stimulating activities. Proc Natl Acad Sci U S A. 1986 Apr;83(7):2061-5.
152. Heussler VT, Eichhorn M, Dobbelaere DA. Cloning of a full-length cDNA encoding bovine interleukin 4 by the polymerase chain reaction. Gene. 1992 May 15;114(2):273-8.
153. Li-Weber M, Krammer PH. Regulation of IL4 gene expression by T cells and therapeutic perspectives. Nat Rev Immunol. 2003 Jul;3(7):534-43.
154. Ansel KM, Djuretic I, Tanasa B, Rao A. Regulation of Th2 differentiation and Il4 locus accessibility. Annu Rev Immunol. 2006;24:607-56.
155. Ohtani M, Hayashi N, Hashimoto K, Nakanishi T, Dijkstra JM. Comprehensive clarification of two paralogous interleukin 4/13 loci in teleost fish. Immunogenetics. 2008 Jul;60(7):383-97.
156. Jaillon O, Aury JM, Brunet F, Petit JL, Stange-Thomann N, Mauceli E, et al. Genome duplication in the teleost fish Tetraodon nigroviridis reveals the early vertebrate proto-karyotype. Nature. 2004 Oct 21;431(7011):946-

57.

157. Li JH, Shao JZ, Xiang LX, Wen Y. Cloning, characterization and expression analysis of pufferfish interleukin-4 cDNA: the first evidence of Th2-type cytokine in fish. Mol Immunol. 2007 Mar;44(8):2078-86.

158. Secrist H, Egan M, Peters MG. Tissue-specific regulation of IL-4 mRNA expression in human tonsil. J Immunol. 1994 Feb 1;152(3):1120-6.

159. Lin AF, Xiang LX, Wang QL, Dong WR, Gong YF, Shao JZ. The DC-SIGN of zebrafish: insights into the existence of a CD209 homologue in a lower vertebrate and its involvement in adaptive immunity. J Immunol. 2009 Dec 1;183(11):7398-410.

160. Hu YL, Xiang LX, Shao JZ. Identification and characterization of a novel immunoglobulin Z isotype in zebrafish: implications for a distinct B cell receptor in lower vertebrates. Mol Immunol. 2010 Jan;47(4):738-46.

161. Mitra S, Alnabulsi A, Secombes CJ, Bird S. Identification and characterization of the transcription factors involved in T-cell development, t-bet, stat6 and foxp3, within the zebrafish, Danio rerio. FEBS J. 2010 Jan;277(1):128-47.

162. Takizawa F, Koppang EO, Ohtani M, Nakanishi T, Hashimoto K, Fischer U, et al. Constitutive high expression of interleukin-4/13A and GATA-3 in gill and skin of salmonid fishes suggests that these tissues form Th2-skewed immune environments. Mol Immunol. 2011 Jul;48(12-13):1360-8.

163. Fry TJ, Mackall CL. The many faces of IL-7: from lymphopoiesis to peripheral T cell maintenance. J Immunol. 2005 Jun 1;174(11):6571-6.

164. Alves NL, Richard-Le Goff O, Huntington ND, Sousa AP, Ribeiro VS, Bordack A, et al. Characterization of the thymic IL-7 niche in vivo. Proc Natl Acad Sci U S A. 2009 Feb 3;106(5):1512-7.

165. Mazzucchelli RI, Warming S, Lawrence SM, Ishii M, Abshari M, Washington AV, et al. Visualization and identification of IL-7 producing cells in reporter mice. PLoS One. 2009;4(11):e7637.

166. Repass JF, Laurent MN, Carter C, Reizis B, Bedford MT, Cardenas K, et al. IL7-hCD25 and IL7-Cre BAC transgenic mouse lines: new tools for analysis of IL-7 expressing cells. Genesis. 2009 Apr;47(4):281-7.

167. Kono T, Bird S, Sonoda K, Savan R, Secombes CJ, Sakai M. Characterization and expression analysis of an interleukin-7 homologue in the Japanese pufferfish, Takifugu rubripes. FEBS J. 2008 Mar;275(6):1213-26.

168. Bei JX, Suetake H, Araki K, Kikuchi K, Yoshiura Y, Lin HR, et al. Two interleukin (IL)-15 homologues in fish from two distinct origins. Mol Immunol. 2006 Mar;43(7):860-9.

169. Fang W, Xiang LX, Shao JZ, Wen Y, Chen SY. Identification and characterization of an interleukin-15 homologue from Tetraodon nigroviridis. Comp Biochem Physiol B Biochem Mol Biol. 2006 Mar;143(3):335-43.

170. Gunimaladevi I, Savan R, Sato K, Yamaguchi R, Sakai M. Characterization of an interleukin-15 like (IL-15L) gene from zebrafish (Danio rerio). Fish Shellfish Immunol. 2007 Apr;22(4):351-62.

171. Wang T, Holland JW, Carrington A, Zou J, Secombes CJ. Molecular and functional characterization of IL-15 in rainbow trout Oncorhynchus mykiss: a potent inducer of IFN-gamma expression in spleen leukocytes. J Immunol. 2007 Aug 1;179(3):1475-88.

172. Parrish-Novak J, Dillon SR, Nelson A, Hammond A, Sprecher C, Gross JA, et al. Interleukin 21 and its receptor are involved in NK cell expansion and regulation of lymphocyte function. Nature. 2000 Nov 2;408(6808):57-63.

173. Asao H, Okuyama C, Kumaki S, Ishii N, Tsuchiya S, Foster D, et al. Cutting edge: the common gamma-chain is an indispensable subunit of the IL-21 receptor complex. J Immunol. 2001 Jul 1;167(1):1-5.

174. Vosshenrich CA, Di Santo JP. Cytokines: IL-21 joins the gamma(c)-dependent network? Curr Biol. 2001 Mar 6;11(5):R175-7.

175. Chtanova T, Tangye SG, Newton R, Frank N, Hodge MR, Rolph MS, et al. T follicular helper cells express a distinctive transcriptional profile, reflecting their role as non-Th1/Th2 effector cells that provide help for B cells. J Immunol. 2004 Jul 1;173(1):68-78.

176. Wurster AL, Rodgers VL, Satoskar AR, Whitters MJ, Young DA, Collins M, et al. Interleukin 21 is a T helper (Th) cell 2 cytokine that specifically inhibits the differentiation of naive Th cells into interferon gamma-producing Th1 cells. J Exp Med. 2002 Oct 7;196(7):969-77.

177. Leonard WJ, Zeng R, Spolski R. Interleukin 21: a cytokine/cytokine receptor system that has come of age. J Leukoc Biol. 2008 Aug;84(2):348-56.

178. Suto A, Kashiwakuma D, Kagami S, Hirose K, Watanabe N, Yokote K, et al. Development and characterization of IL-21-producing CD4+ T cells. J Exp Med. 2008 Jun 9;205(6):1369-79.

179. Frederiksen KS, Lundsgaard D, Freeman JA, Hughes SD, Holm TL, Skrumsager BK, et al. IL-21 induces in vivo immune activation of NK cells and CD8(+) T cells in patients with metastatic melanoma and renal cell carcinoma. Cancer Immunol Immunother. 2008 Oct;57(10):1439-49.

180. Konforte D, Simard N, Paige CJ. IL-21: an executor of B cell fate. J Immunol. 2009 Feb 15;182(4):1781-7.

181. Wang HJ, Xiang LX, Shao JZ, Jia S. Molecular cloning, characterization and expression analysis of an IL-21 homologue from Tetraodon nigroviridis. Cytokine. 2006 Aug;35(3-4):126-34.

182. Wang T, Diaz-Rosales P, Costa MM, Campbell S, Snow M, Collet B, et al. Functional characterization of a nonmammalian IL-21: rainbow trout Oncorhynchus mykiss IL-21 upregulates the expression of the Th cell signature cytokines IFN-gamma, IL-10, and IL-22. J Immunol. 2011 Jan 15;186(2):708-21.

183. Lutfalla G, Roest Crollius H, Stange-Thomann N, Jaillon O, Mogensen K, Monneron D. Comparative genomic analysis reveals independent expansion of a lineage-

specific gene family in vertebrates: the class II cytokine receptors and their ligands in mammals and fish. BMC Genomics. 2003 Jul 17;4(1):29.

184. Fickenscher H, Hor S, Kupers H, Knappe A, Wittmann S, Sticht H. The interleukin-10 family of cytokines. Trends Immunol. 2002 Feb;23(2):89-96.

185. Moore KW, de Waal Malefyt R, Coffman RL, O'Garra A. Interleukin-10 and the interleukin-10 receptor. Annu Rev Immunol. 2001;19:683-765.

186. Zou J, Clark MS, Secombes CJ. Characterisation, expression and promoter analysis of an interleukin 10 homologue in the puffer fish, Fugu rubripes. Immunogenetics. 2003 Aug;55(5):325-35.

187. Savan R, Igawa D, Sakai M. Cloning, characterization and expression analysis of interleukin-10 from the common carp, Cyprinus carpio L. Eur J Biochem. 2003 Dec;270(23):4647-54.

188. Zhang DC, Shao YQ, Huang YQ, Jiang SG. Cloning, characterization and expression analysis of interleukin-10 from the zebrafish (Danio rerion). J Biochem Mol Biol. 2005 Sep 30;38(5):571-6.

189. Inoue Y, Kamota S, Ito K, Yoshiura Y, Ototake M, Moritomo T, et al. Molecular cloning and expression analysis of rainbow trout (Oncorhynchus mykiss) interleukin-10 cDNAs. Fish Shellfish Immunol. 2005 Apr;18(4):335-44.

190. Buonocore F, Randelli E, Bird S, Secombes C, Facchiano A, Constantini S, et al. Interleukin-10 expression by real-time PCR and homology modelling analysis in the European sea bass (Dicentrarchus Labrax L.). Aquaculture. 2007;270(1-4):512 - 22.

191. Pinto RD, Nascimento DS, Reis MI, do Vale A, Dos Santos NM. Molecular characterization, 3D modelling and expression analysis of sea bass (Dicentrarchus labrax L.) interleukin-10. Mol Immunol. 2007 Mar;44(8):2056-65.

192. Zhang Z, Swain T, Bogwald J, Dalmo RA, Kumari J. Bath immunostimulation of rainbow trout (Oncorhynchus mykiss) fry induces enhancement of inflammatory cytokine transcripts, while repeated bath induce no changes. Fish Shellfish Immunol. 2009 May;26(5):677-84.

193. Wegenka UM. IL-20: biological functions mediated through two types of receptor complexes. Cytokine Growth Factor Rev. 2010 Oct;21(5):353-63.

194. Stein C, Caccamo M, Laird G, Leptin M. Conservation and divergence of gene families encoding components of innate immune response systems in zebrafish. Genome Biol. 2007;8(11):R251.

195. Wang T, Diaz-Rosales P, Martin SA, Secombes CJ. Cloning of a novel interleukin (IL)-20-like gene in rainbow trout Oncorhynchus mykiss gives an insight into the evolution of the IL-10 family. Dev Comp Immunol. 2010 Feb;34(2):158-67.

196. Liang SC, Tan XY, Luxenberg DP, Karim R, Dunussi-Joannopoulos K, Collins M, et al. Interleukin (IL)-22 and IL-17 are coexpressed by Th17 cells and cooperatively enhance expression of antimicrobial peptides. J Exp Med. 2006 Oct 2;203(10):2271-9.

197. Cupedo T, Crellin NK, Papazian N, Rombouts EJ, Weijer K, Grogan JL, et al. Human fetal lymphoid tissue-inducer cells are interleukin 17-producing precursors to RORC+ CD127+ natural killer-like cells. Nat Immunol. 2009 Jan;10(1):66-74.

198. Duhen T, Geiger R, Jarrossay D, Lanzavecchia A, Sallusto F. Production of interleukin 22 but not interleukin 17 by a subset of human skin-homing memory T cells. Nat Immunol. 2009 Aug;10(8):857-63.

199. Eyerich S, Eyerich K, Pennino D, Carbone T, Nasorri F, Pallotta S, et al. Th22 cells represent a distinct human T cell subset involved in epidermal immunity and remodeling. J Clin Invest. 2009 Dec;119(12):3573-85.

200. Fujita H, Nograles KE, Kikuchi T, Gonzalez J, Carucci JA, Krueger JG. Human Langerhans cells induce distinct IL-22-producing CD4+ T cells lacking IL-17 production. Proc Natl Acad Sci U S A. 2009 Dec 22;106(51):21795-800.

201. Wolk K, Haugen HS, Xu W, Witte E, Waggie K, Anderson M, et al. IL-22 and IL-20 are key mediators of the epidermal alterations in psoriasis while IL-17 and IFN-gamma are not. J Mol Med (Berl). 2009 May;87(5):523-36.

202. Feinen B, Russell MW. Contrasting Roles of IL-22 and IL-17 in Murine Genital Tract Infection by Neisseria gonorrhoeae. Front Immunol. 2012;3:11.

203. Wolk K, Kunz S, Witte E, Friedrich M, Asadullah K, Sabat R. IL-22 increases the innate immunity of tissues. Immunity. 2004 Aug;21(2):241-54.

204. Aujla SJ, Chan YR, Zheng M, Fei M, Askew DJ, Pociask DA, et al. IL-22 mediates mucosal host defense against Gram-negative bacterial pneumonia. Nat Med. 2008 Mar;14(3):275-81.

205. Andoh A, Zhang Z, Inatomi O, Fujino S, Deguchi Y, Araki Y, et al. Interleukin-22, a member of the IL-10 subfamily, induces inflammatory responses in colonic subepithelial myofibroblasts. Gastroenterology. 2005 Sep;129(3):969-84.

206. Zheng Y, Valdez PA, Danilenko DM, Hu Y, Sa SM, Gong Q, et al. Interleukin-22 mediates early host defense against attaching and effacing bacterial pathogens. Nat Med. 2008 Mar;14(3):282-9.

207. Bird L. Mucosal immunology: IL-22 keeps commensals in their place. Nat Rev Immunol. 2012 Jul 6.

208. Wolk K, Kunz S, Asadullah K, Sabat R. Cutting edge: immune cells as sources and targets of the IL-10 family members? J Immunol. 2002 Jun 1;168(11):5397-402.

209. Wilson NJ, Boniface K, Chan JR, McKenzie BS, Blumenschein WM, Mattson JD, et al. Development, cytokine profile and function of human interleukin 17-producing helper T cells. Nat Immunol. 2007 Sep;8(9):950-7.

210. Manel N, Unutmaz D, Littman DR. The differentiation of human T(H)-17 cells requires transforming growth factor-beta and induction of the nuclear receptor RORgammat. Nat Immunol. 2008 Jun;9(6):641-9.

211. Pene J, Chevalier S, Preisser L, Venereau E, Guilleux

MH, Ghannam S, et al. Chronically inflamed human tissues are infiltrated by highly differentiated Th17 lymphocytes. J Immunol. 2008 Jun 1;180(11):7423-30.

212. Cella M, Fuchs A, Vermi W, Facchetti F, Otero K, Lennerz JK, et al. A human natural killer cell subset provides an innate source of IL-22 for mucosal immunity. Nature. 2009 Feb 5;457(7230):722-5.

213. Hughes T, Becknell B, McClory S, Briercheck E, Freud AG, Zhang X, et al. Stage 3 immature human natural killer cells found in secondary lymphoid tissue constitutively and selectively express the TH 17 cytokine interleukin-22. Blood. 2009 Apr 23;113(17):4008-10.

214. Donnelly RP, Sheikh F, Dickensheets H, Savan R, Young HA, Walter MR. Interleukin-26: an IL-10-related cytokine produced by Th17 cells. Cytokine Growth Factor Rev. 2010 Oct;21(5):393-401.

215. Zou J, Yoshiura Y, Dijkstra JM, Sakai M, Ototake M, Secombes C. Identification of an interferon gamma homologue in Fugu, Takifugu rubripes. Fish Shellfish Immunol. 2004 Oct;17(4):403-9.

216. Igawa D, Sakai M, Savan R. An unexpected discovery of two interferon gamma-like genes along with interleukin (IL)-22 and -26 from teleost: IL-22 and -26 genes have been described for the first time outside mammals. Mol Immunol. 2006 Mar;43(7):999-1009.

217. Corripio-Miyar Y, Zou J, Richmond H, Secombes CJ. Identification of interleukin-22 in gadoids and examination of its expression level in vaccinated fish. Mol Immunol. 2009 Jun;46(10):2098-106.

218. Kolls JK. Th17 cells in mucosal immunity and tissue inflammation. Semin Immunopathol. 2010 Mar;32(1):1-2.

219. Haugarvoll E, Bjerkas I, Nowak BF, Hordvik I, Koppang EO. Identification and characterization of a novel intraepithelial lymphoid tissue in the gills of Atlantic salmon. J Anat. 2008 Aug;213(2):202-9.

220. Moseley TA, Haudenschild DR, Rose L, Reddi AH. Interleukin-17 family and IL-17 receptors. Cytokine Growth Factor Rev. 2003 Apr;14(2):155-74.

221. Yao Z, Timour M, Painter S, Fanslow W, Spriggs M. Complete nucleotide sequence of the mouse CTLA8 gene. Gene. 1996 Feb 12;168(2):223-5.

222. Li H, Chen J, Huang A, Stinson J, Heldens S, Foster J, et al. Cloning and characterization of IL-17B and IL-17C, two new members of the IL-17 cytokine family. Proc Natl Acad Sci U S A. 2000 Jan 18;97(2):773-8.

223. Shi Y, Ullrich SJ, Zhang J, Connolly K, Grzegorzewski KJ, Barber MC, et al. A novel cytokine receptor-ligand pair. Identification, molecular characterization, and in vivo immunomodulatory activity. J Biol Chem. 2000 Jun 23;275(25):19167-76.

224. Hymowitz SG, Filvaroff EH, Yin JP, Lee J, Cai L, Risser P, et al. IL-17s adopt a cystine knot fold: structure and activity of a novel cytokine, IL-17F, and implications for receptor binding. EMBO J. 2001 Oct 1;20(19):5332-41.

225. Lee J, Ho WH, Maruoka M, Corpuz RT, Baldwin DT, Foster JS, et al. IL-17E, a novel proinflammatory ligand for the IL-17 receptor homolog IL-17Rh1. J Biol Chem. 2001 Jan 12;276(2):1660-4.

226. Starnes T, Robertson MJ, Sledge G, Kelich S, Nakshatri H, Broxmeyer HE, et al. Cutting edge: IL-17F, a novel cytokine selectively expressed in activated T cells and monocytes, regulates angiogenesis and endothelial cell cytokine production. J Immunol. 2001 Oct 15;167(8):4137-40.

227. Starnes T, Broxmeyer HE, Robertson MJ, Hromas R. Cutting edge: IL-17D, a novel member of the IL-17 family, stimulates cytokine production and inhibits hemopoiesis. J Immunol. 2002 Jul 15;169(2):642-6.

228. Gunimaladevi I, Savan R, Sakai M. Identification, cloning and characterization of interleukin-17 and its family from zebrafish. Fish Shellfish Immunol. 2006 Oct;21(4):393-403.

229. Wang T, Martin SA, Secombes CJ. Two interleukin-17C-like genes exist in rainbow trout Oncorhynchus mykiss that are differentially expressed and modulated. Dev Comp Immunol. 2010 May;34(5):491-500.

230. Kumari J, Larsen AN, Bogwald J, Dalmo RA. Interleukin-17D in Atlantic salmon (Salmo salar): molecular characterization, 3D modelling and promoter analysis. Fish Shellfish Immunol. 2009 Nov;27(5):647-59.

231. Korenaga H, Kono T, Sakai M. Isolation of seven IL-17 family genes from the Japanese pufferfish Takifugu rubripes. Fish Shellfish Immunol. 2010 May-Jun;28(5-6):809-18.

232. Kono T, Korenaga H, Sakai M. Genomics of fish IL-17 ligand and receptors: a review. Fish Shellfish Immunol. 2011 Nov;31(5):635-43.

233. Huang D, Cancilla MR, Morahan G. Complete primary structure, chromosomal localisation, and definition of polymorphisms of the gene encoding the human interleukin-12 p40 subunit. Genes Immun. 2000 Dec;1(8):515-20.

234. Yoshiura Y, Kiryu I, Fujiwara A, Suetake H, Suzuki Y, Nakanishi T, et al. Identification and characterization of Fugu orthologues of mammalian interleukin-12 subunits. Immunogenetics. 2003 Aug;55(5):296-306.

235. Nascimento DS, do Vale A, Tomas AM, Zou J, Secombes CJ, dos Santos NM. Cloning, promoter analysis and expression in response to bacterial exposure of sea bass (Dicentrarchus labrax L.) interleukin-12 p40 and p35 subunits. Mol Immunol. 2007 Mar;44(9):2277-91.

236. Li MO, Flavell RA. TGF-beta: a master of all T cell trades. Cell. 2008 Aug 8;134(3):392-404.

237. Li MO, Wan YY, Sanjabi S, Robertson AK, Flavell RA. Transforming growth factor-beta regulation of immune responses. Annu Rev Immunol. 2006;24:99-146.

238. Saxena V, Lienesch DW, Zhou M, Bommireddy R, Azhar M, Doetschman T, et al. Dual roles of immunoregulatory cytokine TGF-beta in the pathogenesis of autoimmunity-mediated organ damage. J Immunol. 2008 Feb 1;180(3):1903-12.

239. Zhang L, Yi H, Xia XP, Zhao Y. Transforming growth factor-beta: an important role in CD4+CD25+ regulatory T cells and immune tolerance. Autoimmunity. 2006

Jun;39(4):269-76.

240. Li B, Samanta A, Song X, Furuuchi K, Iacono KT, Kennedy S, et al. FOXP3 ensembles in T-cell regulation. Immunol Rev. 2006 Aug;212:99-113.

241. Wan YY, Flavell RA. 'Yin-Yang' functions of transforming growth factor-beta and T regulatory cells in immune regulation. Immunol Rev. 2007 Dec;220:199-213.

242. Haddad G, Hanington PC, Wilson EC, Grayfer L, Belosevic M. Molecular and functional characterization of goldfish (Carassius auratus L.) transforming growth factor beta. Dev Comp Immunol. 2008;32(6):654-63.

243. Kadowaki T, Yasui Y, Takahashi Y, Kohchi C, Soma G, Inagawa H. Comparative immunological analysis of innate immunity activation after oral administration of wheat fermented extract to teleost fish. Anticancer Res. 2009 Nov;29(11):4871-7.

244. Yang M, Wang Y, Wang X, Chen C, Zhou H. Characterization of grass carp (Ctenopharyngodon idellus) Foxp1a/1b/2: evidence for their involvement in the activation of peripheral blood lymphocyte subpopulations. Fish Shellfish Immunol. 2010 Feb;28(2):289-95.

245. Cai Z, Gao C, Li L, Xing K. Bipolar properties of red seabream (Pagrus major) transforming growth factor-beta in induction of the leucocytes migration. Fish Shellfish Immunol. 2010 Apr;28(4):695-700.

246. Yang M, Wang X, Chen D, Wang Y, Zhang A, Zhou H. TGF-beta1 exerts opposing effects on grass carp leukocytes: implication in teleost immunity, receptor signaling and potential self-regulatory mechanisms. PLoS One. 2012;7(4):e35011.

247. Samuel CE. Antiviral actions of interferons. Clin Microbiol Rev. 2001 Oct;14(4):778-809, table of contents.

248. Robertsen B. The interferon system of teleost fish. Fish Shellfish Immunol. 2006 Feb;20(2):172-91.

249. Bergan V, Steinsvik S, Xu H, Kileng O, Robertsen B. Promoters of type I interferon genes from Atlantic salmon contain two main regulatory regions. FEBS J. 2006 Sep;273(17):3893-906.

250. Kileng O, Brundtland MI, Robertsen B. Infectious salmon anemia virus is a powerful inducer of key genes of the type I interferon system of Atlantic salmon, but is not inhibited by interferon. Fish Shellfish Immunol. 2007 Aug;23(2):378-89.

251. Robertsen B, Bergan V, Rokenes T, Larsen R, Albuquerque A. Atlantic salmon interferon genes: cloning, sequence analysis, expression, and biological activity. J Interferon Cytokine Res. 2003 Oct;23(10):601-12.

252. Rokenes TP, Larsen R, Robertsen B. Atlantic salmon ISG15: Expression and conjugation to cellular proteins in response to interferon, double-stranded RNA and virus infections. Mol Immunol. 2007 Feb;44(5):950-9.

253. Zou J, Tafalla C, Truckle J, Secombes CJ. Identification of a second group of type I IFNs in fish sheds light on IFN evolution in vertebrates. J Immunol. 2007 Sep 15;179(6):3859-71.

254. Robertsen B. Expression of interferon and interferon-induced genes in salmonids in response to virus infection, interferon-inducing compounds and vaccination. Fish Shellfish Immunol. 2008 Oct;25(4):351-7.

255. Milev-Milovanovic I, Long S, Wilson M, Bengten E, Miller NW, Chinchar VG. Identification and expression analysis of interferon gamma genes in channel catfish. Immunogenetics. 2006 Feb;58(1):70-80.

256. Stolte EH, Savelkoul HF, Wiegertjes G, Flik G, Lidy Verburg-van Kemenade BM. Differential expression of two interferon-gamma genes in common carp (Cyprinus carpio L.). Dev Comp Immunol. 2008;32(12):1467-81.

257. Zou J, Carrington A, Collet B, Dijkstra JM, Yoshiura Y, Bols N, et al. Identification and bioactivities of IFN-gamma in rainbow trout Oncorhynchus mykiss: the first Th1-type cytokine characterized functionally in fish. J Immunol. 2005 Aug 15;175(4):2484-94.

258. Grayfer L, Garcia EG, Belosevic M. Comparison of macrophage antimicrobial responses induced by type II interferons of the goldfish (Carassius auratus L.). J Biol Chem. 2010 Jul 30;285(31):23537-47.

259. Arts JA, Tijhaar EJ, Chadzinska M, Savelkoul HF, Verburg-van Kemenade BM. Functional analysis of carp interferon-gamma: evolutionary conservation of classical phagocyte activation. Fish Shellfish Immunol. 2010 Nov;29(5):793-802.

260. Sun B, Skjaeveland I, Svingerud T, Zou J, Jorgensen J, Robertsen B. Antiviral activity of salmonid gamma interferon against infectious pancreatic necrosis virus and salmonid alphavirus and its dependency on type I interferon. J Virol. 2011 Sep;85(17):9188-98.

261. Duguid JR, Dinauer MC. Library subtraction of in vitro cDNA libraries to identify differentially expressed genes in scrapie infection. Nucleic Acids Res. 1990 May 11;18(9):2789-92.

262. Hara E, Kato T, Nakada S, Sekiya S, Oda K. Subtractive cDNA cloning using oligo(dT)30-latex and PCR: isolation of cDNA clones specific to undifferentiated human embryonal carcinoma cells. Nucleic Acids Res. 1991 Dec;19(25):7097-104.

263. Hedrick SM, Cohen DI, Nielsen EA, Davis MM. Isolation of cDNA clones encoding T cell-specific membrane-associated proteins. Nature. 1984 Mar 8-14;308(5955):149-53.

264. Lisitsyn N, Wigler M. Cloning the differences between two complex genomes. Science. 1993 Feb 12;259(5097):946-51.

265. Hubank M, Schatz DG. Identifying differences in mRNA expression by representational difference analysis of cDNA. Nucleic Acids Res. 1994 Dec 25;22(25):5640-8.

266. Liang P, Pardee AB. Differential display of eukaryotic messenger RNA by means of the polymerase chain reaction. Science. 1992 Aug 14;257(5072):967-71.

267. Welsh J, Chada K, Dalal SS, Cheng R, Ralph D, McClelland M. Arbitrarily primed PCR fingerprinting of RNA. Nucleic Acids Res. 1992 Oct 11;20(19):4965-70.

268. Sompayrac L, Jane S, Burn TC, Tenen DG, Danna KJ. Overcoming limitations of the mRNA differential

display technique. Nucleic Acids Res. 1995 Nov 25;23(22):4738-9.

269. Bertioli DJ, Schlichter UH, Adams MJ, Burrows PR, Steinbiss HH, Antoniw JF. An analysis of differential display shows a strong bias towards high copy number mRNAs. Nucleic Acids Res. 1995 Nov 11;23(21):4520-3.

270. Siebert PD, Chenchik A, Kellogg DE, Lukyanov KA, Lukyanov SA. An improved PCR method for walking in uncloned genomic DNA. Nucleic Acids Res. 1995 Mar 25;23(6):1087-8.

271. Diatchenko L, Lau YF, Campbell AP, Chenchik A, Moqadam F, Huang B, et al. Suppression subtractive hybridization: a method for generating differentially regulated or tissue-specific cDNA probes and libraries. Proc Natl Acad Sci U S A. 1996 Jun 11;93(12):6025-30.

272. Wang L, Wu X. Identification of differentially expressed genes in lipopolysaccharide-stimulated yellow grouper Epinephelus awoara spleen. Fish Shellfish Immunol. 2007 Aug;23(2):354-63.

273. Xia JH, Yue GH. Identification and analysis of immune-related transcriptome in Asian seabass Lates calcarifer. BMC Genomics. 2010;11:356.

274. Feng CY, Johnson SC, Hori TS, Rise M, Hall JR, Gamperl AK, et al. Identification and analysis of differentially expressed genes in immune tissues of Atlantic cod stimulated with formalin-killed, atypical Aeromonas salmonicida. Physiol Genomics. 2009 May 13;37(3):149-63.

275. Tsoi SC, Ewart KV, Penny S, Melville K, Liebscher RS, Brown LL, et al. Identification of immune-relevant genes from atlantic salmon using suppression subtractive hybridization. Mar Biotechnol (NY). 2004 May-Jun;6(3):199-214.

276. Li CH, Chen J, Shi YH, Lu XJ. Use of suppressive subtractive hybridization to identify differentially expressed genes in ayu (Plecoglossus altivelis) associated with Listonella anguillarum infection. Fish Shellfish Immunol. 2011 Sep;31(3):500-6.

277. Matsuyama T, Fujiwara A, Takano T, Nakayasu C. Suppression subtractive hybridization coupled with microarray analysis to examine differential expression of genes in Japanese flounder Paralichthys olivaceus leucocytes during Edwardsiella tarda and viral hemorrhagic septicemia virus infection. Fish Shellfish Immunol. 2011 Oct;31(4):524-32.

278. Bo J, Giesy JP, Ye R, Wang KJ, Lee JS, Au DW. Identification of differentially expressed genes and quantitative expression of complement genes in the liver of marine medaka Oryzias melastigma challenged with Vibrio parahaemolyticus. Comp Biochem Physiol Part D Genomics Proteomics. 2012 Jun;7(2):191-200.

279. Purcell MK, Smith KD, Hood L, Winton JR, Roach JC. Conservation of Toll-Like Receptor Signaling Pathways in Teleost Fish. Comp Biochem Physiol Part D Genomics Proteomics. 2006 Mar;1(1):77-88.

280. Trinchieri G, Sher A. Cooperation of Toll-like receptor signals in innate immune defence. Nat Rev Immunol. 2007 Mar;7(3):179-90.

281. Bayne CJ, Gerwick L. The acute phase response and innate immunity of fish. Dev Comp Immunol. 2001 Oct-Dec;25(8-9):725-43.

282. Bayne CJ, Gerwick L, Fujiki K, Nakao M, Yano T. Immune-relevant (including acute phase) genes identified in the livers of rainbow trout, Oncorhynchus mykiss, by means of suppression subtractive hybridization. Dev Comp Immunol. 2001 Apr;25(3):205-17.

283. Holland MC, Lambris JD. The complement system in teleosts. Fish Shellfish Immunol. 2002 May;12(5):399-420.

284. Rise ML, Hall J, Rise M, Hori T, Gamperl AK, Kimball J, et al. Functional genomic analysis of the response of Atlantic cod (Gadus morhua) spleen to the viral mimic polyriboinosinic polyribocytidylic acid (pIC). Dev Comp Immunol. 2008;32(8):916-31.

285. He W, Yinlt ZX, Li Y, Huo WL, Guan HJ, Weng SP, et al. Differential gene expression profile in spleen of mandarin fish Siniperca chuatsi infected with ISKNV, derived from suppression subtractive hybridization. Dis Aquat Organ. 2006 Dec 14;73(2):113-22.

286. Dios S, Poisa-Beiro L, Figueras A, Novoa B. Suppression subtraction hybridization (SSH) and macroarray techniques reveal differential gene expression profiles in brain of sea bream infected with nodavirus. Mol Immunol. 2007 Mar;44(9):2195-204.

287. Poisa-Beiro L, Dios S, Ahmed H, Vasta GR, Martinez-Lopez A, Estepa A, et al. Nodavirus infection of sea bass (Dicentrarchus labrax) induces up-regulation of galectin-1 expression with potential anti-inflammatory activity. J Immunol. 2009 Nov 15;183(10):6600-11.

288. Rise ML, Hall JR, Rise M, Hori TS, Browne MJ, Gamperl AK, et al. Impact of asymptomatic nodavirus carrier state and intraperitoneal viral mimic injection on brain transcript expression in Atlantic cod (Gadus morhua). Physiol Genomics. 2010 Jul 7;42(2):266-80.

289. Xu D, Wei J, Cui H, Gong J, Yan Y, Lai R, et al. Differential profiles of gene expression in grouper Epinephelus coioides, infected with Singapore grouper iridovirus, revealed by suppression subtractive hybridization and DNA microarray. J Fish Biol. 2010 Aug;77(2):341-60.

290. Fisicaro N, Aminian A, Hinchliffe SJ, Morgan BP, Pearse MJ, D'Apice AJ, et al. The pig analogue of CD59 protects transgenic mouse hearts from injury by human complement. Transplantation. 2000 Sep 27;70(6):963-8.

291. Janeway CA, Jr., Medzhitov R. Innate immune recognition. Annu Rev Immunol. 2002;20:197-216.

292. Tasumi S, Ohira T, Kawazoe I, Suetake H, Suzuki Y, Aida K. Primary structure and characteristics of a lectin from skin mucus of the Japanese eel Anguilla japonica. J Biol Chem. 2002 Jul 26;277(30):27305-11.

293. Tsutsui S, Iwamoto K, Nakamura O, Watanabe T. Yeast-binding C-type lectin with opsonic activity from conger eel (Conger myriaster) skin mucus. Mol Immunol. 2007 Feb;44(5):691-702.

294. Zhang H, Robison B, Thorgaard GH, Ristow SS. Cloning, mapping and genomic organization of a fish C-type lectin gene from homozygous clones of rainbow

trout (Oncorhynchus mykiss). Biochim Biophys Acta. 2000 Nov 15;1494(1-2):14-22.

295. Honda K, Yanai H, Takaoka A, Taniguchi T. Regulation of the type I IFN induction: a current view. Int Immunol. 2005 Nov;17(11):1367-78.

296. Krasnov A, Timmerhaus G, Afanasyev S, Jorgensen SM. Development and assessment of oligonucleotide microarrays for Atlantic salmon (Salmo salar L.). Comp Biochem Physiol Part D Genomics Proteomics. 2011 Mar;6(1):31-8.

297. von Schalburg KR, Rise ML, Cooper GA, Brown GD, Gibbs AR, Nelson CC, et al. Fish and chips: various methodologies demonstrate utility of a 16,006-gene salmonid microarray. BMC Genomics. 2005;6:126.

298. Koop BF, von Schalburg KR, Leong J, Walker N, Lieph R, Cooper GA, et al. A salmonid EST genomic study: genes, duplications, phylogeny and microarrays. BMC Genomics. 2008;9:545.

299. Krasnov A, Koskinen H, Pehkonen P, Rexroad CE, 3rd, Afanasyev S, Molsa H. Gene expression in the brain and kidney of rainbow trout in response to handling stress. BMC Genomics. 2005;6:3.

300. Schiotz BL, Jorgensen SM, Rexroad C, Gjoen T, Krasnov A. Transcriptomic analysis of responses to infectious salmon anemia virus infection in macrophage-like cells. Virus Res. 2008 Sep;136(1-2):65-74.

301. Kurobe T, Yasuike M, Kimura T, Hirono I, Aoki T. Expression profiling of immune-related genes from Japanese flounder Paralichthys olivaceus kidney cells using cDNA microarrays. Dev Comp Immunol. 2005;29(6):515-23.

302. Williams TD, Diab AM, George SG, Godfrey RE, Sabine V, Conesa A, et al. Development of the GENIPOL European flounder (Platichthys flesus) microarray and determination of temporal transcriptional responses to cadmium at low dose. Environ Sci Technol. 2006 Oct 15;40(20):6479-88.

303. Park KC, Osborne JA, Montes A, Dios S, Nerland AH, Novoa B, et al. Immunological responses of turbot (Psetta maxima) to nodavirus infection or polyriboinosinic polyribocytidylic acid (pIC) stimulation, using expressed sequence tags (ESTs) analysis and cDNA microarrays. Fish Shellfish Immunol. 2009 Jan;26(1):91-108.

304. Cerda J, Mercade J, Lozano JJ, Manchado M, Tingaud-Sequeira A, Astola A, et al. Genomic resources for a commercial flatfish, the Senegalese sole (Solea senegalensis): EST sequencing, oligo microarray design, and development of the Soleamold bioinformatic platform. BMC Genomics. 2008;9:508.

305. Skugor S, Jorgensen SM, Gjerde B, Krasnov A. Hepatic gene expression profiling reveals protective responses in Atlantic salmon vaccinated against furunculosis. BMC Genomics. 2009;10:503.

306. Tilton SC, Gerwick LG, Hendricks JD, Rosato CS, Corley-Smith G, Givan SA, et al. Use of a rainbow trout oligonucleotide microarray to determine transcriptional patterns in aflatoxin B1-induced hepatocellular carcinoma compared to adjacent liver. Toxicol Sci. 2005 Dec;88(2):319-30.

307. Millan A, Gomez-Tato A, Fernandez C, Pardo BG, Alvarez-Dios JA, Calaza M, et al. Design and performance of a turbot (Scophthalmus maximus) oligo-microarray based on ESTs from immune tissues. Mar Biotechnol (NY). 2010 Aug;12(4):452-65.

308. Peatman E, Baoprasertkul P, Terhune J, Xu P, Nandi S, Kucuktas H, et al. Expression analysis of the acute phase response in channel catfish (Ictalurus punctatus) after infection with a Gram-negative bacterium. Dev Comp Immunol. 2007;31(11):1183-96.

309. MacKenzie S, Balasch JC, Novoa B, Ribas L, Roher N, Krasnov A, et al. Comparative analysis of the acute response of the trout, O. mykiss, head kidney to in vivo challenge with virulent and attenuated infectious hematopoietic necrosis virus and LPS-induced inflammation. BMC Genomics. 2008;9:141.

310. Prieto-Alamo MJ, Abril N, Osuna-Jimenez I, Pueyo C. Solea senegalensis genes responding to lipopolysaccharide and copper sulphate challenges: large-scale identification by suppression subtractive hybridization and absolute quantification of transcriptional profiles by real-time RT-PCR. Aquat Toxicol. 2009 Mar 9;91(4):312-9.

311. Li RW, Waldbieser GC. Production and utilization of a high-density oligonucleotide microarray in channel catfish, Ictalurus punctatus. BMC Genomics. 2006;7:134.

312. Fittipaldi N, Gottschalk M, Vanier G, Daigle F, Harel J. Use of selective capture of transcribed sequences to identify genes preferentially expressed by Streptococcus suis upon interaction with porcine brain microvascular endothelial cells. Appl Environ Microbiol. 2007 Jul;73(13):4359-64.

313. Lun ZR, Wang QP, Chen XG, Li AX, Zhu XQ. Streptococcus suis: an emerging zoonotic pathogen. Lancet Infect Dis. 2007 Mar;7(3):201-9.

314. Rise ML, von Schalburg KR, Brown GD, Mawer MA, Devlin RH, Kuipers N, et al. Development and application of a salmonid EST database and cDNA microarray: data mining and interspecific hybridization characteristics. Genome Res. 2004 Mar;14(3):478-90.

315. Matsumoto M, Tanaka T, Kaisho T, Sanjo H, Copeland NG, Gilbert DJ, et al. A novel LPS-inducible C-type lectin is a transcriptional target of NF-IL6 in macrophages. J Immunol. 1999 Nov 1;163(9):5039-48.

316. MacKenzie S, Iliev D, Liarte C, Koskinen H, Planas JV, Goetz FW, et al. Transcriptional analysis of LPS-stimulated activation of trout (Oncorhynchus mykiss) monocyte/macrophage cells in primary culture treated with cortisol. Mol Immunol. 2006 Mar;43(9):1340-8.

317. Martin SA, Blaney SC, Houlihan DF, Secombes CJ. Transcriptome response following administration of a live bacterial vaccine in Atlantic salmon (Salmo salar). Mol Immunol. 2006 Apr;43(11):1900-11.

318. Ching B, Jamieson S, Heath JW, Heath DD, Hubberstey A. Transcriptional differences between triploid and diploid Chinook salmon (Oncorhynchus tshawytscha) during live Vibrio anguillarum challenge. Heredity (Edinb). 2010 Feb;104(2):224-34.

319. Dumrongphol Y, Hirota T, Kondo H, Aoki T, Hirono I. Identification of novel genes in Japanese flounder

(Paralichthys olivaceus) head kidney up-regulated after vaccination with Streptococcus iniae formalin-killed cells. Fish Shellfish Immunol. 2009 Jan;26(1):197-200.

320. Gerwick L, Corley-Smith G, Bayne CJ. Gene transcript changes in individual rainbow trout livers following an inflammatory stimulus. Fish Shellfish Immunol. 2007 Mar;22(3):157-71.

321. Meijer AH, Verbeek FJ, Salas-Vidal E, Corredor-Adamez M, Bussman J, van der Sar AM, et al. Transcriptome profiling of adult zebrafish at the late stage of chronic tuberculosis due to Mycobacterium marinum infection. Mol Immunol. 2005 Jun;42(10):1185-203.

322. van der Sar AM, Spaink HP, Zakrzewska A, Bitter W, Meijer AH. Specificity of the zebrafish host transcriptome response to acute and chronic mycobacterial infection and the role of innate and adaptive immune components. Mol Immunol. 2009 Jul;46(11-12):2317-32.

323. Purcell MK, Nichols KM, Winton JR, Kurath G, Thorgaard GH, Wheeler P, et al. Comprehensive gene expression profiling following DNA vaccination of rainbow trout against infectious hematopoietic necrosis virus. Mol Immunol. 2006 May;43(13):2089-106.

324. Byon JY, Ohira T, Hirono I, Aoki T. Use of a cDNA microarray to study immunity against viral hemorrhagic septicemia (VHS) in Japanese flounder (Paralichthys olivaceus) following DNA vaccination. Fish Shellfish Immunol. 2005 Feb;18(2):135-47.

325. Jorgensen SM, Afanasyev S, Krasnov A. Gene expression analyses in Atlantic salmon challenged with infectious salmon anemia virus reveal differences between individuals with early, intermediate and late mortality. BMC Genomics. 2008;9:179.

326. Encinas P, Rodriguez-Milla MA, Novoa B, Estepa A, Figueras A, Coll J. Zebrafish fin immune responses during high mortality infections with viral haemorrhagic septicemia rhabdovirus. A proteomic and transcriptomic approach. BMC Genomics. 2010;11:518.

CHAPTER 12

A NOVEL LIPOSOME-BASED NANOCARRIER LOADED WITH AN LPS-DSRNA COCKTAIL FOR FISH INNATE IMMUNE SYSTEM STIMULATION

Angels Ruyra, Mary Cano-Sarabia, Simon A. MacKenzie, Daniel Maspoch, and Nerea Roher

INTRODUCTION

The development of sustainable aquaculture, a strategic sector to feed the ever-increasing human population [1], relies on disease prevention through the implementation of preventive immunostimulation and effective vaccination strategies [2]. With the advent of liposomal vaccines, one can begin to conceive new non-invasive, non-stressful and easy-to-manage methods for administering immunostimulants and vaccines to a large number of cultured fish at any time of their life cycle. Liposomes are hollow spherical, safe and well-tolerated assemblies formed by a single lipid bilayer or multiple concentric bilayers that can be tailored (via selecting their composition, size, charge, etc.) to efficiently entrap a wide variety of immunostimulants and vaccines [3]. This encapsulation provides the obvious potential advantages of increasing their stability and protection, thus enhancing their immune response and disease protection, and opening up the possibility to design more efficient immunostimulant-vaccine cocktails. In addition, liposomes have been proven to act as adjuvants to potentiate immune responses alone and to be rapidly cleared from sites of administration, being preferentially distributed among macrophages [4]. Taking into account these excellent properties and since liposomes can be stable in solution or be dried [5], new opportunities will be available to aquaculture to study such systems as new immunostimulant vehicles, which could be administered either dissolved in water (immersion bath), by injection, or orally via coated-food. Herein, we describe a novel liposomal immunostimulant cocktail (hereafter called liposomal IS-cocktail) composed of two immunostimulants: the bacterial lipopolysaccharide (LPS) and the synthetic analog of dsRNA viruses, poly (I:C). Both bacterial and viral compounds were chosen to stimulate two potent innate immune pathways (TLR3 and TLR4 pathways) virtually present in all fish species[6]. The molecular basis of the immunostimulant action lies in the stimulation of innate immunity through the binding and activation of innate pathogen recognition receptors (PRRs) located on antigen-presenting cells (APCs) [7]. The principal APCs in fish are macrophages, neutrophils, dendritic cells and B cells [8], [9], [10]. Upon immunization, APCs release a variety of cytokines and chemokines regulating both innate and adaptive immunity [11]. Triggering combinations of PRRs on APCs with natural or synthetic ligands can induce synergistic activation and production of cytokines [12], [13]. Indeed, LPS is present in the cell wall of G negative bacteria and signals through TLR4 in mammals. The synthetic analog poly (I:C) (dsRNA) mimics RNA viruses and signals through TLR3 located on endosomal membranes and through RIG-I and MDA5 located in the cytosol [11]. Teleost fish can respond to dsRNA through TLR3, RIG-I and MDA5 [14] and to crude LPS preparations probably through a sensing mechanism not involving TLR4 [15]–[17], but involving peptidoglycan recognition proteins and other intracellular receptors like Nod-like receptor 3 [18]. LPS would be an excellent candidate for immunostimulation purposes, but it has been scarcely used due to its high endotoxic potential in mammals. Fish are much less sensitive to LPS toxic effects [17] and, by encapsulating

Citation: Ruyra A, Cano-Sarabia M, MacKenzie SA, Maspoch D, Roher N (2013). A Novel Liposome-Based Nanocarrier Loaded with an LPS-dsRNA Cocktail for Fish Innate Immune System Stimulation. PLoS ONE, 8(10): e76338. DOI:10.1371/journal.pone.0076338.

LPS, we have assayed a simple way to stimulate fish innate immune system. On the other hand, the addition of dsRNA to the nanocarrier would also target anti-viral response pathways [13].

Prior to this study, some advances have been made on the encapsulation of vaccines for fish vaccination and immunostimulation. Some of these studies have suggested that microencapsulated vaccines significantly enhance the protection and immune response in various fish species [19]–[22]. Thus far, however, no one has demonstrated the ability to simultaneously control the encapsulation of several immunostimulants in unilamellar, biocompatible liposomes. Such capabilities would allow one to construct much more sophisticated and efficient liposomal immunostimulants for aquaculture. The approach used herein relies on the ability of using the surface charge of liposomes, which can be tailored by properly selecting the lipid head-groups, to optimize the encapsulation of both negatively charged LPS and dsRNA. In such design, PEGylated lipids have also been used in liposomal immunostimulant formulations to control the unilamellarity of liposomes and to prolong the plasma half-life of the immunostimulants [23], [24]. This study provides evidence that the optimized multifunctional liposomal IS-cocktail induces a concurrent anti-viral and pro-inflammatory state in zebrafish hepatocytes and trout macrophages. Moreover, insights into the mechanisms controlling the cell interaction and metabolism of the liposomes have demonstrated the possibility to target plasmatic membrane and intracellular compartments essential to achieve an optimum immune response. Our findings have also shown that the designed liposome formulations are safe at therapeutic doses and could be used in future fish health applications.

MATERIALS AND METHODS

Materials

1,2-didodecanoyl-sn-glycero-3-phosphocholine (DLPC), 1,2-dioleoyl-sn-glycero-3-phosphoric acid monosodium salt (DOPA), Cholesterol (Chol), 3β-N-(di-methyl-amino-ethyl) carbamate hydrochloride (Cholesteryl), Cholesterol-PEG$_{600}$ (Chol-PEG), lipopolysaccharides (LPS) from*E. coli* 0111:B4, TriReagent, insulin, EGF, chloroquine and all endocytosis inhibitors were purchased from Sigma-Aldrich. MarinaBlue-DHPE, fluorescein-DHPE, LPS-AlexaFluor594, antibiotic/antimycotic solution, TrypLE Express, Cell Mask Deep Red, Hoechst 33342 and Superscript III reverse transcriptase were purchased from Invitrogen. Poly(I:C) High Molecular Weight, poly (I:C)-Fluorescein and Primocin were purchased from InvivoGen, whereas ZFL cells were purchased from ATCC. Oligo-dT15, GelGreen and SYBR Green I were purchased from Promega, Biotium and Bio-Rad, respectively.

Ethics Statement

All experimental procedures involving rainbow trout (*Onchorynchus mykiss*) and zebrafish (*Danio rerio*) were submitted and authorized by the Ethics Committee of the Autonomous University of Barcelona (CEEH number 1582) who agree with the International Guiding Principles for Research Involving Animals (EU 2010/63).

Preparation and Characterization of Liposomes of Immunoliposomal Formulations

Liposomal formulations were prepared by the thin film hydratation method [25] with some modifications. Briefly, DOPA, DLPC, Chol, Cholesteryl and Chol-PEG$_{600}$ were dissolved in chloroform solutions (100 mg/ml) and mixed at the desired molar ratios (**Table 1**). The organic solvent was then evaporated by rotary evaporation to obtain a lipid film. Later, the film was hydrated with 2 ml of PBS at 0.5 mg/ml poly (I:C) or 1.5 mg/ml LPS. The encapsulation of poly (I:C) or LPS was done with an immunostimulant:lipid ratio of 1:30 and 1:10, respectively. For the preparation of the liposomes that contained a cocktail of immunostimulants (hereafter called liposomal IS-cocktail), the dry lipid film was hydrated with a solution containing 0.5 mg/ml poly (I:C) and 1.0 mg/ml LPS in PBS. The co-encapsulation of poly (I:C) and LPS was done with an immunostimulant:lipid ratio of 1:30 and 1:15, respectively. The resulting lipid suspensions were then vigorously shaken, and the liposomes obtained were homogenized by means of an extruder (Lipex Biomembranes, Canada) through 2 stacked polycarbonate membranes (200 nm pore size, Avanti Polar Lipids) to finally obtain unilamellar liposomes. In all cases, non-encapsulated immunostimulants were removed from liposome preparations by ultracentrifugation at 110000 *xg* for 30 min at 10°C. Liposome integrity was checked by DLS and Cryo-TEM. The particle size distribution and zeta potential (ζ) of the final liposomal formulations were measured by dynamic light scattering (DLS) using a Zetasizer Nano ZS (Malvern Instruments, UK). The morphology was examined using Cryo-Transmission electron microscopy (Cryo-TEM) in a JEOL-JEM 1400 microscope (JEOL Ltd., Japan). Liposome stability was followed (48 h at 28°C) by turbidity measurement in a Turbiscan Lab Expert (Formulaction, France).

Table 1. Composition and characterization of non-loaded liposomal formulations.

Name	Liposome composition		ζ′ potential (mV)	Size (nm)
$NL_{1,n}$	DLPC 50% - Cholesteryl 35% - Cholesterol 10% - PEG5%	++	23.5±0.4	197.3±54.7
$NL_{2,n}$	DLPC 50% - Cholesteryl 10% - Cholesterol 35% - PEG5%	+	10.4±1.8	182.7±8.4
$NL_{3,n}$	DLPC 50% - Cholesteryl 45% - PEG5%		−5.4±1.7	204.5±21.6
$NL_{4,n}$	DLPC 40% - DOPA 10% - Cholesterol 45% - PEG5%	−	−19.0±0.5	185.1±9.5
$NL_{5,n}$	DLPC 15% - DOPA 35% - Cholesterol 45% - PEG5%	− −	−30.9±2.5	161.1±12.6

doi:10.1371/journal.pone.0076338.t001

Encapsulation Efficiency (EE)

Encapsulation efficiencies (EE) were calculated according to the equation $EE(\%)=[(C_{IS,total}-C_{IS,out})/C_{IS,total}]$ x100, where $C_{IS,total}$ is the initial immunostimulant concentration and $C_{IS,out}$ is the concentration of non-encapsulated immunostimulant. To measure the $C_{IS,out}$, all liposome suspensions were centrifuged at 110000 *xg* for 30 min at 10°C. Supernatant aliquots were taken to quantify the concentration of non-encapsulated poly (I:C) and LPS by UV-Vis spectroscopy using a Nanodrop ND-1000 (Thermo Scientific, USA). Poly (I:C) was linearly detected in a range from 2.5 µg/ml to 1 mg/ml (Abs at 250 nm, r^2=0.999), whereas LPS was linearly detected in a range from 4.0 µg/ml to 1 mg/ml (Abs at 269 nm, r^2=0.999). Liposomes that did not contain any encapsulated immunostimulant were also ultracentrifuged and their supernatant quantified (Abs at 220 nm) to verify that liposomes were properly precipitated. To calculate the EE of the liposomal IS-cocktail, the putative non-encapsulated immunostimulants in the supernatant were separated by aqueous Gel Permeation Chromatography (GPC, Ultrahydrogel 120, Waters, USA) and quantified by UV-Vis spectroscopy, where poly (I:C) and LPS were linearly detected. All experiments were done in triplicate.

Localization of Liposome-encapsulated Immunostimulants

Evaluation of the distribution of encapsulated immunostimulants in liposomes was done by confocal microscopy. The liposome bilayer was labeled with MarinaBlue-DHPE (0.005 molar ratio). Fluorescent LPS-AlexaFluor594 and poly (I:C)-Fluorescein were individually or simultaneously encapsulated in liposomes and the resulting liposomal formulations were examined using a Leica TCS SP5 confocal microscope (Leica Microsystems, Germany).

Cell Culture

Zebrafish ZFL cells (CRL-2643, ATCC) were cultured at 28°C, 5% CO_2 in Dulbecco's modified Eagle's medium (DMEM) 4.5 g/l glucose, supplemented with 0.01 mg/ml insulin, 50 ng/ml EGF, 5% (v/v) of antibiotic/antimycotic solution, 10% (v/v) heat-inactivated FBS and 0.5% (v/v) heat-inactivated trout serum (TS). HepG2 cells were grown at 37°C, 5% CO_2 in DMEM 4.5 g/l glucose, supplemented with 5% (v/v) of antibiotic/antimycotic solution and 10% (v/v) heat-inactivated FBS. Adherent trout monocyte/macrophages were isolated as previously described[8]. Before treatments, cells were incubated 3 h in serum free medium.

Cytotoxicity Assays

Two different cell viability assays (MTT and LDH) were simultaneously performed using three cell lines (ZFL, HepG2 and primary trout macrophages). Cells were seeded at 2.5×10^5cells/well. The medium was removed and fresh non-supplemented medium containing the liposome formulation at indicated concentration was added, incubating the cells for 24 h. Lactate dehydrogenase (Cytotoxicity Detection Kit LDH, Roche) activity in the medium and MTT assay on the cells were performed. Cell viability was expressed as a percentage of the control. All the measurements were done in triplicate in 3 independent experiments. Dose-response curves were fitted using a sigmoidal dose-response curve model provided in the GraphPad Prism 5.0 (GraphPad software, USA). EC_{50} value was derived from these fitted curves for single experiments. Differences among data were analyzed using One-way ANOVA followed by Tukey's post test $p<0.001$.

Endocytosis Analysis Using Flow Cytometry

To visualize liposome endocytosis, DHPE-fluorescein was incorporated at a 0.05 molar ratio into the liposomal IS-cocktail. Labeled liposomal IS-cocktail was added to either ZFL or trout macrophages to a final concentration of 750 µg/ml liposomal IS-cocktail (containing 25 µg/ml poly (I:C) and 12.5 µg/ml LPS) and incubated for selected times. After treatment, cells were cooled down, washed 3× with ice-cold PBS, trypsinized and centrifuged at 200 *xg* for 5 min. Pellets were resuspended in ice-cold PBS for FACS analysis using a BD FACSCanto cytometer (Becton Dickinson, USA). Experiments were performed in triplicate (10,000 events for each sample). The internalization of fluorescence was calculated as the mean fluorescence intensity (MFI). To compare membrane-bound versus endocyted liposomes, the medium was removed at different times (5, 15, 30 and 60 min), and the cells were washed either with ice-cold PBS (pH=7.4) or with an ice-cold PBS-acetic acid (pH=4.2) to remove the liposomes attached to the membrane. The remaining (internal) fluorescence of the cells was then analyzed using the PBS washed cells as a total uptake. The uptake of liposomes at long incubation times was also studied. When needed, cells were pretreated for 1 h with 100 µM chloroquine. Then, fluorescent liposomes were added and left to incubate 15 min for the ZFL cells and 30 min for the trout macrophages. After 3× PBS washes, liposome-free medium was added and cells were incubated for 1, 6 or 16 h in the presence of chloroquine, when required. Finally, cells were routinely treated for flow cytometry analysis. To determine the liposome endocytosis pathways, the following inhibitors were used: methyl-β-cyclodextrin (MβCD, 5 mM), 5-(N-Ethyl-N-isopropyl)amiloride (EIPA, 50 mM), sucrose (300 mM for ZFL, 150 mM for trout macrophages) and wortmannin (W, 100 nM). The inhibitor›s toxicity was assessed and working concentrations were selected. Cells were pretreated for 1 h with each inhibitor, and liposomes were added for 15 min (ZFL cells) or 30 min (trout macrophages). Finally, 1 h after adding the liposomes, cells were analysed by flow cytometry.

Endocytosis Analysis Using Confocal Microscopy

Cells were seeded one day before the endocytosis experiments. For short incubation times (from 30 min to 1.5 h), liposomal IS-cocktail was added at 750 µg/ml liposomal IS-cocktail (containing 25 µg/ml poly (I:C) and 12.5 µg/ml LPS). For the 16 h incubation time, liposomal IS-cocktail was added at 375 µg/ml liposomal IS-cocktail (containing 12.5 µg/ml poly (I:C) and 6.25 µg/ml LPS). After 3× PBS washes, cells were stained with CellMask and Hoechst and viewed under a Leica TCS SP5 confocal microscope (Leica Microsystems, Germany). Image analysis was performed using Imaris software and z-stacks were analyzed to visualize the particle contact sites and location.

Gene Expression Studies

Cells were stimulated for 16 h with 750 μg/ml of liposomal IS-cocktail containing 25 μg/ml poly (I:C) and 12.5 μg/ml LPS and 375 μg/ml of liposomal IS-cocktail containing 12.5 μg/ml poly (I:C) and 6.25 μg/ml LPS. Non-loaded liposomes and non-encapsulated IS were used as controls. Total RNA from the ZFL and trout macrophages cell cultures was extracted using TriReagent following manufacturer›s instructions. The RNA quality and concentration was assessed and cDNA was synthesized with 1.0 μg and 0.5 μg of total RNA for ZFL cells and macrophages, respectively, using SuperScript III reverse transcriptase and oligo-dT15 primer. PCR was carried out with 1 μl of cDNA as a template with specific primers and qPCR was carried out using SYBR Green I mix, 500 nM of primers and 5 μl of cDNA. Samples from 3 independent experiments were run in triplicate, and quantification was done according to Livak method [26].

TNF Secretion

Trout macrophages were incubated for 16 h with 375 μg/ml of liposomal IS-cocktail (with 12.5 μg/ml poly (I:C) and 6.25 μg/ml LPS). Non-loaded liposomes and free LPS were used as controls. Supernatants were collected, centrifuged and precipitated with 25% trichloroacetic acid. TNFα secretion was assessed by Western blot as previously described [16].

In Vivo Toxicological Assays

Adult AB zebrafish (*Danio rerio*) were held in tanks with recirculating water under a photoperiod of 14 h light/10 h dark at 28°C. Embryos were obtained from random pair-wise mating collected, rinsed and kept in E3 medium at 28°C. Viable embryos and post-hatching larvae were plated in 96-well plates. Liposomal IS-cocktail (liposome concentrations from 0.75 to 6 mg/ml) were added to the wells (200 μl), and incubated for 120 h. The plate evaporation rate was minimized as previously described [27]. Non-loaded liposomes and non-encapsulated immunostimulants were used as controls, and 24 individuals for each condition were used. Hatching rate, cumulative mortality and malformations of the embryos were recorded every 24 h, and survival curves were plotted using the Kaplan-Meier method and analysed using the log-rank test. Larvae were also frozen at −80°C and total RNA was isolated as indicated before for gene expression evaluation.

RESULTS

Preparation and Characterization of Liposomal Formulations

Series of liposomal formulations with different lipid membrane composition and net surface charges were prepared to determine the optimal liposomal formulation to achieve the maximum encapsulation efficiency of LPS and poly (I:C). Three lipid mixtures were studied, $NL_{1,n}$ and $NL_{2,n}$, formed by the cationic lipid mixture of DLPC- Cholesteryl-Chol-PEG, $NL_{3,n}$, constituted by the neutral mixture DLPC-Chol, and $NL_{4,n}$ and $NL_{5,n}$, formed by the anionic lipid mixture DLPC-DOPA-Chol-PEG (**Table 1**). In all formulations, small unilamellar vesicles (**Figure 1A**) were obtained with a mean size ranging from 161.1±12.6 nm to 204.5±21.6 nm. In all cases, a 5% of Chol-PEG_{600} was included to achieve uniform samples. Encapsulation efficiencies of LPS or poly (I:C) in the different $NL_{1,n}$ to $NL_{5,n}$ formulations were studied, showing that a positively charged liposome surface, like in $NL_{1,n}$ (+23.47±0.40 mV) and $NL_{2,n}$ (+10.43±1.77 mV), favors the encapsulation of LPS and poly (I:C). In contrast, the encapsulation efficiency of both LPS and poly (I:C) in liposomes decreased as the surface charge became more negative like in $NL_{5,n}$ (−30.87±2.53 mV), as previously described by Balazs *et al.* and Nakhla *et al.* [28],[29]. It has been suggested that the attractive interaction between the negative charge of the immunostimulants and the positive charge of the liposome surface results in near-perfect conditions to achieve the highest encapsulation efficiencies [30]. For example, the influence of these interactions to the encapsulation of both LPS and poly (I:C) was further confirmed by a decrease of the positive ζ′ potential value down to −4.34±0,41 and 4.5±1.1 for both $NL_{2,LPS}$ and $NL_{2,poly (I:C)}$, respectively. The maximum loading efficiencies for LPS were 49.6±5.9% and 66.0±0.1% for $NL_{2,LPS}$ and $NL_{1,LPS}$, respectively. Interestingly, loading efficiencies achieved for poly (I:C) were always higher, with values of 95.0±1.4% and 91.2±0.1% for $NL_{2, poly (I:C)}$ and $NL_{1, poly (I:C)}$, respectively (**Table 2**). To further characterize the physicochemical structure of such cationic liposomal formulations, we encapsulated AlexaFluor594-labeled LPS (**Figure 1C**) and fluorescein-labeled poly (I:C) (**Figure 1D**) into Marina Blue-labeled liposomes (**Figure 1B**). Confocal microscope images of non-extruded liposomes demonstrated that both LPS and poly (I:C) were incorporated into their lipidic bilayer. Figures 1C–D show the spatial superimposition between fluorescence intensities of AlexaFluor594-LPS and Marina Blue-liposomes (**Figure 1C**) as well as of fluorescein-poly (I:C) and Marina Blue-liposomes (**Figure 1D**), further confirming that both immunostimulants are localized in the lipidic bilayer of cationic liposomes. Next, we investigated the cytotoxicity of cationic liposomes without encapsulated immunostimulants of both, $NL_{1,n}$ and $NL_{2,n}$ formulations, showing the maximum loading efficiencies. Both types of liposomes were *in vitro* assayed on ZFL and HepG2 cell lines using MTT and LDH assays. Interestingly, $NL_{1,n}$ and $NL_{2,n}$ liposomes showed similar cytotoxicity activity in HepG2 cells. However, the more cationic liposomes ($NL_{1,n}$) clearly showed higher toxicity on ZFL cells (EC_{50}=0.166 mg/ml) than the less cationic one ($NL_{2,n}$). Because of their similar loading efficiencies but different cytotoxicity, the less toxic $NL_{2,n}$ formulation (DLPC 50%-Cholesteryl 10%-Chol 35%-Chol-PEG 5%) was finally chosen as the ideal liposomal composition to co-encapsulate LPS and poly (I:C) (**Figure 1E**). Using these conditions, the resulting liposomal IS-cocktail (hereafter referred to as NL_c formulation) was composed of 125.8±6.6 nm-in-diameter liposomes that entrapped both LPS and poly (I:C) with loading efficiencies of 22.3±2.1% and 99.6±0.1%, respectively. Therefore, the NL_c formulation was composed of a mixture of 15 mg/ml of liposomes containing 250 μg/ml and 500 μg/ml of LPS and poly I:C, respectively. Importantly, after co-encapsulating LPS and poly (I:C), such liposomes exhibited a slightly positive surface charge (1.37±3.58 mV), which was attributed to electrostatic interactions between their positively charged lipidic bilayer and the negatively charged immunostimulants. The occurrence of these attractive interactions was corroborated by co-encapsulating AlexaFluor594-labeled LPS and fluorescein-labeled poly (I:C) into cationic liposomes, from which the localization of both immunostimulants in the lipidic bilayer was observed (**Figure 1F**).

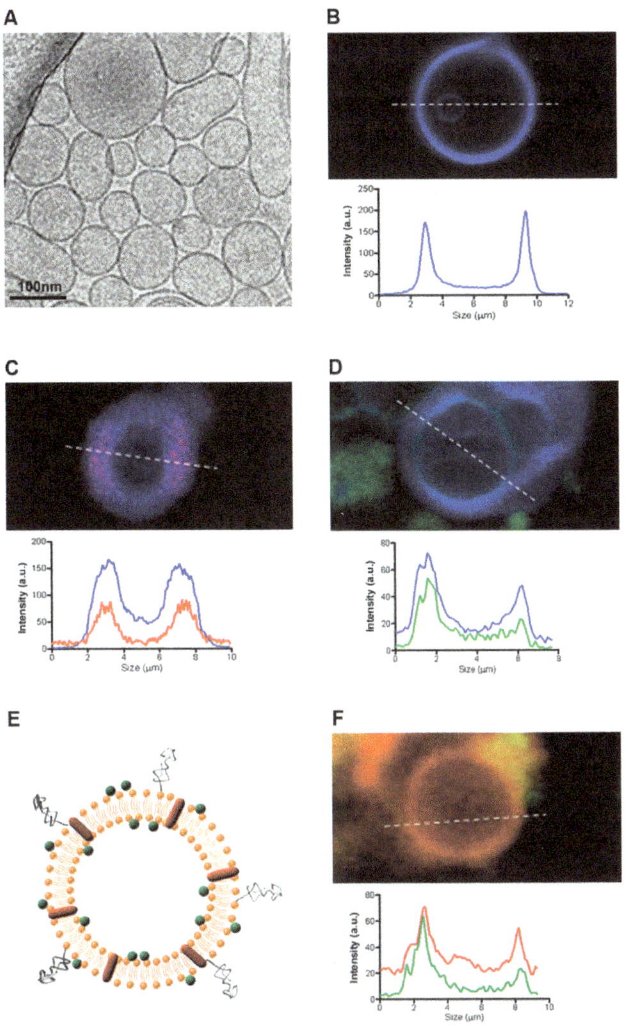

Figure 1. Characterization of liposomal formulations. (A) Representative Cryo-TEM image of DLPC/Chol/Cholesteryl/PEG$_{600}$-Chol (5:3.5:1:0.5) liposomes extruded through a 200 nm pore size membrane. (B) Confocal fluorescence image of a single liposome tagged on its lipid bilayer with Marina Blue-DHPE (blue) and its corresponding fluorescence intensity profile. (C) Confocal fluorescence image of a single Marina Blue-labeled liposome containing AlexaFluor594-labeled LPS (red) and their corresponding fluorescence intensity profiles. (D) Confocal fluorescence image of a single Marina Blue-labeled liposome containing fluorescein-labeled poly (I:C) and their corresponding fluorescence intensity profiles. (E) Schematic representation of the liposomal IS-cocktail (NL$_c$) showing the presence of both encapsulated LPS (red) and poly (I:C) (green) in the lipidic bilayer of liposomes. (F) Confocal fluorescence image of a single liposome containing both fluorescein-labeled poly (I:C) (green) and AlexaFluor594-labeled LPS (red) and their corresponding fluorescence intensity profiles.

Table 2. Efficiencies for the encapsulation of LPS and poly (I:C).

Name	EE LPS (%)	EE poly (I:C) (%)
NL$_{1,LPS}$	66.0±0.1	
NL$_{1,\ poly\ (I:C)}$		91.2±5.9
NL$_{2,LPS}$	49.6±5.9	
NL$_{2,\ poly\ (I:C)}$		95.0±1.4
NL$_{3,LPS}$	6.9±0.4	
NL$_{3,\ poly\ (I:C)}$		25.8±7.6
NL$_{4,LPS}$	5.9±3.2	
NL$_{4,\ poly\ (I:C)}$		38.0±4.5
NL$_{5,LPS}$	2.0±1.3	
NL$_{5,\ poly\ (I:C)}$		12.9±4.3

Encapsulation efficiencies (EE) for separately encapsulating an initial concentration of 1.5 mg/ml of LPS and 0.5 mg/ml of poly (I:C) into 15 mg/ml of the liposomal (NL$_{1-5}$) formulation.
doi:10.1371/journal.pone.0076338.t002

Evaluation of Cell Toxicity of Liposomal NL$_{2,LPS}$, NL$_{2,poly\ (I:C)}$ and NL$_c$ Formulations on Zebrafish Hepatocytes and Trout Macrophages Primary Cultures

To fully characterize the safety of our formulations, we carried out *in vitro* cytotoxic studies (**Figure 2**). The therapeutic immunostimulant doses were chosen according to our previous results on LPS and poly (I:C) responses in different fish species [16], [31]. Based on these results, dose-response experiments were conducted with NL$_{2,n}$, NL$_{2,LPS}$, NL$_{2,poly\ (I:C)}$ and NL$_c$ in ZFL cells at the indicated concentrations (**Figure 2**). None of the encapsulating formulations showed toxicity at potential therapeutic doses in these cells. Moreover, free LPS toxicity (50 μg/ml LPS, 51.8%±17.9 viability and 25 μg/ml LPS, 62.0%±6.01 viability) was avoided by nanoencapsulation. Also, poly (I:C) treatment prompted a slight decrease in viability (50 μg/ml poly I:C, 80.32%±7.01 viability) that was fully reverted when this molecule was encapsulated (**Figure 2B**). Further, empty NL$_{2,n}$ showed low toxicity but higher than NL$_c$ in all cases, which can be attributed to changes suffered by the liposomes after the encapsulation of LPS and poly (I:C) that further improve its biocompatibility. The same results were obtained by using the LDH assay. Finally, toxicity studies were also carried out using trout primary cell cultured APCs. In this cells, we observed low toxicity levels of NL$_c$ formulations (20% over basal mortality), but did not observe a LPS/poly (I:C) mediated toxicity at the indicated doses.

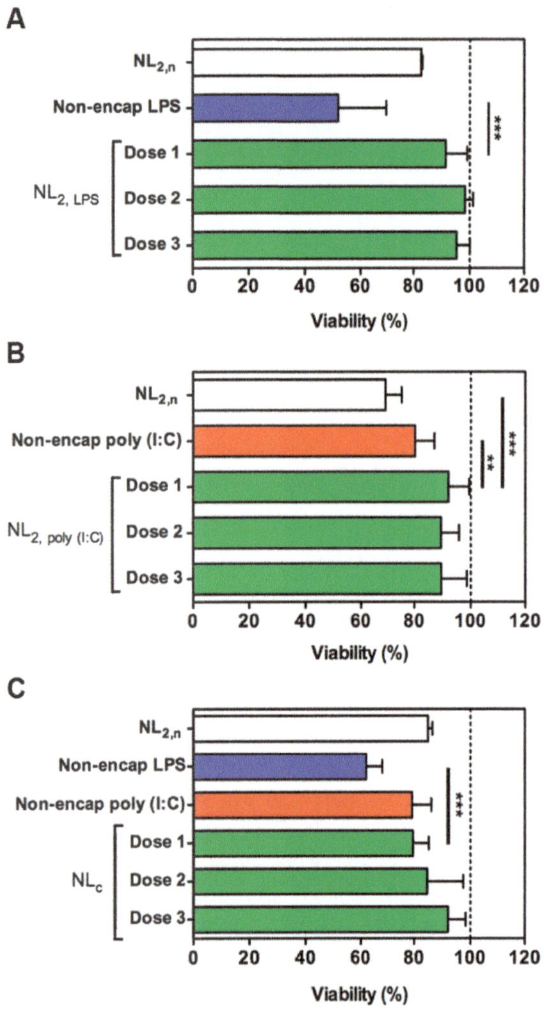

Figure 2. Cytotoxicity of $NL_{2, LPS}$, $NL_{2, poly (I:C)}$, and NL_c formulations in ZFL cells by MTT-based assay. (A) Viability of ZFL after 24 h incubation with liposome-encapsulated LPS ($NL_{2, LPS}$, green bars) at Dose 1=1 mg/ml liposome with 50 μg/ml LPS, Dose 2=0.5 mg/ml liposome with 25 μg/ml LPS and Dose 3=0.20 mg/ml liposome with 10 μg/ml LPS. The white bar is the empty liposome control ($NL_{2,n}$, 1 mg/ml liposome) and the blue bar is the free LPS control (50 μg/ml). (B) Viability of ZFL after 24 h incubation the liposome-encapsulated poly (I:C) ($NL_{2, poly (I:C)}$, green bars) at Dose 1=1.5 mg/ml liposome with 50 μg/ml poly (I:C), Dose 2=0.75 mg/ml liposome with 25 μg/ml poly (I:C) and Dose 3=0.375 mg/ml liposome with 10 μg/ml poly (I:C). The white bar is the empty liposome control treatment ($NL_{2,n}$, 1.5 mg/ml liposome) and the red bar is the non-encapsulated poly (I:C) control (50 μg/ml). (C) Viability of ZFL cells after 24 h incubation with liposomal LPS-poly (I:C) cocktail (NL_c, green bars) at Dose 1=1.5 mg/ml liposome with 50 μg/ml poly (I:C) and 25 μg/ml LPS, Dose 2=0.75 mg/ml liposome with 25 μg/ml poly (I:C) and 12.5 μg/ml LPS and Dose 3=0.375 mg/ml liposome with 12.5 μg/ml poly (I:C) and 6.25 μg/ml LPS. The white bar is the empty liposome control treatment ($NL_{2,n}$, 1.5 mg/ml liposome), the blue bar indicates the free LPS (25 μg/ml) and the red bar is the free (I:C) control (50 μg/ml). Non-treated cells were used as 100% viability control (dotted line). Data represent means ± SD of three independent experiments. Differences were analyzed using One-way ANOVA followed by Tukey's post test. **, $p<0.01$; ***, $p<0.001$.

Endocytosis of NL_c Formulation by ZFL Cells and Trout Macrophages Primary Cultures

Since hepatocytes play a major role in physiological detoxification processes and APCs are the key targets of our liposomes, we next evaluated the liposome uptake in both systems using flow cytometry and confocal microscopy. In ZFL cells, we observed a rapid (5 min) and efficient liposome uptake (**Figure 3A**) that reached a maximum in 1 h, and then started to decrease during the next 16 h, indicating that NL_c were probably metabolized by the endosomal/lysosomal system. Different studies have shown the ability of cationic liposomes to deliver different compounds to endosomal compartments [12], [32]. To further explore this process, we assayed the NL_c endocytosis in the presence of chloroquine (CQ), an inhibitor of lysosomal acidification, and we observed a significant increase of fluorescence in the presence of CQ. This observation confirmed the occurrence of NL_c in the endosomal/lysosomal compartment (55.53±0.83% CQ-dependant endocytosis inhibition at 16 h). To discriminate between membrane-bounded and endocytosed NL_c, we measured the total versus endocytosed fluorescence at different times, observing that around 80% of total fluorescence signal corresponded to endocytosed liposomes (**Figure 3A**) that accumulated intracellularly forming cytosolic agglomerates of 1.13±0.42 μm mean size (**Figure 3C**). To distinguish between the various mechanisms of endocytosis, a series of FITC-labelled NL_cliposome uptake assays were performed in the presence of inhibitors (methyl-β-cyclodextrin, MβCD, sucrose, wortmannin and EIPA) known to block a particular endocytosis pathway (**Figure 3B**). Treatment of cells with MβCD, a caveolae-mediated endocytosis inhibitor, led to a 60±5.9% ($p<0.001$) decrease in liposome uptake, whereas treatment with the macropinocytosis inhibitors wortmannin and EIPA provided contradictory results. While wortmannin inhibited uptake (19±4%; $p<0.01$), EIPA, a more specific macropinocytosis inhibitor, did not. The PI3K inhibitors (e.g wortmannin) have been described to have pleiotropic effects on endocytosis as they can block the internalization of ligands of the clathrin- and caveolae- mediated pathways[33], [34]. Thus, in ZFL cells, wortmannin could affect caveolae-mediated endocytosis instead of macropinocytosis. Finally, treatment with hypertonic medium (sucrose) led to a 15±6% ($p<0.05$) inhibition, indicating that clathrin-mediated endocytosis may also contribute to NL_c uptake. All these data suggested that NL_c could be endocytosed by ZFL cells mainly through the caveolae-dependent endocytosis pathway, but clathrin-mediated endocytosis could also be involved in liposome uptake.

The uptake in differentiated trout macrophages was also evaluated. As shown in Figure 4, these cells were able to efficiently endocyte NL_c. We measured total versus intracellular fluorescence by flow cytometry, and similarly to ZFL cells, macrophages were able to internalize around 80% of fluorescent liposomes in 1 h (**Figure 4A**). In contrast to ZFL cells, however, macrophages did not metabolize liposomes in the endosomal/lysosomal compartment since we could detect the same fluorescence levels even 24 h later. Note that the intracellular liposomes, as in ZFL cells, were present primarily in the cytosol as agglomerates (1.09±0.37 μm), with no fluorescence in the nuclei (**Figure 4C**). Again, we performed a series of liposome uptake assays in the presence of inhibitors, and we observed that in macrophages both MβCD and EIPA were able to inhibit the endocytosis by 31.09±14.52% ($p<0.01$) and 15.57±1.72% ($p<0.05$), respectively (**Figure 4B**). These results indicated that caveolae-mediated endocytosis and macropinocytosis/phagocytosis are the main endocytic pathways for liposome internalization in trout macrophages.

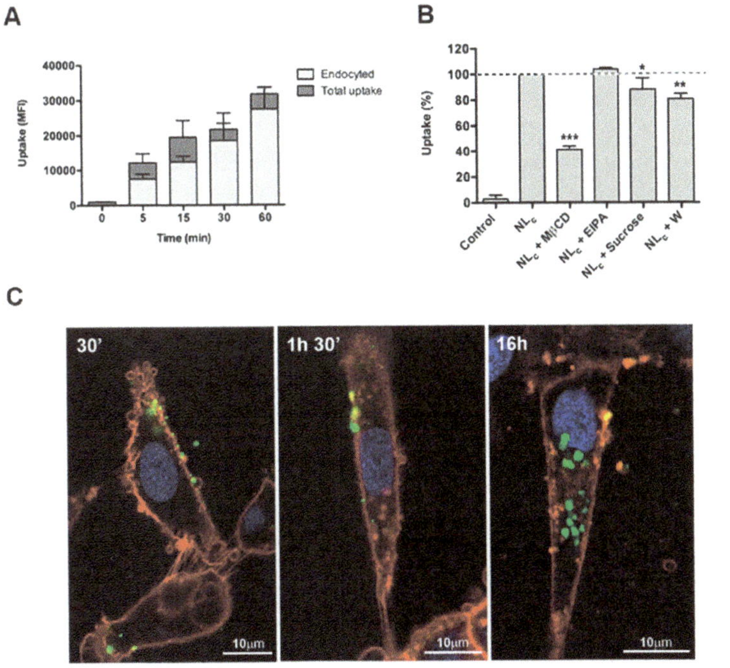

Figure 3. Endocytosis of NL_c formulation by ZFL cells. (A) Flow cytometry time-course comparison of the membrane-bound (dark grey bar) versus the endocyted liposomes (light grey bar) after incubation with NL_c (750 μg/ml liposome, 25 μg/ml poly (I:C) and 12.5 μg/ml LPS) at the indicated times. Data represent means ± SD of three independent experiments. (B) Effect of chemical inhibitors on the endocytosis of the NL_c (750 μg/ml liposome, 25 μg/ml poly (I:C) and 12.5 μg/ml LPS). Inhibitors were used at the following concentrations: MβCD at 5 mM, EIPA at 50 μM, sucrose at 300 mM and W at 100 nM. The uptake of cells without inhibitors (NL_c bar) was used as 100% uptake control and non-treated cells were used as control (control bar). Data represent means ± SD of three independent experiments. Differences were analyzed using One-way ANOVA followed by Tukey's post test. *, $p<0.05$; **, $p<0.01$; ***, $p<0.001$. (C) Confocal microscopy images of fluorescent liposomes (NL_c) endocyted by ZFL cells. Cells were incubated for 30 min, 1.5 h and 16 h with NL_c containing DHPE-Fluorescein (green) at a 0.05 molar ratio. Cell membranes were stained with CellMask (red) and the nucleus was stained with Hoechst (blue).

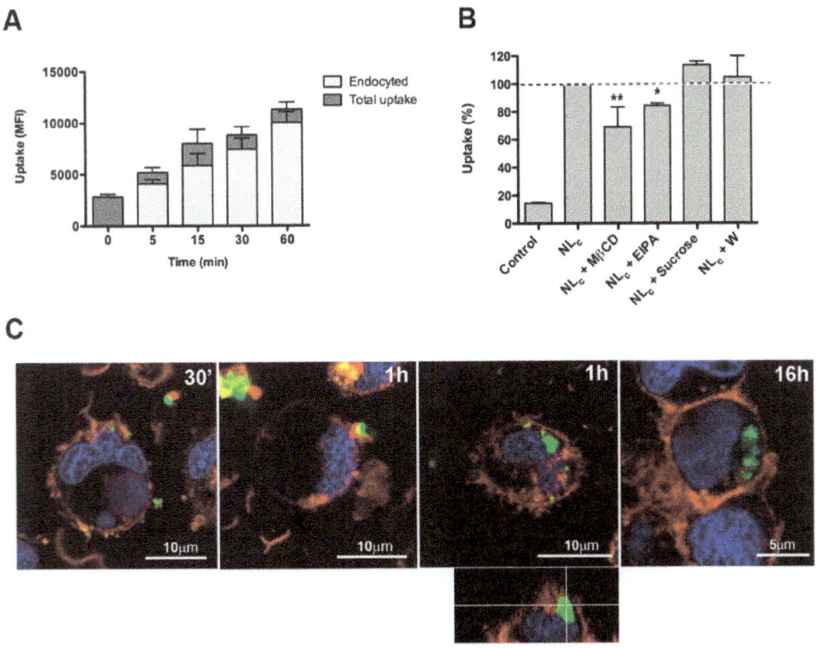

Figure 4. Endocytosis of NL_c formulation by trout macrophages. (A) Flow cytometry time-course comparison of the membrane-bound (dark grey bar) versus the endocyted liposomes (light grey bar) after incubation with 750 μg/ml liposome-encapsulated 25 μg/ml poly (I:C) and 12.5 μg/ml LPS at the indicated times. Data represent means ± SD of three independent experiments. (B) Effect of chemical inhibitors on the endocytosis of NL_c (750 μg/ml liposome-encapsulated 25 μg/ml poly (I:C) and 12.5 μg/ml LPS) macrophages uptake. Inhibitors were used at the following concentrations: MβCD at 5 mM, EIPA at 50 μM, sucrose at 150 mM and W at 100 nM. The uptake of cells not treated with inhibitors (NL_c bar) was used as 100% uptake control and non-treated cells were used as control (control bar). Data represent means ± SD of 3 independent experiments. Differences were analyzed using One-way ANOVA followed by Newman-Keuls post-test. *, $p<0.05$; **, $p<0.01$. (C) Confocal microscopy images of fluorescent liposomes (NL_c) endocyted by macrophages. Cells incubated 30 min, 1 h and 16 h

with NL_c containing DHPE-Fluorescein (green) at a 0.05 molar ratio. Cell membranes were stained with CellMask (red) and nucleus with Hoechst (blue).

The Immunostimulatory Effects of NL_c Formulation on ZFL Cells and Trout Macrophages

We examined the gene expression patterns in response to NL_c treatment in ZFL cells and trout macrophages (**Figure 5A and 5B**) by evaluating the expression of marker genes of pro-inflammatory (TNFα and IL-6) and anti-viral responses (IFNΦ and α, GIG2 and CCL4). Figure 5A shows that IFNΦ and GIG2 gene expression was significantly induced by the NL_c formulation at both doses, but we did not observe significant differences between Dose 1 and 2. Importantly, IFNΦ (NL_c Dose 1: 11±2 -fold change; p<0.01) and GIG2 (NL_c Dose 1: 2250±49 -fold change; p<0.01) had higher expression levels in NL_c formulation than in non-loaded liposomes ($NL_{2,n}$: 5±4 -fold change and 17±1.5 -fold change, respectively). The chemokine CCL4, a chemotactic cytokine that is induced in fish after viral infection [35], was also efficiently induced after NL_c treatment (**Figure 5A**). We also observed that non-loaded liposomes ($NL_{2,n}$) were still able to induce low levels of gene expression (**Figure 5A and 5B**). Several groups have indeed described that cationic liposomes have an immunological adjuvant effect and that they are able to regulate the transcription of several chemokines and cytokines [36].

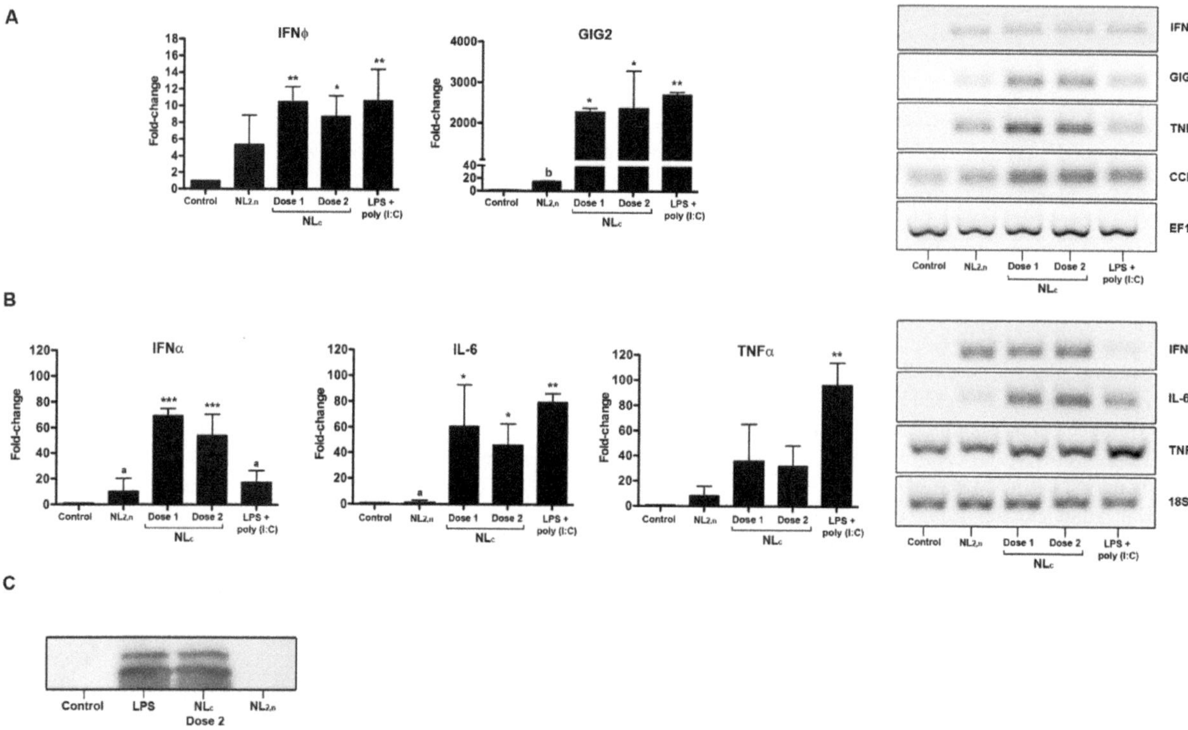

Figure 5. Analysis of gene expression in ZFL cell culture (A) and trout macrophage primary cell culture (B) after 16 h exposure to liposomes. $NL_{2,n}$=liposomes without immunostimulants (750 µg/ml), NL_c Dose 1=liposomes (750 µg/ml) containing 25 µg/ml poly (I:C) and 12.5 µg/ml LPS, NL_c Dose 2=liposomes (375 µg/ml) containing 12.5 µg/ml poly (I:C) and 6.25 µg/ml LPS, and LPS+poly (I:C)=stimulation control (25 µg/ml poly (I:C), 12.5 µg/ml LPS). Elongation factor (EF1) was used as reference gene for ZFL cells and 18S for trout macrophages. IFN (φ for ZFL and α for macrophages), GIG2, CCL4, IL-6 and TNFα abundance was analyzed by Q-PCR (left panel) and conventional PCR (right panel). Data represent means ± SD of 3 independent experiments. Values with asterisk are statistically significant relative to the control (*, p<0.05; **, p<0.01; ***, p<0.001) and values with letters ($_{a}$,$_{b}$) are statistically significant relative to NL_c Dose 1 ($_a$, p<0.001, $_b$, p<0.05). Differences were analyzed using One-way ANOVA and Tukey's post test. (C) TNFα secretion from trout macrophages stimulated with liposomes for 16 h was assessed by Western blot. NL_cDose 2=375 µg/ml liposomes, 12.5 µg/ml poly (I:C), 6.25 µg/ml LPS, $NL_{2,n}$=empty liposomes (375 µg/ml) and LPS=stimulation control (6.25 µg/ml). A representative Western Blot is shown.

We also assessed the IFNα, IL-6 and TNFα expression levels in trout macrophages (**Figure 5B**) to further evaluate the stimulatory ability of NL_c formulation. The IFN expression was significantly induced after NL_c Dose 1 and 2 treatment (68±5 and 50±10.5 -fold change; p<0.001) as compared to non-loaded liposomes ($NL_{2,n}$; 9.2±3.8 -fold change; p<0.001) and to the free LPS/poly (I:C) mixture (12±4 -fold change; p<0.001). The pro-inflammatory cytokines IL-6 and TNFα showed a slightly different pattern, achieving good stimulation levels after NL_ctreatment with respect to non-loaded liposomes, but similar or lower levels when compared to the free-LPS/poly (I:C) mixture (**Figure 5B**). Consistent with gene expression results, TNFα protein secretion was strongly induced by NL_c formulation, and most importantly, it was undetectable after stimulation with non-loaded liposomes $NL_{2,n}$ (**Figure 5C**). TNFα is one of the pivotal early response cytokines that are secreted by macrophages and enters the circulation to exert its systemic action [37].

In Vivo Biocompatibility of the NL_c Formulation

We conducted different dose-response survival experiments with the NL_c formulation and non-loaded liposomes $NL_{2,n}$ in pre- and post-hatching larvae (**Figure 6**). A NL_c concentration range from an extremely high dose (NL_c Dose 4=6 mg/ml) to a putative therapeutic dose (NL_c Dose 1=0.75 mg/ml) was chosen. We did not observe significant differences in survival curves obtained with pre-hatched larvae incubated with NL_c formulation at different doses (**Figure 6A**), and only very high doses (NL_c Dose 4) caused a significant increase in mortality (100% at 72 h, $p<0{,}0001$). In contrast, high LPS toxicity with free-LPS treatment both in pre- and post-hatching larvae was observed. A moderate poly (I:C) toxicity in pre-hatching larvae (62.5% mortality at 120 hpf; $p<0.0001$) versus control (36.12% mortality at 120 hpf; $p<0.0001$) was also recorded. Therefore, and in accordance with our previous *in vitro* toxicity studies (**Figure 2**), the encapsulation of both immunostimulants avoided the embryo/larvae mortality induced by free LPS and poly (I:C) (**Figure 6A**). Importantly, the embryos incubated with NL_c formulations were able to hatch and develop normally until 120 h with no morphological defects. The survival curves in post-hatching larvae incubated with these liposomal formulations were substantially different (**Figure 6B**). In this case, non-loaded liposomes ($NL_{2,n}$ Dose 2, 1.5 mg/ml) showed less toxicity than that of the corresponding liposomal IS-cocktail (NL_c Dose 2, 1.5 mg/ml liposomes, 50 μg/ml poly (I:C), 25 μg/ml LPS). In addition, a dose-dependent toxicity for the NL_c formulation after 48 h incubation was observed (**Figure 6B**). Analysis of gene expression in NL_c challenged larvae at 24, 48 and 72 h showed expression of marker genes of pro-inflammatory (TNFα and iNOS) and anti-viral responses (TLR3 and GIG2), indicating a stimulation of the zebrafish larvae immune system. Finally, DLS measurements done using NL_c and $NL_{2,n}$ formulations after 5 days incubation in E3 medium indicated a good stability after the *in vivo* challenge. We also characterized the NL_c stability in*in vivo* experimental conditions by Turbiscan, and we found that the NL_c stability index was not significantly changed after incubation in aquarium water or in E3 medium at 28°C for 2 days (stability indexes of 6.16 and 3.8, respectively). These data further confirm that this liposomal IS-cocktail might be used for future *in vivo* immunization in aquatic species.

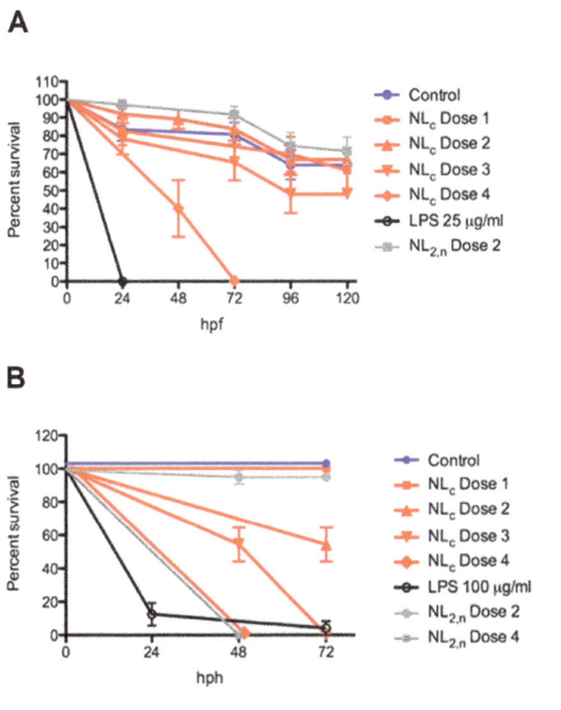

Figure 6. *In vivo* NL_c formulation toxicities. Survival of zebrafish embryos was recorded every 24(hpf) (A) and 72 h post-hatching (hph) (B) after exposure to four concentrations of liposomal IS cocktail (red, NL_c Dose 1=750 μg/ml liposomes, 25 μg/ml poly (I:C) and 12.5 μg/ml LPS; NL_c Dose 2=1.5 mg/ml liposomes, 50 μg/ml poly (I:C) and 25 μg/ml LPS; NL_c Dose 3=3 mg/ml liposomes, 100 μg/ml poly (I:C) and 50 μg/ml LPS; NL_c Dose 4=6 mg/ml liposomes, 200 μg/ml poly (I:C) and 100 μg/ml LPS). Liposomes without encapsulated immunostimulants (grey, $NL_{2,n}$Dose 2=1.5 mg/ml, $NL_{2,n}$ Dose 4=6 mg/ml) and non-treated embryos (blue) were used as controls. Non-encapsulated LPS (black, 25 μg/ml and 100 μg/ml) was used as mortality control. Differences were analyzed using log rank test. *, $p<0.05$; **, $p<0.01$; ***, $p<0.001$.

DISCUSSION

Vaccination and preventive immunostimulation has become the principal prophylactic tool for disease control in aquaculture. Some conventional vaccines made with inactivated bacteria (e.g. *Listonella anguillarum* causing vibriosis) have achieved good protection levels against different fish infections [38]. However, most diseases have no prevention tools, causing massive mortalities in fish farms and generating important economic losses. It is still unclear whether teleost fish have immunological memory but the secondary humoral responses are by far less prominent than in mammals [9], [38]. Thus, the activation of the innate immune system seems the most effective way for the initiation of an efficient immune response in fish. The binding of antigens to the innate pathogen receptors (PRRs) located on antigen-presenting cells (APCs) is critical for developing an effective immune response. Fish have a powerful innate immune system with a high molecular diversity and complexity [39], being APCs (especially the macrophages and dendritic cells) the main players of the innate immune response and responsibles for the activation of adaptive immunity [40]. With these specific premises, we have designed a nanosized and non-toxic unilamellar liposomal formulation loaded with TLR ligands (LPS and poly (I:C)) which was able to induce a potent anti-viral and pro-inflammatory response *in vitro* and *in vivo* in fish. As far as we know, this study is the first attempt to co-encapsulate two model immunostimulants specifically designed to target fish APCs in nanosized liposomes. To date, the unique attempt to vaccinate fish using liposomes was done by Irie *et al.*, who explored the use of microsized liposomes containing *A. salmonicida* total extracts in carp [22]. Recently, Fredriksen *et al.* have also shown that a combination of poly(lactic-*co*-glycolic acid) microparticles loaded with β-glucan and human γ-globulins were able to target head kidney macrophages inducing an adaptive *in vivo* immune response in salmon [41]. The LPS would be an excellent candidate for immunostimulation purposes, but it has been scarcely used due to its high endotoxic potential in mammals. Fish are less sensitive to LPS toxic effects [17], and by encapsulating LPS we have assayed a simple way to stimulate fish innate immune system. On the other hand, we also target anti-viral response pathways by adding dsRNA to the nanocarrier [13]. We have achieved high co-encapsulation efficiencies by using liposomes with positive charge that can easily incorporate LPS and poly (I:C) into the lipid bilayer and become neutral liposomes. Although liposomes are in principle highly biocompatible, *in vitro* toxicity of cationic liposomes has been reported by several groups [42], [43]. Thus, the observed charge neutralization has been an advantage, making our formulation highly biocompatible. Another advantage of this encapsulation system has been the elimination of the free

LPS associated toxicity observed in cells and larvae (**Figures 2 and 6**). The LPS toxicity *in vitro* and *in vivo* has been well documented in different vertebrates [15], and it has also been demonstrated that encapsulation of LPS into liposomes decreased its toxicity compared to the free form [29]. Our system minimizes the detrimental effects of LPS while maintaining the immune system activation potency.

By developing an *in vitro* endocytosis assay with fish cells, we have also demonstrated that NL_c liposomes contact with plasma membranes and they are efficiently internalized by fish macrophages and zebrafish hepatocytes. Different studies in rodents and humans have shown the ability of liposomes to deliver different compounds to endosomal compartments [12], [32]. The liposomes developed in this study are 125 nm in size and its endocytosis is inhibited mainly by MβCD and sucrose, which indicates that they likely utilize the caveolae-mediated and the clathrin-mediated endocytosis pathways to reach intracellular compartments. The fact that the NL_c liposomes accumulate in endosomal-lysosomal compartments is a potential advantage since TLR3 is located in endosomal membranes, and antigen processing for MHCII presentation takes place in this compartment [3]. In addition, this simple and active formulation designed for virtually all fish species vaccination could be upgraded with specific pathogenic antigens of any particular fish species.

In recent years, health and environmental safety of nanoparticle-based therapeutics is a major concern for nanotechnology that has to be carefully addressed [44]. The zebrafish embryos and larvae have become a reference model for *in vivo* toxicological studies due to its sensitivity and logistic convenience [45]–[47]. Zebrafish embryos are protected from the environment with the chorion, a rigid but permeable membrane, which embryos lose after 48 h (hatching) to become free-swimming larvae [48], [49]. We have taken advantage of the zebrafish model to demonstrate the biocompatibility of our formulation at therapeutic doses and also the ability of NL_c to target innate immune system. The activation of the innate immune system in trout macrophages and in zebrafish larvae can be assessed by following the expression of key cytokines [16], [50]. Our study demonstrates that NL_c formulation stimulates the expression of several cytokines involved in anti-viral and bacterial response, and in some cases, the treatment with empty NL formulations also stimulates cytokine gene expression. Importantly, TNFα secretion by trout macrophages is potently and specifically stimulated by the liposomal IS-cocktail and not by the non-loaded liposomes. However, several groups have indeed described that cationic liposomes had an immunological adjuvant effect and that they were able to regulate the transcription of different chemokines and cytokines [36]. The induction of specific immune responses with liposomal immunostimulant formulations should be a promising strategy to improve disease control in fish farms.

ACKNOWLEDGMENTS

We thank Dr. Piferrer and Dr. Ribas for providing zebrafish AB embryos, Dr. Itarte for providing HepG2 cell line and Carlos Carbonell for the artwork in Figure 1F.

AUTHOR CONTRIBUTIONS

Conceived and designed the experiments: AR MCS DM NR. Analyzed the data: AR MCS SAM DM NR. Wrote the paper: AR MCS DM NR.

REFERENCES

1. Khan MA, Khan S, Miyan K (2011) Aquaculture as a food production system: A review. Biol Med 3: 291–302.
2. Evensen O (2009) Development in fish vaccinology with focus on delivery methodologies, adjuvants and formulations. Options Mediterraneennes 86: 177–186.
3. Nordly P, Madsen HB, Nielsen HM, Foged C (2009) Status and future prospects of lipid-based particulate delivery systems as vaccine adjuvants and their combination with immunostimulators. Expert Opin Drug Deliv 6: 657–672 doi:10. 1517/17425240903018863.
4. Watson DS, Endsley AN, Huang L (2012) Design considerations for liposomal vaccines: Influence of formulation parameters on antibody and cell-mediated immune responses to liposome associated antigens. Vaccine 30: 2256–2272 doi:10. 1016/j. vaccine. 2012. 01. 070.
5. Wieber A, Selzer T, Kreuter J (2011) Characterisation and stability studies of a hydrophilic decapeptide in different adjuvant drug delivery systems: A comparative study of PLGA nanoparticles versus chitosan-dextran sulphate microparticles versus DOTAP-liposomes. International Journal of Pharmaceutics 421: 151–159 doi:10. 1016/j. ijpharm. 2011. 09. 011.
6. Zhu L-Y, Nie L, Zhu G, Xiang L-X, Shao J-Z (2013) Advances in research of fish immune-relevant genes: A comparative overview of innate and adaptive immunity in teleosts. Dev Comp Immunol 39: 39–62 doi:10. 1016/j. dci. 2012. 04. 001.
7. Coffman RL, Sher A, Seder RA (2010) Vaccine adjuvants: putting innate immunity to work. Immunity 33: 492–503 doi:10. 1016/j. immuni. 2010. 10. 002.
8. MacKenzie S, Planas JV, Goetz FW (2003) LPS-stimulated expression of a tumor necrosis factor-alpha mRNA in primary trout monocytes and in vitro differentiated macrophages. Dev Comp Immunol 27: 393–400 doi:10. 1016/S0145-305X(02)00135-0.
9. Whyte SK (2007) The innate immune response of finfish – A review of current knowledge. Fish Shellfish Immunol 23: 1127–1151 doi:10. 1016/j. fsi. 2007. 06. 005.
10. Bassity E, Clark TG (2012) Functional identification of dendritic cells in the teleost model, rainbow trout (Oncorhynchus mykiss). PLoS ONE 7: e33196 doi:10. 1371/journal. pone. 0033196. g008.
11. Takeuchi O, Akira S (2010) Pattern recognition receptors and inflammation. Cell 140: 805–820 doi:10. 1016/j. cell. 2010. 01. 022.
12. Kasturi SP, Skountzou I, Albrecht RA, Koutsonanos D, Hua T, et al. (2011) Programming the magnitude and persistence of antibody responses with innate immunity. Nature 470: 543–547 doi:10. 1038/nature09737.
13. Thim HL, Iliev DB, Christie KE, Villoing S, McLoughlin MF, et al. (2012) Immunoprotective activity of a Salmonid Alphavirus Vaccine: Comparison of the immune responses induced by inactivated whole virus antigen

formulations based on CpG class B oligonucleotides and poly I:C alone or combined with an oil adjuvant. Vaccine 30: 4828–4834 doi:10. 1016/j. vaccine. 2012. 05. 010.

14. Palti Y (2011) Toll-like receptors in bony fish: From genomics to function. Dev Comp Immunol 35: 1263–1272 doi:10. 1016/j. dci. 2011. 03. 006.
15. Sepulcre MP, Alcaraz-Perez F, Lopez-Munoz A, Roca FJ, Meseguer J, et al. (2009) Evolution of lipopolysaccharide (LPS) recognition and signaling: Fish TLR4 does not recognize LPS and negatively regulates NF- B activation. J Immunol 182: 1836–1845 doi:10. 4049/ jimmunol. 0801755.
16. Roher N, Callol A, Planas JV, Goetz FW, Mackenzie SA (2011) Endotoxin recognition in fish results in inflammatory cytokine secretion not gene expression. Innate Immun 17: 16–28 doi:10. 1177/1753425909348232.
17. Iliev DB, Roach JC, MacKenzie S, Planas JV, Goetz FW (2005) Endotoxin recognition:*In fish or not in fish ?*. FEBS Letters 579: 6519–6528 doi:10. 1016/j. febslet. 2005. 10. 061.
18. Mackenzie SA, Roher N, Boltaña S, Goetz FW (2010) Peptidoglycan, not endotoxin, is the key mediator of cytokine gene expression induced in rainbow trout macrophages by crude LPS. Mol Immunol 47: 1450–1457 doi:10. 1016/j. molimm. 2010. 02. 009.
19. Wang K-Y, Deng L-J, Huang J-L, Fu X, Chen D-F, et al. (2011) Study on the immunogencity of poly(D,L-lactide-co- glycolide) (PLGA) microspheres-encapsulated vaccine preparation against *Stenotrophomonas maltophilia* infection in channel catfish (*Ictalurus punctatus*). Afr J Biotech 10: 2751–2761 doi:10. 5897/ AJB10. 175.
20. Tian J, Yu J, Sun X (2008) Chitosan microspheres as candidate plasmid vaccine carrier for oral immunisation of Japanese flounder (Paralichthys olivaceus). Vet Immunol Immunopathol 126: 220–229 doi:10. 1016/j. vetimm. 2008. 07. 002.
21. Rodrigues AP, Hirsch D, Figueiredo HCP, Logato PVR, Moraes ÂM (2006) Production and characterisation of alginate microparticles incorporating Aeromonas hydrophila designed for fish oral vaccination. Process Biochem 41: 638–643 doi:10. 1016/j. procbio. 2005. 08. 010.
22. Irie T, Watarai S, Iwasaki T, Kodama H (2005) Protection against experimental Aeromonas salmonicida infection in carp by oral immunisation with bacterial antigen entrapped liposomes. Fish Shellfish Immunol 18: 235–242 doi:10. 1016/j. fsi. 2004. 07. 006.
23. Klibanov AL, Maruyama K, Torchilin VP, Huang L (1990) Amphipathic polyethyleneglycols effectively prolong the circulation time of liposomes. FEBS Lett 268: 235–237 doi:10. 1016/0014-5793(90)81016-H.
24. Milla P, Dosio F, Cattel L (2012) PEGylation of proteins and liposomes: a powerful and flexible strategy to improve the drug delivery. Curr Drug Metab 13: 105–119 doi:10. 2174/138920012798356934.
25. Bennett-Guerrero E, McIntosh TJ, Barclay GR, Snyder DS, Gibbs RJ, et al. (2000) Preparation and preclinical evaluation of a novel liposomal complete-core lipopolysaccharide vaccine. Infection and Immunity 68: 6202–6208 doi:10. 1128/IAI. 68. 11. 6202-6208. 2000.
26. Livak KJ, Schmittgen TD (2001) Analysis of relative gene expression data using real-time quantitative PCR and the 2−ΔΔCT method. Methods. METHODS 25: 402–408 doi:10. 1006/meth. 2001. 1262.
27. Ali S, Mil HGJV, Richardson MK (2011) Large-scale assessment of the zebrafish embryo as a possible predictive model in toxicity testing. PLoS ONE 6: e21076 doi:10. 1371/journal. pone. 0021076. t003.
28. Balazs DA, Godbey WT (2011) Liposomes for use in gene delivery. J Drug Deliv 2011: 326497 doi:10. 1155/2011/326497.
29. Nakhla AN, Banoub JH, Hernandez-Borrell J, Keough KMW (1996) Incorporation of the lipopolysaccharide and polysaccharide from Aeromonas salmonicida into liposomes. Journal of Liposome Research 6: 141–154 doi:10. 3109/08982109609037209.
30. Puyal C, Milhaud P, Bienvenüe A, Philippot JR (1995) A new cationic liposome encapsulating genetic material. A potential delivery system for polynucleotides. Eur J Biochem 228: 697–703 doi:10. 1111/j. 1432-1033. 1995. tb20312. x.
31. Doñate C, Roher N, Balasch JC, Ribas L, Goetz FW, et al. (2007) CD83 expression in sea bream macrophages is a marker for the LPS-induced inflammatory response. Fish Shellfish Immunol 23: 877–885 doi:10. 1016/j. fsi. 2007. 03. 016.
32. Dow S (2008) Liposome–nucleic acid immunotherapeutics. Expert Opin Drug Deliv 5: 11–24 doi:10. 1517/17425247. 5. 1. 11.
33. Ivanov AI (2008) Pharmacological inhibition of endocytic pathways: Is it specific enough to be useful? Methods Mol Biol 440: 15–33 doi:_10. 1007/978-1-59745-178-9_2.
34. Jess TJ, Belham CM, Thomson FJ, Scott PH, Plevin RJ, et al. (1996) Phosphatidylinositol 3′-kinase, but not p70 ribosomal S6 kinase, is involved in membrane protein recycling: wortmannin inhibits glucose transport and downregulates cell-surface transferrin receptor numbers independently of any effect on fluid-phase endocytosis in fibroblasts. Cell Signal 8: 297–304 doi:10. 1016/0898-6568(96)00054-X.
35. Wang W-L, Liu W, Gong H-Y, Hong J-R, Lin C-C, et al. (2011) Activation of cytokine expression occurs through the TNFα/NF-κB-mediated pathway in birnavirus-infected cells. Fish Shellfish Immunol 31: 10–21 doi:10. 1016/j. fsi. 2011. 01. 015.
36. Lonez C, Vandenbranden M, Ruysschaert J-M (2012) Cationic lipids activate intracellular signaling pathways. Adv Drug Deliv Rev 64: 1749–1758 doi:10. 1016/j. addr. 2012. 05. 009.
37. Goetz F (2004) Tumor necrosis factors. Dev Comp Immunol 28: 487–497 doi:10. 1016/j. dci. 2003. 09. 008.
38. Sommerset I, Krossoy BBE, Frost P (2005) Vaccines for fish in aquaculture. Expert Rev Vaccines 4: 89–101 doi:10. 1586/14760584. 4. 1. 89.

39. Lieschke GJ, Trede NS (2009) Fish immunology. Curr Biol 19: R678–R682 doi:10. 1016/j. cub. 2009. 06. 068.

40. Zhu L-Y, Nie L, Zhu G, Xiang L-X, Shao J-Z (2012) Advances in research of fish immune-relevant genes: A comparative overview of innate and adaptive immunity in teleosts. Dev Comp Immunol 1–24 doi:10. 1016/j. dci. 2012. 04. 001.

41. Fredriksen BN, Grip J (2012) PLGA/PLA micro- and nanoparticle formulations serve as antigen depots and induce elevated humoral responses after immunization of Atlantic salmon (Salmo salar L.). Vaccine 30: 656–667 doi:10. 1016/j. vaccine. 2011. 10. 105.

42. Filion MC, Phillips NC (1997) Toxicity and immunomodulatory activity of liposomal vectors formulated with cationic lipids toward immune effector cells. Biochim Biophys Acta 1329: 345–356 doi:10. 1016/S0005-2736(97)00126-0.

43. Takano S, Aramaki Y, Tsuchiya S (2003) Physicochemical properties of liposomes affecting apoptosis induced by cationic liposomes in macrophages. Pharm Res 20: 962–968 doi: 10. 1023/A:1024441702398.

44. Moghimi SM, Peer D, Langer R (2011) Reshaping the future of nanopharmaceuticals: Ad Iudicium. ACS Nano 5: 8454–8458 doi:10. 1021/nn2038252.

45. Harper SL, Carriere JL, Miller JM, Hutchison JE, Maddux BLS, et al. (2011) Systematic evaluation of nanomaterial toxicity: utility of standardized materials and rapid assays. ACS Nano 5: 4688–4697 doi:10. 1021/nn200546k.

46. George S, Xia T, Rallo R, Zhao Y, Ji Z, et al. (2011) Use of a high-throughput screening approach coupled with in vivo zebrafish embryo screening to develop hazard ranking for engineered nanomaterials. ACS Nano 5: 1805–1817 doi:10. 1021/nn102734s.

47. Peterson RT, Macrae CA (2012) Systematic approaches to toxicology in the zebrafish. Annu Rev Pharmacol Toxicol 52: 433–453 doi:10. 1146/annurev-pharmtox-010611-134751.

48. Lee KJ, Nallathamby PD, Browning LM, Osgood CJ, Xu X-HN (2007) In vivo imaging of transport and biocompatibility of single silver nanoparticles in early development of zebrafish embryos. ACS Nano 1: 133–143 doi:10. 1021/nn700048y.

49. Fent K, Weisbrod CJ, Wirth-Heller A, Pieles U (2010) Assessment of uptake and toxicity of fluorescent silica nanoparticles in zebrafish (Danio rerio) early life stages. Aquat Toxicol 100: 1–11 doi:10. 1016/j. aquatox. 2010. 02. 019.

50. Galindo-Villegas J, García-Moreno D, de Oliveira S, Meseguer J, Mulero V (2012) Regulation of immunity and disease resistance by commensal microbes and chromatin modifications during zebrafish development. Proc Natl Acad Sci USA 109: E2605–E2614 doi:10. 1073/pnas. 1209920109.

CHAPTER 13

COMPLETE GENOME SEQUENCE OF YERSINIA RUCKERI STRAIN CSF007-82, ETIOLOGIC AGENT OF RED MOUTH DISEASE IN SALMONID FISH

Michael C. Nelson, Scott E. LaPatra, Timothy J. Welch, Joerg Graf

GENOME ANNOUNCEMENT

Yersinia ruckeri, a Gram-negative gammaproteobacterium, is the etiologic agent of enteric red mouth disease (ERM) a hemorrhagic septicemia of farmed salmonid fish species worldwide (1, 2). Y. ruckeri strains include several O-serotypes, yet nearly all outbreaks are caused by a genetically homogenous group of serotype 01 strains. Successful bacterin vaccines for ERM were developed in the 1970s; however, outbreaks in vaccinated fish have recently occurred and have been associated with the emergence of novel variants of this pathogen (3, 4). The reemergence of *Y. ruckeri* mediated ERM has renewed interest in the virulence and evolution of this pathogen.

Presently, only two draft *Y. ruckeri* genomes are available; the genome of the type strain, ATCC 29473^T, consists of 174 contigs, while the genome of a strain from Chile consists of 75 contigs (5). An Illumina-based assembly of the pathogenic Y. ruckeri strain CSF007-82 (6) yielded 63 contigs with N_{50} of 160,889 bp. Because of this, we resequenced strain CSF007-82 using PacBio in order to obtain a complete, circularized genomic reference for use in future studies examining determinants of *Y. ruckeri* pathogenicity.

Large insert libraries were prepared and sequenced on a Pacific Biosciences RS II instrument using the P5-C3 sequencing chemistry at the Yale Center for Genomic Analysis, generating over 1.64 Gbp of sequencing data from ~200,000 raw reads. The resulting data was assembled using the HGAP version 2 assembler (7) with standard parameters, yielding two unique contigs ~100 kbp and ~3.8 Mbp in size. Overlapping nucleotides were manually edited and removed from the contig ends to create circularized replicons, which were then used as reference sequences for remapping of the raw sequencing reads to identify and correct any remaining errors. Interestingly, a contig corresponding to the plasmid pYR2 identi- fied in the Illumina assembly was not generated, however, sequencing reads did map to pYR2 during the reference assembly. The largest contig, representing the chromosome, is 3,799,036 bp with a G + C content of 47.5% while the second contig is 103,917 bp with a G + C content of 48.4%. pYR2 is 16,923 bp with a G + C content of 47.2%.

The three finished contigs were functionally annotated by RAST (8). The chromosome encodes 3,517 predicted coding sequences (CDS), 22 rRNAs organized into 7 ribosomal operons, and 80 tRNAs covering all standard amino acids. Numerous phage and transposable element proteins were identified. The ~100 kbp contig encoded 99 predicted CDS, 16 of which belonged to the IncI1 class of replication proteins. No RNAs were predicted for this contig, indicating that it represents a previously unknown plasmid we named pYR3.

The availability of a complete, finished *Y. ruckeri* genome sequence, in combination with additional genome sequences should allow for more thorough analysis and identification of novel virulence factors associated with ERM. Additionally, the availability of a closed genome sequence should enable further studies of the evolution of the *Yersinia* genus as well as an examination of the transmission and acquisition of mobile genetic elements.

Citation: Nelson MC, LaPatra SE, Welch TJ, and Graf J. 2015. Complete Genome Sequence of Yersinia ruckeri Strain CSF007-82, Etiologic Agent of Red Mouth Disease in Salmonid Fish. Genome Announcements, 2015; Volume 3, Issue 1, e01491-14. DOI:10.1128/genomeA.01491-14.

Nucleotide Sequence Accession Numbers

The sequencing data is available under the GenBank/ENA/DDBJ accession no. PRJEB6967. The annotated genome is available under accession numbers LN681229 to LN681231.

ACKNOWLEDGMENT

This research was supported through USDA ARS agreement 58-1930-4- 002.

REFERENCES

1. Busch RA. 1978. Enteric redmouth disease (Hagerman strain). Mar Fish Rev 40:467–472.
2. Ewing WH, Ross AJ, Brenner DJ, Fanning GR. 1978. Yersinia ruckeri sp. nov., the redmouth (RM) bacterium. Int J Syst Bacteriol 28:37–44. http:// dx.doi.org/10.1099/00207713-28-1-37.
3. Welch TJ, Verner-Jeffreys DW, Dalsgaard I, Wiklund T, Evenhuis JP, Cabrera JAG, Hinshaw JM, Drennan JD, Lapatra SE. 2011. Independent emergence of Yersinia ruckeri biotype 2 in the United States and Europe. Appl Environ Microbiol 77:3493–3499. http://dx.doi.org/10.1128/ AEM.02997-10.
4. Wheeler RW, Davies RL, Dalsgaard I, Garcia J, Welch TJ, Wagley S, Bateman KS, Verner-Jeffreys DW. 2009. Yersinia ruckeri biotype 2 isolates from mainland Europe and the UK likely represent different clonal groups. Dis Aquat Org 84:25–33. http://dx.doi.org/10.3354/dao02039.
5. Navas E, Bohle H, Henríquez P, Grothusen H, Bustamante F, Bustos P, Mancilla M. 2014. Draft genome sequence of the fish pathogen Yersinia ruckeri strain 37551, serotype O1b, isolated from diseased, vaccinated Atlantic salmon (Salmo salar) in Chile. Genome Announc 2(4):e00858-14. http://dx.doi.org/10.1128/genomeA.00858-14.
6. Evenhuis JP, Lapatra SE, Verner-Jeffreys DW, Dalsgaard I, Welch TJ. 2009. Identification of flagellar motility genes in Yersinia ruckeri by transposon mutagenesis. Appl Environ Microbiol 75:6630 –6633. http:// dx.doi.org/10.1128/AEM.01415-09.
7. Chin C-S, Alexander DH, Marks P, Klammer AA, Drake J, Heiner C, Clum A, Copeland A, Huddleston J, Eichler EE, Turner SW, Korlach J. 2013. Nonhybrid, finished microbial genome assemblies from long-read SMRT sequencing data. Nat Methods 10:563–569. http://dx.doi.org/ 10.1038/nmeth.2474.
8. Aziz RK, Bartels D, Best AA, DeJongh M, Disz T, Edwards RA, Formsma K, Gerdes S, Glass EM, Kubal M, Meyer F, Olsen GJ, Olson R, Osterman AL, Overbeek RA, McNeil LK, Paarmann D, Paczian T, Parrello B, Pusch GD, Reich C, Stevens R, Vassieva O, Vonstein V, Wilke A, Zagnitko O. 2008. The RAST server: Rapid Annotations using Subsystems Technology. BMC Genomics 9:75. http://dx.doi.org/10.1186/1471-2164-9-75.

CHAPTER 14

IMPROVEMENT OF INNATE IMMUNE RESPONSES AND DEFENSE ACTIVITY IN TIGER SHRIMP (PENAEUS MONODON FAB.) BY INTRAMUSCULAR ADMINISTRATION OF THE OUTER MEMBRANE PROTEIN VIBRIO ALGINOLYTICUS

Maftuch, E Prasetio, A Sudianto, M Rozik, R Nurdiyani, E Sanusi, H Nursyam, F Fariedah, Marsoedi, and Murachman

Keywords: Outer membrane protein, Immunomodulation, Tiger shrimp, *Vibrio harveyi*

INTRODUCTION

Tiger shrimp (*Penaeus monodon* Fabricus) are one of the most important aquaculture products in several countries, especially in South-East Asia. But in Indonesia, tiger shrimp have almost entirely disappeared because of diseases. Disease problems in aquaculture are caused by environmental problems such as poor water conditions, leading to stress and disease outbreaks (Sadovy, 2000). The big disease outbreaks in tiger shrimp aquaculture are caused by viruses and bacteria e.g., *Vibrio* spp. Therefore, control of diseases is an important task for the further development of tiger shrimp culture.

There are many techniques widely used to control the diseases, such as using antibiotics and chemotherapy. However, experience demonstrates that there are several problems associated with the use of antibiotics and chemotherapy in clinical treatment of shrimp diseases. Such as, potential environmental hazards, the spread of antibiotic-resistant bacteria and associated stress (Jin et al. 2008). Thus, researchers have focused on finding alternative methods of disease control such as the application of vaccines and immunostimulants of microbial origin (Sakai, 1999). The OMP is an effective microbial origin immunostimulant and has been successfully used to strengthen the non-specific defense systems of shrimp and also non-specific and specific defense systems of fish (Maftuch, 2006). The OMP was isolated from the *Vibrio alginolyticus* cell wall.

Many reports showed that OMP contain LPS, beta-glucan and proteins that can enhance macrophage phagocytic activity, which is part of the non-specific defense system (Jin et al. 2008). However, there are a few reports about the application of OMP in Shrimp. The purpose of the study was to explore the effect of OMP on the immune response in tiger shrimp, as well as on resistance against *Vibrio harveyi.*

MATERIALS AND METHODS

Preparations of Outer Membrane (OMP) from *V. Alginolyticus*

V. alginolyticus strains were isolated from pathogenic shrimp (*Penaeus monodon* Fab.) that cultured in marine ponds at

Citation: Maftuch, E Prasetio, A Sudianto, M Rozik, R Nurdiyani, E Sanusi, H Nursyam, F Fariedah, Marsoedi, and Murachman. Improvement of innate immune responses and defense activity in tiger shrimp (Penaeus monodon Fab.) by intramuscular administration of the outer membrane protein Vibrio alginolyticus, SpringerPlus, 2013; 2:432. DOI:10.1186/2193-1801-2-432.

the Research Centre of Brackish Water, East Java Province, Indonesia. The isolate was cultured on thiosulfate-citrate-bile salt-sucrose (TCBS) agar plates. Therefore the *V. alginolyticus* was grown in 1000 ml Brain Heart Infusion Broth (BHIB) at 30°C for 24 hours, and then were separated into 100 bottles containing Thiaproline Carbonate Glutamate (TCG) medium at 30°C, for 48 hours to enrich growth pili of *V. alginolyticus*. The medium contains 0.02% Thioproline, 0.3% NaHCO3, 0.1% monosodium l-glutamate, 1% bactotryptone, 0.2% yeast extract, 0.5% NaCl, 2% bacto agar and 1 mM Beta-amino ethyl ether N,N,N,"n"-tetra acetic acid (EGTA).

OMP was isolated from both pili (growth medium) and cell membrane (cell pellet), following (Garavito and Rosenbusch 1986). The pili were isolated using micro centrifugation (3,500 g, 15 min, 4°C) six times. The Pili's content was put in the solution. The cell pellet (1 gram) was resuspended in 15 ml of lysis buffer. After incubation for 1 min with gentle shaking and centrifugation at 12,000 g and 4°C for 30 min, the solution containing the membrane fraction was dialyzed for 24 hours by dH_2O and following with Phosphate Buffer Saline (PBS) pH 7.4 for 24 hours. The determination of protein concentration was estimated using the BCA assay kit (Pierce, Rockford, IL, USA) according to the manufacturer's instructions. A protein sample (12.5 μl) was mixed with 100 μl of the BCA working reagent. After the reaction mixture was incubated at 37°C for 30 min, absorbance at 540 nm was measured with a microplate reader spectrophotometer (Labsystem, Finland). BSA at various concentrations, ranging from 0.025 to 1.0 mg/ml, was used to construct a standard calibration curve and to determine protein concentrations of the samples.

Preparation of *V. Harveyi*

V. harveyi strain was isolated from pathogenic shrimp (*P. monodon* Fab.) from marine ponds at the Research Centre of Brackish Water, East Java Province, Indonesia. The isolate was cultured on thiosulfate-citrate-bile salt-sucrose (TCBS) agar plates, and then grown in 1000 ml of Brain Heart Infusion Broth (BHIB) at 30°C for 24 hours. The harvested bacteria were washed 3 times in 1.5% physiological saline by centrifugation at 1,000 g for 10 min. The concentration of bacterial suspension was determined by spectrophotometer and diluted to 10^9 cell/ml in 1.5% sterile physiological saline for preparation.

Shrimp Immunization and Bacterial Challenge

A total two hundred and fourty tiger srimps were divided into four groups, each group comprised 60 tiger shrimps were immunized by OMP 10, 20, 30 μg/kg bw, and control (without immunized). The Immunizations were done by injecting OMP directly in the first swimming leg of the abdomen using a 1 mL syringe equipped with a 22 gauge needle. Second injection or boostering was done after 7 days first immunization. After 14 days immunization, the srimps were chalanged with *V. harveyi* for 24 hours. Sixty tiger shrimps for each treatment were cultured at 10 liters of water diluted with *V. harveyi* 10^7 cell/ml in culture media for 24 hours. Another 60 tiger shrimps without immunized with OMP were cultured in standard media as control. The mortality was monitored for 1 day after infection (Jin et al.2008). Haemolymph was collected after 24 hours infection, from abdomen sides in the second legs of shrimp by using a 1 ml syringe equipped with a 22 gauge needle. The Haemolymph was stored at room temperature and ready to be used for sample of each parameters assay (Johansson et al. 2000).

Total Haemocyte Count (THC)

The THC was measured according to modified method from (Wootton et al. 2003). A hundred μl of haemolymph was transferred into a tube which contained 900 μl of Natt-Herricks's stain solution. A drop of haemolymph collected from each individual was introduced into an improved Neubauer hemocytometer and the number of haemocytes was determined microscopically. THC was calculated using the following formulation (Wootton et al.2003):

$$\text{THC} = \frac{\text{Total hemolymph counted} \times \text{dilution factor} \times 2 \times 10^4}{\text{Number of square counted}}$$

Differential Haemocyte Counts (DHCs)

The DHCs was performed according to (Wootton et al.2003). The haemolymph of shrimp was collected (100 μl) and diluted with 100 μl of TBS buffer, then a total volume of 0.2 ml (200 μl). The cells solution was mixed well and 200 μl was spread on a glass slide and placed at room temperature for 60 minutes. Then, the cells were fixed using Baker's formol calcium (+ 2% NaCl). After 30 minutes, the glass slide was rinsed with absolute methanol for 5 minutes. The cells were stained with H&E staining cells (pH 6.6) for 5 minutes. Afterward, the slide was washed with dH_2O. The cell was observed under light microscope using 100× objective lenses. A minimum of 100 haemocytes per slide were examined to determine the DHC.

The Amount of Protein Plasma Assay

The *in vitro* amount of TPP of the haemocytes of *P. monodon* was assessed by following the modified procedure of (Van de Braak, 2002). A hundred microliters of haemolymph were centrifuged at 1500 rpm, at 4°C, for 10 minutes. Then 0.05 ml supernatant was mixed by adding of 1 mg/ml Bovine Serum Albumin (BSA), aquades 4.75 ml, 0.5 ml Alk.$CuSO_4$ and 0.05 ml Lowry Reagent medias, and by following the modification procedure of (Lowry et al. 1951). The values were expressed as O.D. at 660 nm/15 min.

Superoxide Dismutase (SOD) Activity Assay

Superoxide Dismutase activity Assay, considered as its ability to inhibit superoxide radical, was done by previously method (Chandra et al. 2006). One unit of SOD activity was defined as the amount required to inhibit the rate of xanthine reduction by 50% (Jin et al. 2008). Specific activity was expressed as SOD unit/mg protein.

Protease Enzyme Activity Assay

Protease enzyme activity assay was assessed by tyrosine reduction following the modification procedure of (Celis-Guerrero et al. 2004). One unit protease activity was counted by converting the value of absorbance to tyrosine concentration (Celis-Guerrero et al. 2004).

Statistical Analysis

All the immune parameters analysis was performed with the SPSS 15 statistical software. One-way analysis of variance (One-way ANOVA) was applied to analyze the differences among treatments and control. Then, a post hoc test (Duncan) was conducted to display the mean for groups in the homogenous subset.

RESULTS AND DISCUSSION

The Tiger shrimp was administered by OMP intramuscularly (IM) at 10, 20, 30 µg/kg bw. The THC was collected every two days from day 2 until day 14 after imunization. The data showed that after day 2 administered with OMP decreased THCs compared to control (day 0) and increased again at day 4 and day 6. Futhermore THCs decreased again in day 8, after second imunization, and then increased at day 10, day 12 and day 14. This result indicated that administration OMP 20 µg/kg bw stimulated the highest THC in tiger shrimps compare to other doses (Figure 1). This phenomenone was similar to Johansson statement, stated that ß-glucan treatment as immunostimulant could decrease haemocyte at the moment and then increased dramatically (Johansson et al. 2000). (Yeh et al. 2005) reported that administration of *Sargassum duplicatum* extract could increase haemocyte of *Litopenaeus vannamei*. Haemocyte is part of cells in haemolymph, it has function to protect infectious disease as an immune response of the invertebrate organism (Van de Braak, 2002).

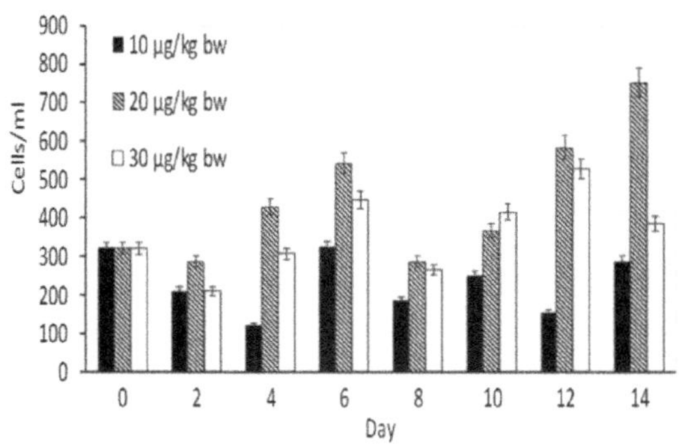

Figure 1. Effects of different dosages of intramuscular administration of OMP on total haemocyte count (THC) of tiger shrimp (×10 ⁴ sel/mm ³) for fourteen days. Data are represented as mean±S.D. (n=10).

Total Haemocyte Count

The total haemocyte count (THC) of tiger shrimp was significantly different with intramuscular administration of OMP and *V. harveyi* infections (P<0.01). Greatly improved THC was observed in the treatment of OMP 20 µg with THC value of 317.5 × 10^4 cells/ml (Figure 2).

The haemocyte circulation system may cause modulation of THC. Indeed, invertebrates (including shrimp) blood circulation system is open while haemocytes are distributed in both the vascular system and tissues. Consequently, an increase of THC values was assumed from either proliferation or movement of cells from tissues into haemolymphs. The concentration of cells varied as function of the development stage of an intermoult cycle of *P. japonicus* post larvae (Tsing et al. 1989). A decrease of THC may be due to cell lysis or increased movement of cells from haemolymph to tissues (Pipe and Cole, 1995). The decreasing of THC is also presumably due to the altered physiological conditions of the shrimp, as enhanced susceptibility to bacterial infection (Ford et al. 1993).

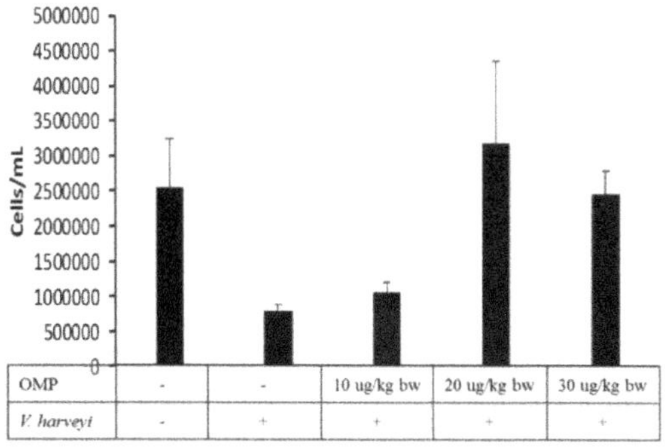

Figure 2. Effects of different dosages of intramuscular administration of OMP and *V. harveyi* infections on total haemocyte count (THC). Data are represented as mean±S.D. (n=10). It indicates significant differences (P<0.05) among five treatments determined by one-way ANOVA.

Differential Haemocyte Count

The DHC (hyaline, semigranular, granular) of tiger shrimp were given intramuscular administration of OMP and *V. harveyi* infections, where hyaline, semigranular, and granular were significantly different among five treatments determined by one-way ANOVA (Figure 3). Phagocytosis is believed to be one of the major cellular defence mechanisms in crustaceans. The semigranular haemocytes are the primary cells involved in the phagocytosis of foreign particles in shrimp (Bachère et al. 1995). The similar morphological haemocytes were also reported as haemoblasts in *Tapes philippinarum* (Matozzo et al. 2008) and *Saccostrea glomerata* (Aladaileh et al. 2007b). This cells were identified with different functions which do not play a role in defensive responses like phagocytosis or encapsulation, and they lack the common intracellular enzyme systems associated with host defense, unlike hyaline and granular (Aladaileh et al. 2007b).

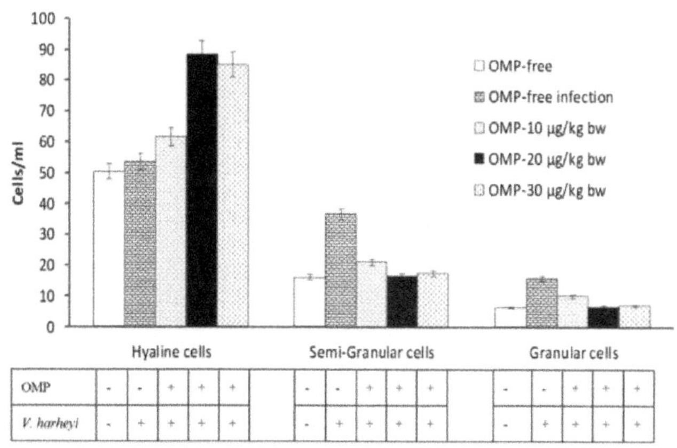

Figure 3. Effects of different dosages of intramuscular administration of OMP and *V. harveyi* infections on differential haemocyte count (DHC). Data are represented as mean±S.D. (n=10). It indicates significant differences (P<0.05) among five treatments determined by one-way ANOVA.

Granular haemocytes are also capable of phagocytosis of foreign material but with less frequency than the smaller ones (Hose and Martin, 1989). Granular cells have been proven to play a significant role in the shrimp defence system because of their antibacterial activity (Chisholm and Smith, 1995). Granular played a role in phagocytosis, which may be correlated by their capacity for intracellular killing. Moreover, several studies showed that granular contain high levels of acid phosphatase and phenol oxidase enzymatic activities, as well as the ability to form superoxides and peroxides (Xue et al. 2000). The other haemocytes are the hyaline cells. They are also considered as phagocytes and superoxide anions production (Söderhäll and Cerenius, 1992).

The results of the study showed that treatment significantly increased the percentages of hyaline and decreased the percentages of semigranular and granular cells in the haemolymph. It was presumed that the young are generally known as the hyaline cells and that those cells gave rise to two haemocytic developmental series, i.e. the large- and small-granular cell line. Granular cells of the large-granular cell line mature and accumulate in the connective tissue, and many cells of the small-granular cell line were located in the lymphoid organ (Van de braak, 2002). Decreases in the percentages of semigranular and granular cells were compensated for by a proportional increase in the percentages of hyaline population. In the present study, haemocytes may display recruitment of immature cells from haematopoietic tissues (Hine, 1999). (Aladaileh et al. 2007a) predicted that the increased percentages of particular haemocyte types were due to induced cellular proliferation, recruitment of cells from non-circulating compartments of the haemolymph, or rapid cellular differentiation in response to antigenic challenge.

Total Plasma Protein

The total plasma protein (TPP) or haemolymph protein activity of black tiger shrimp was significantly different with intramuscular administration of OMP and *V. harveyi* infections ($P<0.01$) (Figure 4). A Great improved TPP was observed in treatment of OMP 20 µg with TPP value of 16.56 mg/ml (Figure 4).

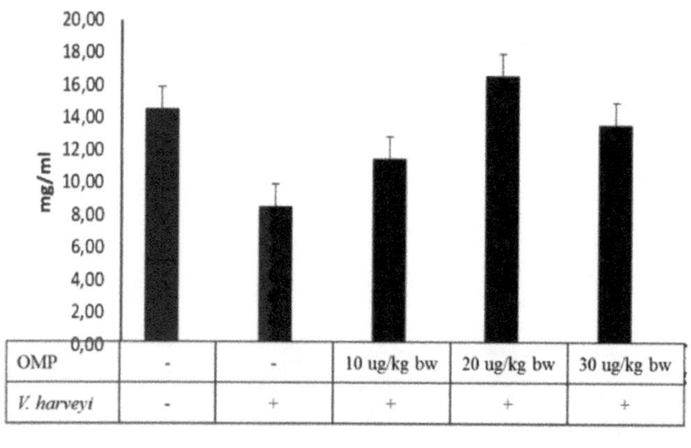

Figure 4. The Effects of different dosages of intramuscular administration of OMP and *V. harveyi* infections on total plasma protein (TPP). Data are represented as mean±S.D. (n=10). Different letters indicate significant differences ($P<0.05$) among five treatments determined by one-way ANOVA.

The TPP has been observed in previous studies on shrimp. The changing of TPP composition in the blood was affected with animal size, sex, nutritional state, environmental factors (such as temperature, salinity) and the moult cycle but this was inversely related to haemolymph volume (Chen and Cheng, 1993) and a decrease of TPP during infection as well as during repeated haemolymph sampling (Ford et al. 1993). In this study, TPP increased with treatment of OMP-10 and 20 µg/kg bw but decreased during infection of bacteria and treatment OMP-30 µg/kg bw. Those were similar to (Chen et al. 1994), the TPP increased with treatment of 10 and 20 ppt and decreased with treatment of 30 ppt.

Superoxide Dismutase and Protease Enzyme Activity

The humoral parameters are SOD and protease activities of tiger shrimp. The SOD and protease activities were significantly different with intramuscular administration of OMP and *V. harveyi* infections ($P<0.05$) (Figure 5). Greatly improved SOD activity was observed in the treatment of OMP 10 µg with SOD value of 139,09 unit /mg protein and protease activity of OMP 20 µg with protease activity value of 130,17 unit/mg protein (Figure 5).

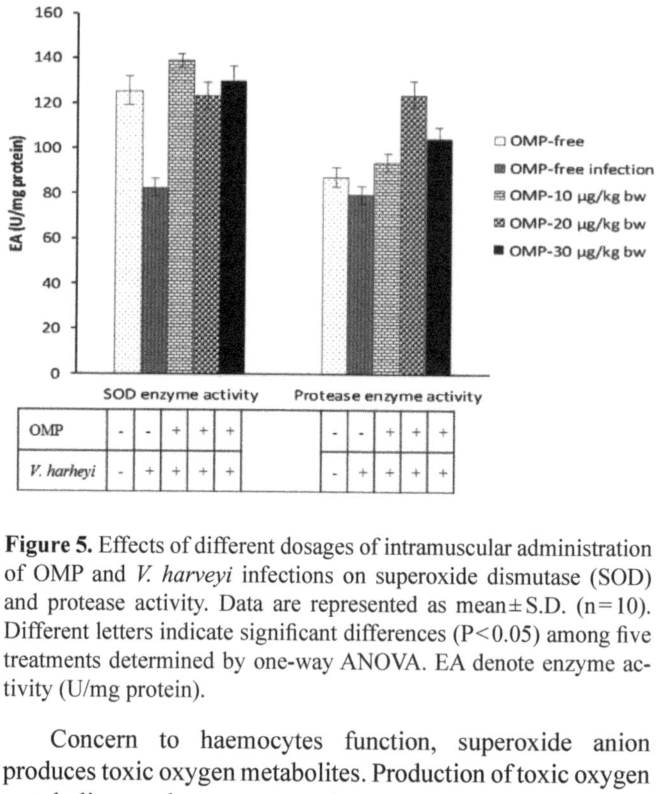

Figure 5. Effects of different dosages of intramuscular administration of OMP and *V. harveyi* infections on superoxide dismutase (SOD) and protease activity. Data are represented as mean±S.D. (n=10). Different letters indicate significant differences ($P<0.05$) among five treatments determined by one-way ANOVA. EA denote enzyme activity (U/mg protein).

Concern to haemocytes function, superoxide anion produces toxic oxygen metabolites. Production of toxic oxygen metabolites, such as protease and *reactive oxygen species* (ROS) activation anion superoxide/$O_2^{\cdot-}$, hydrogen peroxide/H_2O_2, ion hydroxide/OH^- and oxygen/O_2 (Campa-Cordova et al. 2002) is believed to be mediated through phagocytosis, encapsulation, aggregate nodulation, melanocytes and cytotoxicity which destroy invasive pathogens (Rodriguez and Moullac, 2000), thus providing an explanation for the combined suppression of phagocytosis and superoxide anion (Anderson et al. 1992).

The levels of protease activity were enhanced, caused by intramuscular administration of OMP and *V. harveyi*infections in tiger shrimp. The protease enzyme has a role to play as a lysosomal enzyme and prophenoloxidase (proPO) activator to become a phenol oxidase (PO) enzyme (Van de Braak, 2002). During that process, it was followed by phenol oxidation to become quinone that is antibacterial molecules (Smith et al. 2003). (Yeh et al. 2005) said that immunostimulants could enhance Phenol oxidase enzyme activity. Similar study in *Biomphalaria glabrata*reported that serine protease enzyme activity enhancement was followed by phenol oxidase enzyme activity enhancement (Bahgat et al. 2002).

On the other hand, an alteration in the levels of superoxide anion caused by contaminant exposure has been well studied in invertebrates (Wootton et al. 2003). A study on bivalves reported that bivalves' superoxide generation is generally enhanced with low concentrations of contaminant exposure, but inhibited at higher concentrations (Dyrynda et al. 2000). In the present study, intramuscular administration of OMP and *V. harveyi* infections presented significant differences from control to produce protease and superoxide anion.

Mortality

Initial mortality tiger shrimp occurred on the next day after challenge. The highest cumulative mortality (46%) was observed in tiger shrimp with the control infection (OMP-free infection) (Figure 6). While shrimp with various OMP injection had reduced mortality until of 6-13%. Similar results had been reported in previous studies on shrimp when using immunostimulant (beta glucan) (Huang et al. 2006). In this study, treatment of OMP-20 μg/kg bw by intramuscular injection showed the most efficient and desirable influence on resistance of tiger shrimp to *V. harveyi.*However, treatments of OMP-10 μg/kg bw and OMP-30 μg/kg bw reduced of mortality of tiger shrimp but they were not more efficient than treatment of OMP-20 μg/kg bw. It probably concluded that the beneficial effect of OMP of disease resistance of tiger shrimp is dose-dependent and OMP-20 μg/kg bw is an optimal dose after two times of boosters of OMP for 14 days against *V. harveyi* infection for 24 hours.

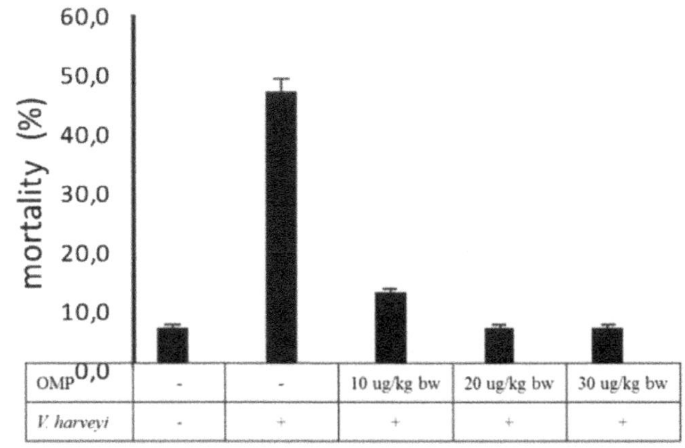

Figure 6. Cumulative mortality (%) of tiger shrimp after two times of boosters of OMP for 14 days against *V. harveyi*infection for 24 hours.

CONCLUSION

The research demonstrated that OMP *V. Alginolyticus* administrated intramuscularly enhance immune responses as well as the resistance to *V. harveyi* in tiger shrimp. In this research, the optimal immune parameters were observed in tiger shrimp induced by OMP-20 μg/kg bw.

ACKNOWLEDGEMENTS

The research was supported by a grant from National Education Ministry of Indonesia. We thanks to Dr. Sumarno, Dr. Widodo, Dr. Supanjani, Larry Cotreen, Ph.D for their technical assistance in the experiment and publication.

AUTHORS' CONTRIBUTIONS

Maf, EP, MR, FF, RN, ES, and AS participated study design and performed the data analysis in the Experiments. HN, Mar, and Mu participated in drafting the manuscript. All authors have read and approved the final manuscript.

REFERENCES

1. Aladaileh S, Nair VS, Birch D, Raftos AD: Sydney rock oyster (*Saccostrea glomerata*) haemocytes. Morphology and function. *J Invertebr Pathol* 2007, 96: 48-63. 10.1016/j.jip.2007.02.011
2. Aladaileh S, Rodney P, Nair VS, Raftos AD: Characterization of phenoloxidase activity in Sydney rock oysters (*Saccostrea glomerata*). *Comp Biochem Physiol* 2007b, 148(b):470-480.
3. Anderson RS, Oliver LM, Brubacher LL: Superoxide anion generation by Crassostrea virginica haemocytes as measured by nitroblue tetrazolium reduction. *J Invertebr Pathol* 1992, 59: 303-307. 10.1016/0022-2011(92)90137-S
4. Bachere E, Mialhe E, Notl D, *et al.*: Knowledge and research prospects in marine mollusc and crustacean immunology.*Aquaculture* 1995, 132: 17-32. 10.1016/0044-8486(94)00389-6
5. Bahgat M, Doenhoff M, Kirschfink M, Ruppel A: Serine protease and phenoloksidase activities in haemocytes of*biomphalaria glabrata* snails with varying susceptibility to infection with the parasite *schistosoma mansoni* . *Parasitol Res*2002, 88: 489-494.
6. Campa-Cordova AI, Hernandez-Saavedra NY, De Philippis R, Ascencio F: Generation of superoxide anion and SOD activity in haemocytes and muscle of American white shrimp (*Litopenaeus vannamei*) as a response to β-glucan and sulphated polysaccharide. *Fish Shellfish Immunol* 2002, 12: 353-366. 10.1006/fsim.2001.0377
7. Celis-Guerrero LE, Fernando LGC, Angeles MNT: Characterization of proteases in the digestive system of spiny lobster (Panulirus interruptus). *Marine Biotechnol* 2004, 6: 262-269.
8. Chandra KM, Basanta KD, Subhsah CM: Effect of multiple injections of β-glucan on non-spesifik immune respone and disease resistance in *Labeo rohita* fingerlings. *Fish Shellfish Immunol* 2006, 20: 305-319. 10.1016/j.fsi.2005.05.007
9. Chen JC, Cheng SY: Studies on hemocyanin and haemolymph protein levels of *Penaeus japonieus* based on sex, size and moulting cycle. *Comp Biochem Physiol* 1993, 106B: 293-296.
10. Chen JC, Chen CT, Cheng SY: Nitrogen excretion and changes of hemocyanin, protein and free amino acid levels in the haemolymph of *Penaeus monodon* exposed to different concentrations of ambient ammonia-N at different salinity levels.*Mar Ecol Prog Ser* 1994, 110: 85-94.
11. Chisholm JRS, Smith VJ: Comparison of antibacterial activity in the haemocytes of different crustacean species. *Comp Bioehem Physiol* 1995, 110A: 39-45.
12. Dyrynda EA, Law RJ, Dyrynda PEJ, Kelly CA, Pipe RK,

Ratcliffe NA: Changes in immune parameters of natural mussel*Mytilus edulis* populations following a major oil spill ('Sea Empress', Wales, UK). *Mar Ecol Prog Ser* 2000, 206: 155-170.

13. Ford SE, Kanaley SA, Littlewood DTJ: Cellular responses of oysters infected with haplosporidium nelsoni: changes in circulating and tissue-infiltrating haemocytes. *J Invertebr Pathol* 1993, 61: 49-57. 10.1006/jipa.1993.1009

14. Garavito RM, Rosenbusch JP: Isolation and crystallization of bacterial porin. *Methods Enzymol* 1986, 125: 309-328.

15. Hine PM: The inter-relationships of bivalve haemocytes. *Fish Shellfish Immunol* 1999, 9: 367-385. 10.1006/fsim.1998.0205

16. Hose JE, Martin GG: Defense functions of granulocytes in the ridgeback prawn *Sicyonia ingentis* . *J Invert Pathol* 1989, 53: 335-346. 10.1016/0022-2011(89)90097-9

17. Huang X, Zhou H, Zhang H: The effect of sargassum fusiforme polysaccharide extract on vibriosis resistance and immune activity of shrimp, Fenneropenaeus chinensis. *Fish Shelfish Immunol* 2006, 20: 750-757. 10.1016/j.fsi.2005.09.008

18. Jin Z, Wang YH, Chu J, Luo L-Z: Improvement of innate immune responses and defense activity in mitten crab (*Eriocheir sinensis*) by oral administration of ß glucan. *Biotechnol Lett* 2008, 30: 1721-1725. 10.1007/s10529-008-9765-7

19. Johansson MW, Keyser P, Sritunyalucksana K: Crustacean haemocytes and haematopoiesis. *Aquaculture* 2000, 191: 45-52. 10.1016/S0044-8486(00)00418-X

20. Lowry OH, Rosenbrough NJ, Farr AL, Randall RJ: Protein measurement with the follin phenol reagent. *J Biol Chem* 1951, 193: 265-275.

21. Maftuch : *Outer membrane protein (Omp) Vibrio alginolyticus vaccine against vibrio disease on grouper. Disertation*. Malang: Brawijaya University; 2006:174.

22. Matozzo V, Monari M, Foschi J, Papi T, Cattani O, Marin GM: Exposure to anoxia of the clam Chamelea gallina: I. Effects on immune responses. *J Exp Mar Biol Ecol* 2008, 325: 163-174.

23. Pipe RK, Coles JA: Environmental contaminants influencing immune function in marine bivalve mollusks. *Fish Shellfish Immunol* 1995, 5: 581-595. 10.1016/S1050-4648(95)80043-3

24. Radriguez J, Moullac GL: State of the art of immunological tools and health control of penaied shrimp. *Aquaculture* 2000, 191: 109-119. 10.1016/S0044-8486(00)00421-X

25. Sadovy Y: *Regional survey for fry and fingerling supply and current practices for grouper mariculture. Evaluating current status and long term prospects for grouper mariculture in south east Asia. Final report to the collaborative APEC grouper research and development network.* 2000. (FWG 01/99)

26. Sakai M: Current research status of fish immunostimulants. *Aquaculture* 1999, 172: 63-92. 10.1016/S0044-8486(98)00436-0

27. Smith VJ, Brown JH, Hautona C: Immunostimulation in crustaceans: does it really protect against infection? *Fish Shellfish Immunol* 2003, 15: 71-90. 10.1016/S1050-4648(02)00140-7

28. Söderhäll K, Cerenius L: Crustacean immunity. *Annu Rev Fish Dis* 1992, 2: 3-23.

29. Tsing A, Arcier JM, Brehélin M: Haemocytes of penaeid and palaemonid shrimps;morphology, cytochemistry, and hemograms. *J Invertebr Pathol* 1989, 53: 64-77. 10.1016/0022-2011(89)90075-X

30. Van de Braak K: *Haemocytic defence in balck tiger shrimp (Penaeus monodon). PhD Degree*. Netherland: Wageningen University; 2002. 159

31. Wootton EC, Dyrinda EA, Pipe RK, Ratcliffe NA: Bivalve immunity:comparisons between the marine mussel (*Mytilus edulis*), the edible cokle (*Cerastoderma edule*), and the razor shell (*Ensis siliqua*). *Fish Shellfish Immunol* 2003, 15: 195-210. 10.1016/S1050-4648(02)00161-4

32. Xue Q, Renault T, Cochennec N, Gerard A: Separation of European flat oyster, Ostrea edulis, haemocytes by density gradient centrifugation and SDS-PAGE characterisation of separated haemocyte sub-populations. *Fish Shellfish Immunol*2000, 10: 155-165. 10.1006/fsim.1999.0234

33. Yeh ST, Chiu S, Lee JCC: Administration of hot-water extract of brown seaweed *Sargassum duplicatum* via immersion and injection enhances the immune resistance of white shrimp *Litopenaeus vannamei* . *Fish Shellfish Immunol* 2005, 20: 332-345.

CHAPTER 15

MOLECULAR CLONING, EXPRESSION AND THE ADJUVANT EFFECTS OF INTERLEUKIN-8 OF CHANNEL CATFISH (ICTALURUS PUNCTATUS) AGAINST STREPTOCOCCUS INIAE

Erlong Wang, Jun Wang, Bo Long, Kaiyu Wang, Yang He, Qian Yang, Defang Chen, Yi Geng, Xiaoli Huang, Ping Ouyang, and Weimin Lai

INTRODUCTION

Aquaculture known as one of the most vibrant sector of the global food system has made a significant contribution in the production of protein-rich food for human consumption[1]. However, aquaculture practices encounter many challenges in recent decades, and one of the most devastating problems is disease outbreaks caused by microbial pathogens. Channel catfish (*Ictalurus punctatus*) as the most extensively cultured food-fish species especially in the United States and China, suffers from widespread disease outbreaks due to various bacterial pathogens[2,3]. *Streptococcus iniae*, a Gram-positive bacterium and one of the β-hemolytic Streptococcus species, has caused serious damage to fish farming in recent years[4,5,6,7], thereby becoming an important fish pathogen with a broad host range, including Asian seabass[8], Nile tilapia[9], zebrafish[10], Japanese flounder[11], rainbow trout[12] and channel catfish[13]. In addition to fish, *S. iniae* is also known to be an opportunistic human pathogen and causes diseases in human[14,15,16].

Traditionally, controlling of *S. iniae* infection in aquaculture relies mainly on the use of antibiotics and antimicrobial compounds. However, due to the indiscriminate use of antibiotics, several problems have arisen such as the development of resistance against antibiotics and safety concerns associated with antibiotic residues. Fortunately, vaccine has been suggested as an effective method and a green intervention to prevent *S. iniae* infection. Compared with inactivated and/or attenuated live vaccines, subunit vaccines composed of conserved proteins are safer and more serotype-independent[17]. Many *S. iniae*–derived proteins were identified as immunogenic[18], and α-enolase (ENO) as one of the surface-exposed and highly conserved proteins between Streptococcus spp[19], had been reported to be used as an efficient protective antigen candidate against *S. iniae* infection in mice according our previous study[20]. It was also confirmed that recombinant ENO (rENO) induced cross-protective immunity against *S. iniae* and*Streptococcus parauberi* in zebrafish[10]. However, poor immunogenicity has been the main constraint to subunit vaccine development. Thus, the use of adjuvants in subunit vaccines is exceedingly necessary and important. Cytokines are a family of low molecular weight and secretory proteins which involved in various pro-inflammatory functions against

Citation: Erlong Wang, Jun Wang, Bo Long, Kaiyu Wang, Yang He, Qian Yang, Defang Chen, Yi Geng, Xiaoli Huang, Ping Ouyang, and Weimin Lai (2016).. Molecular cloning, expression and the adjuvant effects of interleukin-8 of channel catfish (Ictalurus Punctatus) against Streptococcus iniae. Scientific Reports 6, Article number: 29310; DOI:10.1038/srep29310.

invasive pathogens including bacteria, virus or parasite. Chemokines as a kind of cytokines are among the adjuvants more widely used for vaccination against various pathogens in mammals[21,22,23]. To date, many cytokines and chemokines have been found in bony fish, and their functions and signaling are being explored with great progress[24,25,26]. Moreover, several cytokines have already been shown to be efficient adjuvants which are crucial for vaccine effectiveness to obtain the appropriate immune response and ensure a protective outcome in fish[24,27,28]. Interleukin-8 (IL-8) is one of the first CXC chemokines to be discovered in fish[29]. Some studies have focused on the potential use of IL-8 as an adjuvant for genetic engineering vaccine in mammals, while in fish IL-8 has just been cloned and characterized in few species including lamprey[30], Japanese flounder[29], rainbow trout[31], and channel catfish just with a report of an IL-8 like gene[32], and an understanding of the biological role of IL-8 has not been achieved in any of these species.

In the current study, we cloned the IL-8 coding sequence of channel catfish, selected the truncate IL-8 (tIL-8, mature IL-8 regions) to express based on the sequence analysis, and evaluated its adjuvant effects on the protective immunity of a subunit vaccine encoding α-enolase (rENO) of *Streptococcus iniae* from several aspects including lysozyme activity, alternative haemolytic complement activity (ACH50), antibody titers, the expression of immune-related genes and relative percent survival (RPS) in channel catfish.

RESULTS

Cloning, Sequence and Phylogenetic Analyses of IL-8 Coding Regions

The full-length of IL-8 coding sequence of channel catfish was 336 bp (GenBank No. KP701473) (Fig. 1A), which contained a complete open reading frame (ORF) and encoded a 111 a.a. with the predicted molecular weight of 12.20 kDa and the theoretical pI of 10.08. The a.a. sequence was predicted to contain a 23 a.a. signal peptide (Fig. 1C). Thus, the mature peptide of channel catfish IL-8 (tIL-8) was speculated to consist of 88 a.a. There was a Chemokine_CXC domain which was conserved in Chemokine superfamily locating at N-terminal of 30~88 a.a residuals of the IL-8. Three consensus motifs (ELR motif, CXC motif and GPH motif) existing in Chemokine_CXC domain were also found in channel catfish IL-8 at the N-terminal of 30~31, 32~34 and 56~58 a.a. residuals, respectively (Fig. 1C).

The a.a. sequence identity between channel catfish IL-8 and 15 different fish IL-8s were calculated using Clustal W and MegAlign, which suggested that IL-8 in this paper shared high identity of 96.9~99.2% to channel catfish IL-8 (AAN60105 and AAN41455) with only 1 a.a. residual difference from AAN41455, and low identity ranging from 21.1 to 31.2% compared with other fish species (Fig. 2A). The phylogenetic tree constructed with Mega 5.1 showed that IL-8 in this study was clustered together with channel catfish IL-8 (AAN60105 and AAN41455) displaying the close relationship between them, with 100 bootstrap support (Fig. 2B), and a distant relationship existed between channel catfish IL-8 with other fish species.

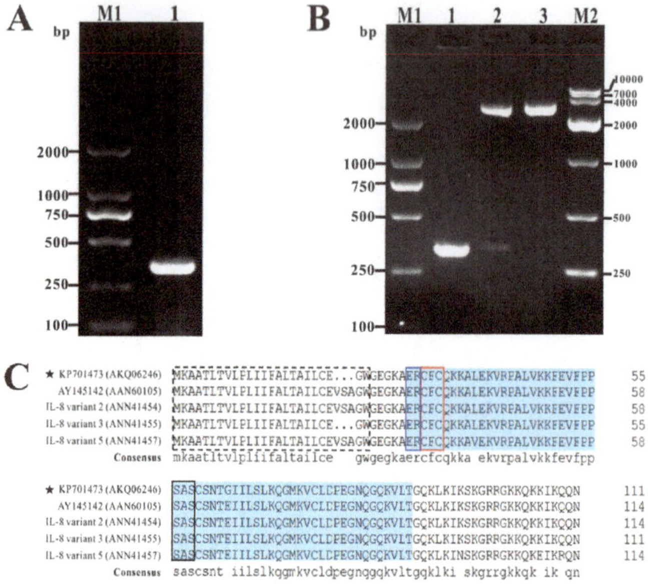

Figure 1: PCR amplification of IL-8 gene, identification of recombinant plasmid pMD19-T-IL-8 and sequence analysis of IL-8 amino acid sequence. (A) PCR amplification of IL-8 gene. M1: DNA marker (DL2000); lane 1: PCR product of IL-8 gene (336 bp). (B) Identification of recombinant plasmid pMD19-T- IL-8 with PCR and enzyme digestion. M1: DNA marker (DL2000); lane 1: PCR product of plasmid pMD19-T-IL-8; lane 2: digestion of plasmid with EcoR I and Xho I; lane 3: digestion of plasmid with XhoI; M2: DNA marker (DL10000). (C) Amino acid sequence analysis of IL-8 sequence with other channel catfish IL-8s. Dotted box presented signal peptides, shadow regions presented the conserved Chemokine_CXC domain. The ELR motif, CXC motif and GPH motif were indicated by blue, red and black box, respectively.

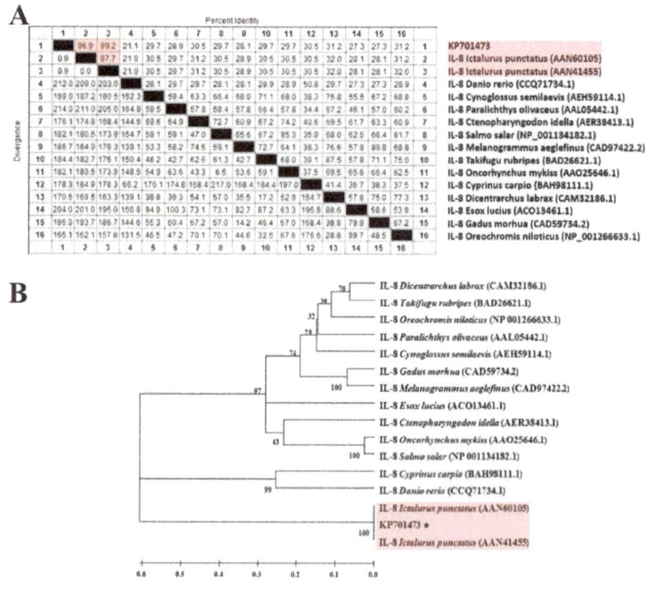

Figure 2: Multiple sequences alignment and phylogenetic tree of IL-8 amino acid sequences with 15 different fish IL-8s. (A) Multiple sequences alignment of IL-8 a.a. sequences. (B) Phylogenetic tree of IL-8 a.a. sequences. The numbers at each branch represent the bootstrap values obtained with 1000 replicates. "KP701473" represents the a.a. sequence of channel catfish IL-8 in this study.

Molecular Cloning, Expression and Purification of rtIL-8

To determine the role of mature peptide of channel catfish IL-8 in the immune regulation, tIL-8 gene (not containing signal peptide) was successfully amplified with PCR and cloned into pMD19-T (Fig. 3A,B), followed by expectedly expressed with vector pET32a (Fig. 3C) in induced BL21(P-tIL-8) sediment and purified by Ni-NTA metal affinity chromatography. Single bands of about 29 kDa of recombinant proteins rtIL-8 were observed after overexpression and purification with SDS-PAGE (Fig. 3D), which was close to the predicted molecular weight. The protein concentration of the rtIL-8 solution was determined as 2.4 mg/ml. Western blot analysis showed that the 29 kDa band of recombinant proteins rtIL-8 reacted specifically with the rabbit anti-6-Histidine antiserum (Fig. 3E).

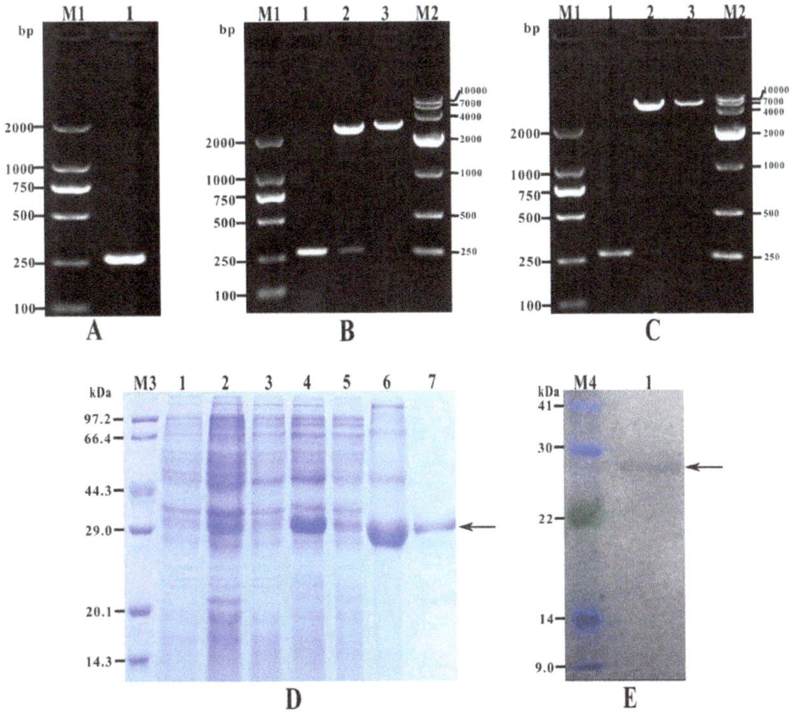

Figure 3: Molecular cloning, expression, purification and western blot analysis of rtIL-8. (A) M1: DNA Marker (DL2000); lane 1: PCR product of tIL-8 gene with 264 bp. (B) M1: DNA Marker (DL2000); lane 1: PCR product of recombinant plasmid T-tIL-8; lane 2: digestion of T-tIL-8 with EcoR I and Xho I; lane 3: digestion of T-tIL-8 with Xho I; M2: DNA Marker (DL10000). (C) M1: DNA Marker (DL2000); lane 1: PCR product of plasmid P-tIL-8; lane 2: digestion of P-tIL-8 with EcoR I and Xho I; lane 3: digestion of P-tIL-8 with Xho I; M2: DNA Marker (DL10000). (D) M3: protein marker; lane 1~6: uninduced BL21(pET32a), induced BL21 (pET32a), uninduced BL21 (P-tIL-8), induced BL21(P-tIL-8), induced BL21(P-tIL-8) supernatant, induced BL21(P-tIL-8) sediment, lane 7: purification of recombinant rtIL-8. (E) M4: protein marker; lane 1: Specific binding between recombinant protein rtIL-8 and rabbit anti-6 × Histidine antibody.

Immunological Analysis

Serum Lysozyme Activity

The serum lysozyme activity of vaccinated fish was assessed by a turbidimetric assay from 1 to 8 week post-secondary vaccination (s.v.) (Fig. 4A). In general, the four vaccinated groups (rtIL-8, rENO, rENO+rtIL-8 and rENO+ISA763) revealed significantly higher levels of lysozyme activity compared with those of control group from 1 to 5 week s.v. with the peak value occurring at 3-week, 4-week, 4-week and 3-week s.v. respectively, and the highest lysozyme activity located in rENO+ISA763 group at 3-week s.v. Furthermore, the lysozyme activity of rENO groups were remarkably higher than that in rtIL-8 group from 1 to 4 week s.v., and the combination of rENO with rtIL-8 enhanced more markedly the action of lysozyme activity than did rENO from 1 to 5 week s.v. Although the difference between rENO+rtIL-8 and rENO+ISA763 were not significant in 1~4 week s.v., it was significant from 5 to 8 week s.v.

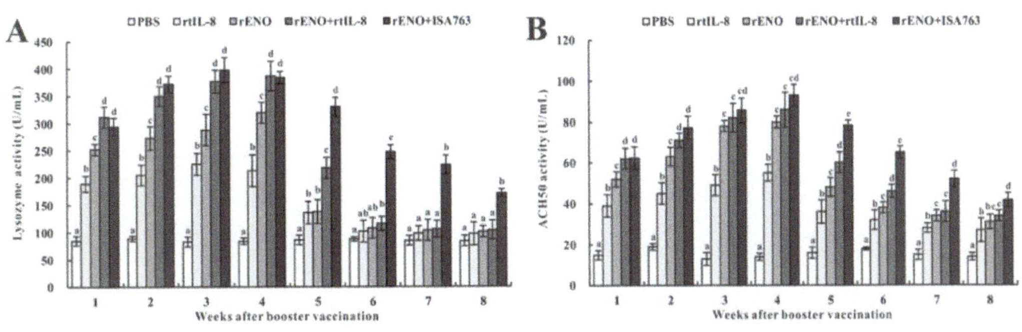

Figure 4. Serum lysozyme activity (A) and ACH50 activity (B) of vaccinated fish. Channel catfish were vaccinated twice

at 2-week intervals, with PBS (control), rtIL-8, rENO, rENO+rtIL-8 and rENO+ISA763. Sera were collected from the fish at 1~8 week s.v. Data are presented as means±SD (n=5). Different letters above a bar denote significant difference (P<0.05).

Serum ACH50 Activity

During all experimental time points, the serum ACH50 activity of four vaccinated groups increased significantly compared with control group (Fig. 4B), and the highest ACH50 activity were all observed at 4-week s.v. in four vaccinated groups. rENO induced remarkably higher ACH50 activity than rtIL-8 from 1 to 7 week s.v., while the ACH50 activity of rENO+rtIL-8 group was significantly higher than that in rENO group at 1, 2, 5 and 6 week s.v. respectively. Moreover, the ACH50 activity in rENO+ISA763 group was higher than that of rENO+rtIL-8 group at 1~8 week s.v. with significant difference of two groups from 5 to 8 week s.v.

Specific Serum Antibody Response

The specific antibody titers in serum of vaccinated fish were detected continuously by ELISA from week 1 to 8 s.v. (Fig. 5). Generally, rENO, rENO+rtIL-8 and rENO+ISA763 stimulated significantly higher levels of antibodies than PBS and rtIL-8 from 2 to 7 week s.v. with the highest antibody level both peaking at 4-week s.v. The antibody levels of rENO+rtIL-8-vaccinated fish were remarkably higher than that in rENO-vaccinated fish from 2 to 5 week s.v., while rENO+ISA763 stimulated the production of antibody significantly than rENO+rtIL-8 at 2, 3, 5~8 week s.v. respectively. Moreover, the antibody levels induced by rtIL-8 alone was just higher than that in PBS-vaccinated fish at each of the examined time point except 3, 4 and 5 week s.v. (P<0.05).

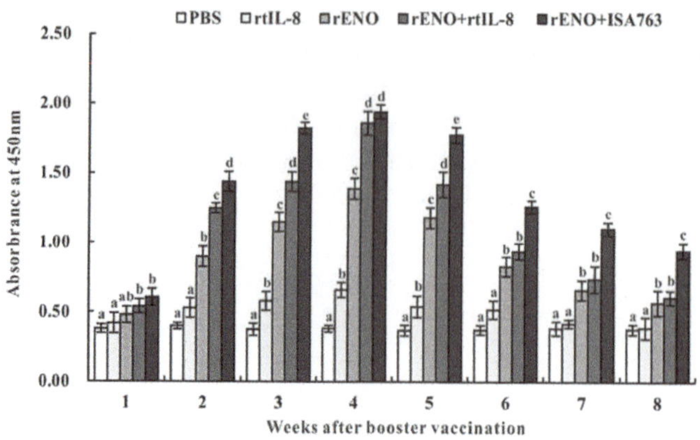

Figure 5: Specific serum antibody detection in vaccinated fish by ELISA. Channel catfish were vaccinated twice at 2-week intervals, with PBS, rtIL-8, rENO, rENO+rtIL-8 and rENO+ISA763. Sera were collected from the fish at 1~8 weeks s.v. Data are presented as means±SD (n=5). Different letters above a bar denote significant difference (P<0.05).

Immunoprotective Efficacy against S. Iniae

Fish vaccinated with PBS, rtIL-8, rENO, rENO+rtIL-8 and/or rENO+ISA763. exhibited accumulated survivals of 6.67%, 13.33%, 53.33%, 73.33% and 80.00% respectively (Fig. 6A) after challenging with the pathogenic *S. iniae* DGX07 at 4 week s.v. The survival of rENO+rtIL-8 and rENO+ISA763 was significant higher than rENO by log-rank test, while the difference of survival of two adjuvant groups was not significant (Fig. 6B). Furthermore, the immunoprotective efficacies, in terms of RPS, of rtIL-8, rENO, rENO+rtIL-8 and rENO+ISA763 were 7.14%, 50.00%, 71.43% and 78.57% respectively compared with PBS.

A

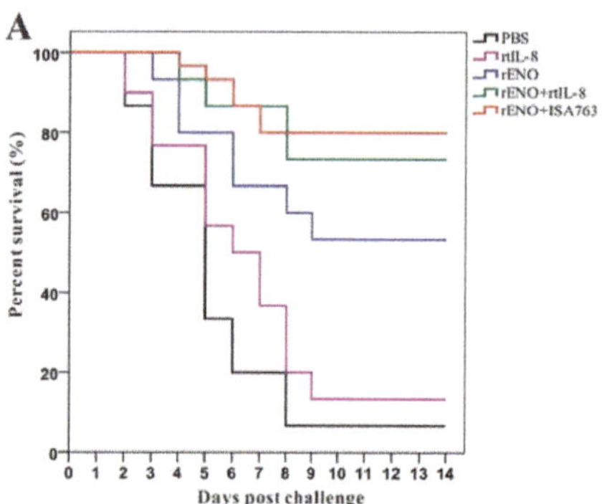

B

Groups	Significance			
	PBS	rtIL-8	rENO	rENO+rtIL-8
PBS				
rtIL-8	0.095^{ns}			
rENO	0.000*	0.001*		
rENO+rtIL-8	0.000*	0.000*	0.036*	
rENO+ISA763	0.000*	0.000*	0.029*	0.592^{ns}

C

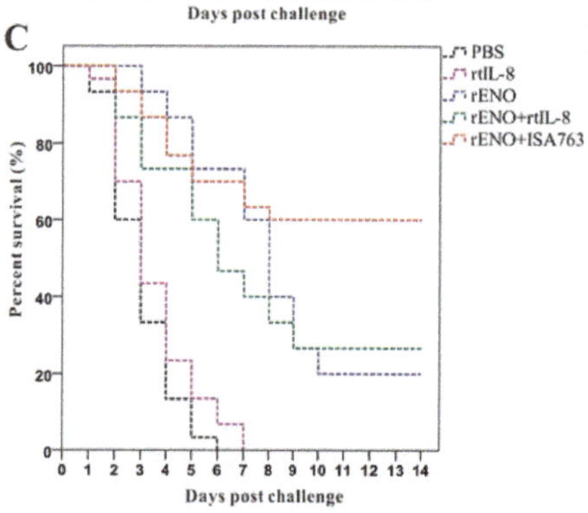

D

Groups	Significance			
	PBS	rtIL-8	rENO	rENO+rtIL-8
PBS				
rtIL-8	0.147^{ns}			
rENO	0.000*	0.000*		
rENO+rtIL-8	0.000*	0.000*	0.605^{ns}	
rENO+ISA763	0.000*	0.000*	0.022*	0.019*

Figure 6. Percent survivals (Kaplan-Meier) of vaccinated fish during the challenges tests of 4 week s.v. (A) and 8 week s.v. (C). Differences among groups were tested using log-rank test shown in (B) (4 week s.v.) and (D) (8 week s.v.). "*"Denotes significant difference (P<0.05), "ns" means not significant.

To examine the duration of protection, fish were also challenged at 8-week s.v., and the results suggested that the cumulative survivals of fish vaccinated with PBS, rtIL-8, rENO, rENO+rtIL-8 and rENO+ISA763 were 0%, 0%, 20.00%, 26.67% and 60% respectively (Fig. 6C), which correspond to an RPS of 20.00% for rENO group, 26.67% for rENO+rtIL-8 group and 60% of rENO+ISA763 group respectively. In addition, the results of log-rank test showed that the survival of rENO+ISA763 was significant higher than rENO+rtIL-8 and rENO, and the difference of survival of rENO+rtIL-8 and rENO was not significant (Fig. 6D). Furthermore, *S. iniae* DGX07 was the only type of bacterial strain recovered from the liver, and kidney of moribund fish during two challenge tests, suggesting that mortality was caused by *S. iniae* DGX07 infection.

Expression of Immune-related Genes

To investigate the effect of vaccination on the expression of immune-related genes, qRT-PCR was carried out to analyze the expression of the genes encoding interleukin 1β (IL-1β), tumor necrosis factor-α (TNF-α), CXC Chemokine Ligand 10 (CXCL10), major histocompatibility complex class Iα (MHC Iα) and IIβ (MHC IIβ), CD4-L2, CD8α and interferon-γ (IFN-γ) with ACTB as an internal control. The results showed that except for IFN-γ, all other investigated genes were induced significantly in fish of rENO, rENO+rtIL-8 and rENO+ISA763 compared with the expression in PBS and rtIL-8 group, especially MHC II β and CD4-L2 with more than 4-fold expression (Fig. 7). The expression of CXCL10, MHC Iα, MHC IIβ, CD4-L2 and CD8α in rtIL-8-vaccinated fish were upregulated significantly compared with the control (PBS) and a slightly higher expression of IL-1β, TNF-α and IFN-γ. In addition, both rENO+rtIL-8 and rENO+ISA763 induced a significantly higher expression of TNF-α, MHC IIβ and CD4-L2, and stimulated a slightly higher expression of IL-1β, MHC Iα, CD8α and IFN-γ than that of rENO group. Moreover, vaccination with rENO+ISA763 upregulated the expression of CD8α and IFN-γ notably, stimulated a higher expression of MHC Iα and a lower expression of other investigated genes especially CXCL10 than that of rENO+rtIL-8.

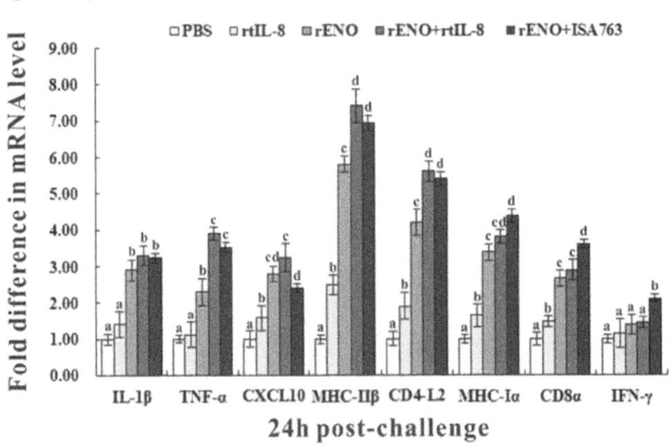

Figure 7: Expression of immune-related genes in vaccinated fish determined by qRT-PCR. Channel catfish were vaccinated twice at 2-week intervals, with PBS (control), rtIL-8, rENO, rENO+rtIL-8 and rENO+ISA763. Total RNA was extracted from the head-kidney at 24 h post-challenge of 4-week s.v. and used for qRT-PCR. For each gene, the mRNA level of the PBS-vaccinated fish was set as 1. Data are presented as means±SD (n=5). Different letters above a bar denote significant difference ($P<0.05$).

DISCUSSION

An effective vaccine candidate usually consists of a strong immunogen (antigen) and a potent adjuvant (immunoenhancer) that can trigger early innate defense mechanisms to aid in the generation of robust and long-lasting immune responses[33]. However, most known effective adjuvants such as Freund's complete adjuvant (FCA) and Freund's incomplete adjuvant (FIA) are unsuitable for human and animal use owing to their toxicity and side effects[24,34]. It has been reported that co-vaccination of antigens with cytokines is a novel and effective strategy to enhance immunization and regulate cellular and humoral immune responses to promote protective immunity against pathogens. Hence, cytokines used as the potential adjuvants have been studied increasingly[22,23,24,35,36,37]. IL-8 as a CXC chemokine produced by numerous cell types, was known to recruit neutrophils to inflammatory sites. In mammals, it had been proved IL-8 was an important immune mediator for innate immunity involved in recruitment and activation of various leukocytes through modulating cytokine production[38], and a potential immune adjuvant to promote cytokines immune response and increase the antibody response by co-administration with antigen[39,40]. However, there were few reports of the potential use of cytokines as an adjuvant in fish especially IL-8. Channel catfish as an important economic fish species, has served as a classical model for the study of comparative immunology. Research on the characterization and biological role of its cytokines should help in elucidation of innate immunity in catfish[32].

In the present study, molecular cloning and recombination expression of mature IL-8 of channel catfish were performed, and rENO-vaccination alone and co-vaccination of rENO with rtIL-8 or commercial adjuvant Montanide™ ISA763 were conducted to evaluate the protective efficacy of rENO and the adjuvant efficacy of rtIL-8 against *S. iniae* in channel catfish. The results showed that the coding sequence of channel catfish IL-8 with 336 bp full-length encoded 111 a.a. belonging to the shorter alternatively spliced transcripts of IL-8 according to Chen *et al.*[32], which existed 3 a.a. (9 bp) difference with the longer transcript. However, the 3-a.a. difference was located in the signal peptides based on sequence analysis (Fig. 1C), leading that both transcripts have the same mature peptide. Thus, we selected the truncate IL-8 (without signal peptides) to express the mature IL-8. In addition, it was interesting that ELR motifs lacking of leucine residue were found to be located immediately before the CXC motif of channel catfish IL-8 in our study, which was different from other fish IL-8 lacking of ELR motifs, indicating that the similarities among the fish IL-8 genes were relatively low (Fig. 2A,B) and the molecular evolution of channel catfish IL-8 occurred during the process of evolution and phylogeny to adapt to the special environment or living way.

Innate immunity as a fundamental defence mechanism of fish, plays an instructive role in the acquired immune response and homeostasis[25,41]. Lysozyme activity and alternative haemolytic complement activity (ACH50) are two important parameters to evaluate the innate immune response[41]. Both are involved in activating the complement system and play a critical role in the innate immune defence against bacterial and fungal pathogens[42,43,44]. In order to stimulate an optimal immune response, we boosted the immune system with the same method and dosage as first immunization at 2 week post-vaccination. The results of the present study showed that both rENO and rtIL-8 enhanced serum lysozyme activity and ACH50 activity, and resulted in stimulating the innate immune responses and activating the protective mechanisms. The lysozyme activity in rENO+rtIL-8-vaccinated fish increased significantly at 1~5 week s.v. and slightly increased at 6~8 week s.v. compared with that of rENO group, while lysozyme activity induced by rENO+ISA763 was remarkably higher than rENO+rtIL-8 at 5~8 week s.v. and the peak value in rENO+ISA763 group was also higher than that of rENO+rtIL-8 group. Similar trends were also observed on the ACH50 activity of rENO and rENO+rtIL-8 and two adjuvant groups, respectively.

Studies have suggested that production of antibodies in response to protein antigens is the basic mechanism of immunoprotection against *S. iniae* infection[7]. Antibody responses induced by subunit or DNA vaccines against *S. iniae* have been observed in various fish models[10,45,46,47,48]. In Japanese flounder vaccinated with recombinant Sip11 subunit vaccine against *S. iniae* SF1, serum specific antibodies were produced at 5~8 weeks post-vaccination and highest antibody titers were detected at the 6 week post-vaccination[46]. Similarly, specific antibody productions in turbot immunized with DNA vaccine pSia10 against *S. iniae* were detected at 4~7 weeks after immunization with the peak of antibody titers appearing at 5 and 6 weeks post-vaccination[45]. In the case of rENO, rENO+rtIL-8 and rENO+ISA763, we found that serum specific antibodies were both detected at 1~8 week s.v. and peaked at 4 week s.v., and the antibody levels of rENO+rtIL-8 group were remarkably higher than that of rENO group from 2 to 5 week s.v., while rENO+ISA763 induced significant higher specific antibody than rENO+rtIL-8 at 2, 3 and 5~8w s.v. respectively, which suggested that rtIL-8 enhanced the antibody levels induced by rENO but failed to hold this enhancement for longer time, and ISA763 not only increased the antibody levels but also maintained the increasement for long time compared with rtIL-8 adjuvant. Comparing with the above other vaccines against the same pathogen[45,46], the time of antibody production and the peak in our study brought forward, mainly due to the difference of vaccines and the host fish, as well as different immune schedules.

Although specific antibodies are important for the immune response against *S. iniae* infection, they are not sufficient for complete protection which involves both humoral and cellular immunity[49]. Therefore, transcriptional analysis of immune-related genes were conducted by qRT-PCR in our study. The results showed that vaccination with rtIL-8, rENO, rENO+rtIL-8 and rENO+ISA763 exhibited comparable inductions in all the examined genes that related to inflammatory response (IL-1β, TNF-α and CXCL10), humoral immunity (MHC II β and CD4-L2) and cellular immunity (MHC Iα, CD8α and IFN-γ) compared with the control, especially MHC IIβ and CD4-L2 which participated in extracellular antigen presentation and the activation of T helper (Th) cells[50], with more than 4-fold expression levels in rENO, rENO+rtIL-8 and rENO+ISA763 group. Compared with rENO, both rENO+rtIL-8 and rENO+ISA763 upregulated the expression of all genes with significantly higher expression of TNF-α, MHC IIβ and CD4-L2, which was, to a certain extent, in line with the protection efficacies exhibited by these vaccines. Besides, rENO+ISA763 upregulated a higher expression of MHC Iα, CD8α and IFN-γ, and a significantly lower expression of CXCL10 than that of rENO+rtIL-8. These results suggested that rENO induced a systemic immune response in channel catfish against *S. iniae* infection, and rtIL-8 activated inflammatory response and strengthened the immune response and ISA763 could promote innate immune reactions and cellular immunity[51], which likely resulted in the higher level protection of rENO+rtIL-8 and rENO+ISA763.

RPS is not only one of the most visual indices to evaluate the immune effect in challenge test, but also a measure of vaccine efficacy against pathogen infection[52]. Our results suggested that rENO provided a moderate immune protection for channel catfish against *S. iniae*infection with the RPS of 50%, which was only half of the RPS (100%) of rENO in zebrafish against same pathogen[10], chiefly because of the different fish species and infection dosages. Besides, the RPS of rENO alone in our study was also lower than the protection of other subunit vaccine against *S. iniae* including RPS (53.8%) of Sia10 in turbot[45], RPS (69.7%) of Sip11 in Japanese flounder[46] and RPS (66.9%) of mtsB in tilapia[47], mostly due to the difference of vaccines, fish and immune schedules. However, adjuvants rtIL-8 and ISA763 enhanced the immune protection with RPS of 71.43% and 78.57% respectively at 4 week s.v., which were not significant each other but were significant higher than that of rENO alone. While rtIL-8 failed to maintain long period of immunity against *S. iniae* infection, only increased the RPS of rENO to 26.67% compared with rENO alone with RPS of 20.00% at 8 week s.v, and both were significant lower than rENO+ISA763 with RPS of 60.00%. In view of this, more effective way such as DNA vaccines like pSia10 with RPS of 73.9% in turbot against *S. iniae*[45] and pSagF with RPS of 78% in Japanese flounder against *S. iniae*[48] should be taken into account.

In conclusion, this study suggested that rtIL-8 hold promise for use as a potent immunopotentiator in subunit vaccines against *S. iniae*infections in channel catfish, while it is insufficient to extend the protective effects of rENO such as two month compared with the commercial adjuvant ISA763. Thus, our study supports the development of IL-8 as an effective adjuvant for subunit vaccine, enhancing the immune response and protection against bacterial infections in fish. However, further studies are required to understand the mechanisms of IL-8 used as an adjuvant and seek for more effective way to strengthen the adjuvanticity of IL-8 such as co-vaccination or fusion in DNA vaccines.

MATERIALS AND METHODS

Bacterial Strains, Plasmids, Reagents and Growth Conditions

S. iniae DGX07 is a pathogenic isolate from diseased channel catfish in China and stored at our laboratory[13], it was cultured in Brain-Heart Infusion (BHI) medium at 37 °C. *E. coli* DH5α and *E. coli* BL21 (DE3) were used as the host strains for cloning and protein expression, respectively. Both were routinely grown in Luria-Bertani (LB) medium containing 100 μg/ml of ampicillin at 37 °C. The plasmids pMD19-T simple (Takara, Japan) and pET32a (+) (Merck, Germany) were used for T-A cloning and protein expression, respectively. Recombinant protein rENO was constructed, expressed and stored at our laboratory[20]. Montanide™ ISA763 A VG (Seppic, France) was selected as a commercial adjuvant for the experiment.

Cloning of IL-8 Complete Coding Regions

Total RNA was extracted from spleen of healthy channel catfish with RNAiso Plus Kit (TaKaRa, Dalian, China) and was reverse-transcribed into first-strand cDNA using PrimeScrip t™ RT reagent Kit with gDNA Eraser (Perfect Real Time) (TaKaRa) according to the manufacturer's instructions. A channel catfish IL-8 complete coding sequence was cloned from the cDNA template using primers IL-8-F1/IL-8-R1 (Table 1), which were designed based on sequence published in GenBank (AY145142, corresponding protein GenBank: AAN60105) and by Primer Premier 5.0 software under default parameters. PCR amplification was performed under the following conditions: 1 cycle of 94 °C for 5 min, 30 cycles of 94 °C for 1 min, 55 °C for 30 s, and 72 °C for 30 s, followed by a final extension of 72 °C for 10 min. The PCR products were purified using the Agarose Gel DNA Extraction Kit (TaKaRa) and cloned into the pMD19-T vector, followed by transformation into *E. coli* DH5α. The positive recombinant clone was then selected using an Amp/IPTG/X-Gal agar plate. The recombinant plasmid was identified by PCR under the aforementioned conditions, digested with restriction enzymes *EcoR*I and *Xho* I, and fractionated on 1% agarose gels. DNA sequencing was also conducted by TaKaRa Bio Inc.

Table 1: Primers used for cloning, expression and qRT-PCR of genes in this paper.

Primers	Sequences (5′→3′)[a]	Target gene
IL-8-F1	CGGAATTCATGAAGGCTGCAACT (EcoRI)	IL-8
IL-8-R1	CCCTCGAGTCAGTTTTGCTGTTTG (XhoI)	
IL-8-F2	CGGAATTCGGAGAAGGAAAAGCAGAG (EcoRI)	tIL-8
IL-8-R2	CCCTCGAGTCAGTTTTGCTGTTTGATCT (XhoI)	
ACTB-F	CCCATCTATGAGGGTTATGCTCTG	ACTB
ACTB-R	GCTCGGTCAGGATCTTCATCAG	
IL-1β-F	GCCATGTTGCTAATGTTGTAATCG	IL-1β
IL-1β-R	TGTCTTGCAGGCTGTAACTCTTG	
TNF-α-F	CGCACAACAAACCAGACGAGAC	TNF-α
TNF-α-R	ACCACTGCATAGATACGCTCGAA	
CXCL10-F	ACAGGCCAGGACCAGTGTAAGG	CXCL10
CXCL10-R	CAAGTTGCACTCGCAGGATGAA	
MHC Iα-F	GGTATCATCGTTGGTGTAGCCG	MHC Iα
MHC Iα-R	GGACAGGTTTGAAGCCAGAGTT	
MHC IIβ-F	CGGGAAGGAGATTAAAGGAGGT	MHC IIβ
MHC IIβ-R	GTTTGGTGAAGCTGGCGTGT	
CD4-L2-F	GCAGGGCACGGATAGATGGA	CD4-L2
CD4-L2-R	TGGGTTCGCAGAGGCTGATAC	
CD8α-F	CCGACAGTGCCTACGACTAAAGC	CD8α
CD8α-R	CCAGCAGCCAAAGGAATGAAG	
IFN-γ-F	TGCACGAAGTGAAAGACCAAA	IFN-γ
IFN-γ-R	TTAAGGTCCAGCAGCTCAGTGA	

[a]Underlined nucleotides are restriction sites of the enzymes indicated in the brackets at the ends.

Sequence and Phylogenetic Analyses

The amino acid (a.a.) sequence was derived from nucleotide sequence using the translate tool, the molecular weight and the isoelectric point (pI) of the peptide were calculated using ProtParam tool, the signal peptide was predicted with SignalP 4.1 program, both available on the ExPASy molecular biology server (http://www.expasy.org/tools)53. Multiple sequence alignments and the identities between each pair of the a.a. sequences were calculated by Clustal W method[54] using MegAlign programe[55], and the phylogenetic tree was constructed using the Neighbor-Joining (N-J) method in the MEGA 5.1 software[56], with 1,000 bootstrap replicates.

Cloning of Truncate *IL-8* Gene (*tIL-8*)

Based on the sequence analysis, the signal peptide was removed in order to obtain the mature IL-8 by prokaryotic expression with *E. coli*BL21 (DE3). The truncate IL-8 gene (tIL-8) was cloned from the cDNA template using degenerate primers IL-8-F2/IL-8-R2 (Table 1) which containing *EcoR* I and *Xho* I restriction enzyme sites. PCR amplification was performed as described above. The resultant amplicons were purified, cloned, transformed and sequenced, and the positive recombinant cloning plasmid named by T-tIL-8.

Expression and Purification of Recombinant Truncate IL-8 (rtIL-8)

The expression and purification of recombinant truncate IL-8 was conducted as described in our previous study[57]. Briefly, the plasmid T-tIL-8 was digested with EcoRI and XhoI and the resultant products were inserted into the EcoRI/XhoI-digested pET32a (+) vector to construct the recombinant expression plasmid, named as P-tIL-8. The plasmid was then transformed into *E. coli* BL21 and induced by adding 1.0 mM IPTG at 37 °C for 4 h. the cells were centrifuged at 8000× g for 10 min at 4 °C and suspended with sterile phosphate buffer saline (PBS), followed by ultrasonicating with an ultrasonic cell disrupter (JY92-IIDN, Ningbo Scientz, China) and examined by 12.5% SDS-PAGE. The inclusion bodies from the insoluble fractions were purified by Ni-NTA-Sefinose Column (Sangon Biotech, Shanghai, China) after dissolution in 8 M urea solutions and filtration with 0.22-μm filters. The refolding of the purified proteins was conducted by dialyzing gradiently from 6 M urea solutions to PBS at 4 °C, and then analyzed by 12.5% SDS-PAGE. The protein was quantified using the Bradford assay with bovine serum albumin (BSA) as standard and a NanoDrop spectrophotometer (Thermo Scientific) according to the manufacturer's instructions. Purified proteins were named as rtIL-8 and stored at 20 °C.

Western-blot Analysis

The werstern-blot analysis of rtIL-8 was performed as previously described[57]. Briefly, the purified proteins were separated by 12.5% SDS-PAGE and transferred to a PVDF membrane at 150 V for 2 h. Non-specific binding sites of the membranes were blocked by incubating for 1 h at 37 °C in TBST (containing 3% BSA). Then the membrane was incubated with rabbit anti-6-Histidine antibody (Sangon Biotech, Shanghai, China) diluted 1:100 in TBST (containing 0.5% BSA) for 12 h at 4 °C. After washing 3 times with TBST, the membrane was incubated with goat-anti-rabbit IgG (H+L)-HRP (Sigma, St. Louis, MO, USA) diluted 1:5000 in TBST (containing 0.5% BSA) at 37 °C for 1 h. After washing off unbound secondary antibody, the specific antigen-bound antibody was visualized using DAB (Sigma) for 5 to 15 min, and terminated by rinsing with distilled water.

Preparation of Fish and Vaccine

All animal experiments were approved by the Committee of the Ethics on Animal Care and Experiments at Sichuan Agricultural University. All experimental procedures were carried out in accordance with the approved guidelines.

Channel catfish (50.0 g±5.0 g) were purchased from a fish farm in Chengdu (Sichuan, China) and acclimatized in the laboratory for 2 weeks before experimental manipulation. The fish were fed a commercial diet daily, and water was partly replaced every day, maintaining a temperature of 28 °C± 1 °C. Before experiments, fish were randomly sampled from blood, liver, kidney, and spleen, the examination of bacterial recovery indicated that no bacteria could be detected and agglutination test showed no reaction between the serum and *S. iniae* DGX07. Fish were anaesthetized with MS222 (Sigma, Beijing, China) prior to the experiments, which involved manipulations such as injections and serum collection. The purified protein rtIL-8 and recombinant protein rENO were diluted in PBS to 0.5 mg/ml and 1.0 mg/ml, respectively. To obtain rENO+rtIL-8, the recombinant protein rENO was mixed with an equal volumes of purified protein rtIL-8. To obtain rENO+ISA763, the recombinant protein rENO was mixed with adjuvant Montanide™ ISA763 A VG (Seppic, France) at a ratio of 3:7 by ultrasonic disruption (JY92-IIDN, Ningbo Scientz, China).

Vaccination and Bacterial Challenge

Five hundred channel catfish were divided randomly into five groups (100 fish/group) and injected intraperitoneally (i.p.) with 0.2 ml of PBS (control group), rtIL-8, rENO, rENO+rtIL-8 and rENO+ISA763. The secondary vaccination were performed to obtain an optimal immune response with the same method and dosage at 2 week post-vaccination. At 4 and 8 weeks post secondary vaccination (s.v.), 30 fish from each group were randomly selected and challenged by i.p. injection with 0.2 ml of *S. iniae* DGX07 that resuspended in PBS to 6×10^7 CFU/ml[57]. Mortality was monitored over a period of 14 days after the challenge, and dying fish were randomly selected for the examination of bacterial recovery from liver, kidney, and spleen. Relative percent of survival (RPS) was calculated according to the following formula: RPS=[1− (% mortality of vaccinated fish/% mortality of control fish)] × 100[58]. Serum samples of five fish in each group were collected for assessment of immune related indexes from 1 to 8 week s.v. and head-kidney of five fish were taken for qRT-PCR at 24 h post-challenge of 4 week s.v. All vaccination trials were repeated once.

Detection of Innate Immune Parameters

Serum lysozyme activity was measured by the turbidimetric method described by Hutchinson[59]. Briefly, 150 μL Micrococcus lysodeikticus with a concentration of 0.2 mg/mL (in 0.04 M PBS, pH=6.2) was added to 15 μL sera in a 96-well U-bottom microtiter plate, quickly mixed by vortex, and the OD_{450} (UV absorption value at 450 nm) (A_0) was assayed at 0.5 and 4.5 min after reaction, respectively. Each test was conducted in triplicate. One activity unit of lysozyme (U) was defined as the amount of serum lysozyme that caused a decrease in absorbancy of 0.001 per min at 450 nm.

Alternative haemolytic complement activity (ACH50) was determined following the method described by Sunyer and Tort[60] based on haemolysis of rabbit red blood cells (RaRBC). The volume of serum yielding 50% haemolysis (ACH50) was determined and used to calculate the complement activity of the samples (value of ACH50 is in units per ml).

Enzyme-linked Immunosorbent Assay (ELISA)

Specific antibody titers for rENO were determined by ELISA as described previously with some modification[57]. Briefly, the rENO was diluted to a concentration of 50 μg/mL in a carbonate buffer (pH=9.6). Each well of the 96-well plate was covered with 100 μL diluted rENO overnight at 4 °C followed by washing with PBST (0.1% Tween-20 in PBS) and blocking with 3% BSA in PBST for 2 h at 37 °C. Serial 2-fold dilutions of sera were added to the wells in triplicate and subsequently incubated for 2 h at 37 °C. Rabbit anti-channel catfish IgM antibody (prepared in our laboratory) (1:2000) and goat-anti-rabbit IgG (H+L)-HRP (1:2000) were used as the secondary and tertiary antibodies, respectively. Color development was performed with the TMB kit (Tiangen, Beijing, China). The plates were read at 450 nm with a microplate reader (Bio-Rad, Hercules, USA).

qRT-PCR Analysis of the Expression of Immune-related Genes

Head-kidney samples were taken from the fish (five in each group) at 24 h post-challenge of 4 week s.v. Total RNA extraction and cDNA synthesis were carried out as described above. qRT-PCR was performed with the SYBR® Premix Ex Taq™ II (Tli RNaseH Plus) (TaKaRa) in an ABI StepOnePlus™ System (Applied Biosystems, USA) as described previously. Each assay was performed in triplicate with beta actin (ACTB) as an internal control. The primers used to amplify the immune-related genes are listed in Table 1. All data are given in terms of relative mRNA.

Statistical Analysis

All statistical analyses were performed with SPSS 19.0 software (SPSS Inc., USA). Mortality data from the challenge experiments were analyzed by the Kaplan-Meier methods and differences among groups were tested using log-rank test. The difference significance of other data were determined using a one-way analysis of variance (ANOVA). In all cases, the significance level was defined as $P<0.05$ and the results were presented as means ± SD (standard deviation).

ACKNOWLEDGEMENTS

This work was supported by the Program for Changjiang Scholars and Innovative Research Team in University (No. IRT0848), Sichuan Technology Support Planning (No. 2014NZ0003) and (No. 2014JY0143).

REFERENCES

1. Troell, M. *et al.* Does aquaculture add resilience to the global food system? *Proc Natl Acad Sci* 111, 13257–13263 (2014).
2. Dong, X. *et al.* Expression profile analysis of two cathepsin S in channel catfish (*Ictalurus punctatus*) mucosal tissues following bacterial challenge. *Fish Shellfish Immun.* 48, 112–118 (2016).
3. Peterson, B. C., Peatman, E., Ourth, D. D. & Waldbieser, G. C.Effects of a phytogenic feed additive on growth performance, susceptibility of channel catfish to *Edwardsiella ictaluri* and levels of mannose binding lectin. *Fish Shellfish Immun.* 44, 21–25 (2015).
4. Shiau, S., Gabaudan, J. & Lin, Y. Dietary nucleotide supplementation enhances immune responses and survival to*Streptococcus iniae* in hybrid tilapia fed diet containing low fish meal. *Aquac Rep.* 2, 77–81 (2015).
5. Zhu, J. *et al.* Trancriptomic profiling revealed the signatures of acute immune response in tilapia (*Oreochromis niloticus*) following *Streptococcus iniae* challenge. *Fish Shellfish Immun.* 46, 346–353 (2015).
6. El Aamri, F. *et al.* Differential innate immune response of European seabass (*Dicentrarchus labrax*) against *Streptococcus iniae. Fish Shellfish Immun.* 46, 436–441 (2015).
7. Agnew, W. & Barnes, A. C. *Streptococcus iniae*: An aquatic pathogen of global veterinary significance and a challenging candidate for reliable vaccination. *Vet Microbiol.* 122, 1–15 (2007).
8. Kayansamruaj, P. *et al.* Susceptibility of freshwater rearing Asian seabass (*Lates calcarifer*) to pathogenic *Streptococcus iniae. Aquac Res.* (2015).
9. LaFrentz, B. R. *et al.* Controlled challenge experiment demonstrates substantial additive genetic variation in resistance of Nile tilapia (*Oreochromis niloticus*) to *Streptococcus iniae.Aquaculture.* 458, 134–139 (2016).
10. Membrebe, J. D. *et al.* Protective efficacy of*Streptococcus iniae*derived enolase against Streptococcal infection in a zebrafish model. *Vet Immunol Immunop.* (2016).
11. Sun, Y. & Hu, Y. Cell-penetrating peptide-mediated subunit vaccine generates a potent immune response and protection against *Streptococcus iniae in Japanese flounder (Paralichthys olivaceus). Vet Immunol Immunop.* 167, 96–103 (2015).
12. Safari, R. *et al.* Host-derived probiotics Enterococcus casseliflavus improves resistance against *Streptococcus iniae*infection in rainbow trout (*Oncorhynchus mykiss*) via immunomodulation. *Fish Shellfish Immun.* 52, 198–205 (2016).
13. Chen, D. *et al. Streptococcus iniae* isolated from channel catfish (*Ictalurus punctatus*) in China. *Isr J Aquacult-Bamid.* 63 (2011).
14. Goh, S. H. *et al. Streptococcus iniae*, a human and animal pathogen: Specific identification by the chaperonin 60 gene identification method. *J Clin Microbiol.* 36, 2164–2166 (1998).
15. Facklam, R., Elliott, J., Shewmaker, L. & Reingold, A. Identification and characterization of sporadic isolates of *Streptococcus iniae*isolated from humans. *J Clin Microbiol.* 43, 933–937 (2005).
16. Lau, S. K. *et al.* Invasive *Streptococcus iniae* infections outside north america. *J Clin Microbiol.* 41, 1004–1009 (2003).
17. He, Y. *et al.* A recombinant truncated surface immunogenic protein (tSip) plus adjuvant FIA confers active protection against Group B streptococcus infection in tilapia. *Vaccine.* 32, 7025–7032 (2014).
18. Baiano, J. C. & Barnes, A. C. Towards control of *Streptococcus iniae. Emerg Infect Dis.* 15, 1891 (2009).
19. Kim, M. S. *et al.* α-enolase, a plasmin (ogen) binding and cell wall associating protein from a fish pathogenic *Streptococcus iniae*strain. *Aquaculture.* 265, 55–60 (2007).
20. Wang, J. *et al.* Cloning and characterization of Surface-Localized α-Enolase of *Streptococcus iniae*, an effective protective antigen in mice. *Int J Mol Sci.* 16, 14490–14510 (2015).
21. Kutzler, M. A. *et al.* Chemokine-adjuvanted electroporated DNA vaccine induces substantial protection from simian immunodeficiency virus vaginal challenge. *Mucosal Immunol.*(2015).

22. Bobanga, I. D., Petrosiute, A. & Huang, A. Y. Chemokines as cancer vaccine adjuvants. *Vaccines.* 1, 444–462 (2013).

23. Xie, X. *et al.* Co-administration of antigen with chemokine MCP-3 or MDC/CCL22 enhances DNA vaccine potency. *Invest New Drug.*33, 810–815 (2015).

24. Tafalla, C., Bøgwald, J. & Dalmo, R. A. Adjuvants and immunostimulants in fish vaccines: Current knowledge and future perspectives. *Fish Shellfish Immun.* 35, 1740–1750 (2013).

25. Zhu, L. *et al.* Advances in research of fish immune-relevant genes: A comparative overview of innate and adaptive immunity in teleosts. *Dev Comp Immunol.* 39, 39–62 (2013).

26. Savan, R. & Sakai, M. Genomics of fish cytokines. *Comp Biochem Physiol D.* 1, 89–101 (2006).

27. Sanchez, E., Coll, J. & Tafalla, C. Expression of inducible CC chemokines in rainbow trout (*Oncorhynchus mykiss*) in response to a viral haemorrhagic septicemia virus (VHSV) DNA vaccine and interleukin 8. *Dev Comp Immunol.* 31, 916–926 (2007).

28. Jimenez, N., Coll, J., Salguero, F. J. & Tafalla, C. Co-injection of interleukin 8 with the glycoprotein gene from viral haemorrhagic septicemia virus (VHSV) modulates the cytokine response in rainbow trout (*Oncorhynchus mykiss*). *Vaccine.* 24, 5615–5626 (2006).

29. Lee, E. *et al.* Cloning and sequence analysis of the interleukin-8 gene from flounder (*Paralichthys olivaceous*). *Gene.* 274, 237–243 (2001).

30. Najakshin, A. M., Mechetina, L. V., Alabyev, B. Y. & Taranin, A. V.Identification of an IL-8 homolog in lamprey(*Lampetra fluviatilis*): Early evolutionary divergence of chemokines. *Eur J Immunol.* 29, 375–382 (1999).

31. Harun, N. O. *et al.* The biological effects of rainbow trout (*Oncorhynchus mykiss*) recombinant interleukin-8. *Dev Comp Immunol.* 32, 673–681 (2008).

32. Chen, L. *et al.* Analysis of a catfish gene resembling interleukin-8: CDNA cloning, gene structure, and expression after infection with *Edwardsiella ictaluri. Dev Comp Immunol.* 29, 135–142 (2005).

33. Kaur, G. & Bansal, A. RIL-22 as an adjuvant enhances the immunogenicity of rGroEL in mice and its protective efficacy against *S. Typhi* and *S. Typhimurium. Cell Mol Immunol.* 12, 96–106 (2015).

34. Gjessing, M. C. *et al.* A sequential study of incomplete Freund's adjuvant-induced peritonitis in Atlantic cod. *Fish Shellfish Immun.*32, 141–150 (2012).

35. Kalams, S. A. *et al.* Safety and immunogenicity of an HIV-1 gag DNA vaccine with or without IL-12 and/or IL-15 plasmid cytokine adjuvant in healthy, HIV-1 uninfected adults. *PLos One.* 7, e29231 (2012).

36. Kaur, G., Chitradevi, C. & Bansal, A. Adjuvant efficacy of IL22 in modulating the immune response and protection accorded by *S. Typhi* rGroEL against multiple pathogens in mice. *Cell Mol Immunol.* 12, 96–106 (2015).

37. Passalacqua, R. *et al.* Adjuvant low-dose interleukin-2 (IL-2) plus interferon-α (IFN-α) in operable renal cell carcinoma (RCC): A Phase III, randomized, multicentre trial of the Italian Oncology Group for Clinical Research (GOIRC). *J Immunother.* 37, 440–447 (2014).

38. Sin, J. *et al.* DNA vaccines encoding interleukin-8 and RANTES enhance antigen-specific Th1-type CD4+ T-cell-mediated protective immunity against herpes simplex virus type 2 *in vivo*. *J Virol.* 74, 11173–11180 (2000).

39. Kim, J. J. *et al.* Chemokine gene adjuvants can modulate immune responses induced by DNA vaccines. *Interferon Cytokine Res.* 20, 487–498 (2000).

40. Min, W. *et al.* Adjuvant effects of IL-1β, IL-2, IL-8, IL-15, IFN-α, IFN-γ, TGF-β4 and lymphotactin on DNA vaccination against*Eimeria acervulina. Vaccine.* 20, 267–274 (2001).

41. Magnadóttir, B. Innate immunity of fish (overview). *Fish Shellfish Immun.* 20, 137–151 (2006).

42. Swain, P. *et al.* Non-specific immune parameters of brood Indian major carp *Labeo rohita* and their seasonal variations. *Fish Shellfish Immun.* 22, 38–43 (2007).

43. Wang, Z. & Zhang, S. The role of lysozyme and complement in the antibacterial activity of zebrafish (*Danio rerio*) egg cytosol. *Fish Shellfish Immun.* 29, 773–777 (2010).

44. Wang, Z. *et al.* Maternal transfer and protective role of the alternative complement components in zebrafish *Danio rerio.PLos One.* 4, e4498 (2009).

45. Sun, Y., Hu, Y., Liu, C. & Sun, L. Construction and analysis of an experimental *Streptococcus iniae* DNA vaccine. *Vaccine.* 28, 3905–3912 (2010).

46. Cheng, S., Hu, Y., Jiao, X. & Sun, L. Identification and immunoprotective analysis of a *Streptococcus iniae* subunit vaccine candidate. *Vaccine.* 28, 2636–2641 (2010).

47. Zou, L. *et al.* MtsB, a hydrophobic membrane protein of*Streptococcus iniae*, is an effective subunit vaccine candidate.*Vaccine.* 29, 391–394 (2011).

48. Sun, Y., Hu, Y., Liu, C. & Sun, L. Construction and comparative study of monovalent and multivalent DNA vaccines against*Streptococcus iniae. Fish Shellfish Immun.* 33, 1303–1310 (2012).

49. Wilson, R. *et al.* Protection against *Streptococcus pneumoniae* lung infection after nasopharyngeal colonization requires both humoral and cellular immune responses. *Mucosal Immunol.* 8, 627–639 (2015).

50. Levine, T. P., Chain, B. M. & Brodsky, F. The cell biology of antigen processing. *Crit Rev Biochem Mol.* 26, 439–473 (1991).

51. Jaafar, R. M. *et al.* Effects of adjuvant Montanide™ ISA 763 a VG in rainbow trout injection vaccinated against *Yersinia ruckeri. Fish Shellfish Immun.* 47, 797–806 (2015).

52. Costa, A. A. *et al.* Effect of vaccination against *yersiniosis* on the relative percent survival, bactericidal and lysozyme response of Atlantic salmon, *Salmo salar. Aquaculture.* 315, 201–206 (2011).

53. Artimo, P. *et al.* ExPASy: SIB bioinformatics resource portal.*Nucleic Acids Res.* s400 (2012).

54. Larkin, M. A. *et al.* Clustal W and Clustal X version 2.0.*Bioinformatics.* 23, 2947–2948 (2007).

55. Burland, T. G. DNASTAR's Lasergene sequence analysis software in *Bioinformatics Methods and Protocols* (eds Stephen, M. *et al.*) 71–91 (1999).

56. Tamura, K. *et al.* MEGA5: Molecular evolutionary genetics analysis using maximum likelihood, evolutionary distance, and maximum parsimony methods. *Mol Biol Evol.* 28, 2731–2739 (2011).

57. Jiang, J. *et al.* Adjuvant immune enhancement of subunit vaccine encoding pSCPI of *Streptococcus iniae* in channel catfish (*Ictalurus punctatus*). *Int J Mol Sci.* 16, 28001–28013 (2015).

58. Amend, D. F. Potency testing of fish vaccines In *Fish biologics: serodiagnostics and vaccines.* (eds Anderson, D. P. *et al.*) 447–454 (1981).

59. Hutchinson, T. H. & Manning, M. J. Seasonal trends in serum lysozyme activity and total protein concentration in dab (*Limanda limandaL*) sampled from Lyme Bay, UK. *Fish Shellfish Immun.* 6, 473–482 (1996).

60. Sunyer, J. O. & Tort, L. Natural hemolytic and bactericidal activities of sea bream *Sparus aurata* serum are effected by the alternative complement pathway. *Vet Immunol Immunop.* 45, 333–345 (1995).

INDEX